D1219401

Interfacial Nanochemistry

Nanostructure Science and Technology

Series Editor: David J. Lockwood, FRSC
National Research Council of Canada
Ottawa, Ontario, Canada

Current volumes in this series:

Alternative Lithography: Unleashing the Potentials of Nanotechnology
Edited by Clivia M. Sotomayor Torres

Interfacial Nanochemistry: Molecular Science and Engineering at Liquid–Liquid Interfaces
Edited by Hitoshi Watarai

Nanoparticles: Building Blocks for Nanotechnology
Edited by Vincent Rotello

Nanostructured Catalysts
Edited by Susannah L. Scott, Cathleen M. Crudden, and Christopher W. Jones

Nanotechnology in Catalysis, Volumes 1 and 2
Edited by Bing Zhou, Sophie Hermans, and Gabor A. Somorjai

Polyoxometalate Chemistry for Nano-Composite Design
Edited by Toshihiro Yamase and Michael T. Pope

Self-Assembled Nanostructures
Jin Z. Zhang, Zhong-lin Wang, Jun Liu, Shaowei Chen, and Gang-yu Liu

Semiconductor Nanocrystals: From Basic Principles to Applications
Edited by Alexander L. Efros, David J. Lockwood, and Leonid Tsybeskov

Interfacial Nanochemistry

Molecular Science and Engineering at Liquid–Liquid Interfaces

Edited by

Hitoshi Watarai

Osaka University
Osaka, Japan

Norio Teramae

Tohoku University
Tohoku, Japan

and

Tsuguo Sawada

Tokyo University
Tokyo, Japan

Kluwer Academic / Plenum Publishers
New York, Boston, Dordrecht, London, Moscow

Library of Congress Cataloging-in-Publication Data

Watarai, Hitoshi.
Interfacial nanochemistry : molecular science and engineering at liquid-liquid interfaces / Professor Hitoshi Watarai.
p. cm.—(Nanostructure science and technology)
Includes bibliographical references and index.
ISBN 0-306-48527-3
1. Liquid-liquid interfaces. I. Title. II. Series.
QD509.L54W37 2005
541′.33—dc22 2004055312

ISBN 0-306-48527-3

233 Spring Street, New York, New York 10013

http://www.kluweronline.com

10 9 8 7 6 5 4 3 2 1

A C.I.P. record for this book is available from the Library of Congress

Printed in the United States of America.

Contributors

Raymond R. Dagastine
Department of Chemical and Biomolecular Engineering
University of Melbourne
Melbourne, Victoria 3010
Australia

Jeremy G. Frey
Department of Chemistry
University of Southampton
Southampton SO17 1BJ
UK

Masahiro Goto
Department of Applied Chemistry
Graduate School of Engineering
Kyushu University
Fukuoka 812-8581
Japan

Fumio Hirata
Institute for Molecular Science
Okazaki National Research Institutes
Myodaiji, Okazaki 444-8585
Japan

Hiroki Hotta
Department of Chemistry
Kobe University
Nada, Kobe 657-8501
Japan

Yasuhiro Ikezoe
Department of Advanced Materials Science
Graduate School of Frontier Sciences
The University of Tokyo
Hongo, Tokyo 113-8656
Japan

Shoji Ishizaka
Department of Chemistry
Graduate School of Science
Hokkaido University
Sapporo 060-0810
Japan

Takashi Kakiuchi
Department of Energy and Hydrocarbon Chemistry
Graduate School of Engineering
Kyoto University
Kyoto 606-8501
Japan

Sorin Kihara
Department of Chemistry
Kyoto Institute of Technology
Matsugasaki
Sakyo, Kyoto 606-8585
Japan

Noboru Kitamura
Department of Chemistry
Graduate School of Science
Hokkaido University
Sapporo 060-0810
Japan

Shū Kobayashi
Graduate School of Pharmaceutical Sciences
The University of Tokyo
Hongo
Bunkyo-ku, Tokyo 113-0033
Japan

Andriy Kovalenko
National Institute for Nanotechnology
National Research Council of Canada
University of Alberta
9107-116 Str.
Edmonton, AB T6G 2V4
Canada

Kei Manabe
Graduate School of Pharmaceutical Sciences
The University of Tokyo
Hongo
Bunkyo-ku, Tokyo 113-0033
Japan

Kiyoharu Nakatani
Department of Chemistry
University of Tsukuba
Tennoudai, Tsukuba 305-8571
Japan

Takayuki Negishi
Department of Chemistry
University of Tsukuba
Tennoudai, Tsukuba 305-8571
Japan

Seiichi Nishizawa
Department of Chemistry
Graduate School of Science
Tohoku University
Aoba-ku, Sendai 980-8578
Japan

Tsutomu Ono
Department of Applied Chemistry
Graduate School of Engineering
Kyushu University
Fukuoka 812-8581
Japan

Toshiyuki Osakai
Department of Chemistry
Kobe University
Nada, Kobe 657-8501
Japan

Geraldine L. Richmond
Department of Chemistry and Materials Science Institute
University of Oregon
Eugene, OR 97403
USA

Tsuguo Sawada
Department of Advanced Materials Science
Graduate School of Frontier Sciences
The University of Tokyo
Hongo, Tokyo 113-8656
Japan

Geoffery W. Stevens
Department of Chemical and Biomolecular Engineering
University of Melbourne
Melbourne, Victoria 3010
Australia

Norio Teramae
Department of Chemistry
Graduate School of Science
Tohoku University
Aoba-ku, Sendai 980-8578
Japan

Satoshi Tsukahara
Department of Chemistry
Graduate School of Science
Osaka University
Toyonaka, Osaka 560-0043
Japan

Tatsuya Uchida
Tokyo University of Pharmacy and Life Science
Hachioji 192-0392
Japan

Hitoshi Watarai
Department of Chemistry
Graduate School of Science
Osaka University
Toyonaka, Osaka 560-0043
Japan

Mark R. Watry
Rocky Mountain College
Billings, MT 59102
USA

Akira Yamaguchi
Department of Chemistry
Graduate School of Science
Tohoku University
Aoba-ku, Sendai 980-8578
Japan

Hiroharu Yui
Department of Advanced Materials Science
Graduate School of Frontier Sciences
The University of Tokyo
Hongo, Tokyo 113-8656
Japan

Preface

The history of the liquid–liquid interface on the earth might be as old as that of the liquid. It is plausible that the generation of the primitive cell membrane is responsible for an accidental advent of the oldest liquid interfaces, since various compounds can be concentrated by an adsorption at the interface. The presence of liquid–liquid interface means that real liquids are far from ideal liquids that must be miscible with any kinds of liquids and have no interface. Thus it can be said that the non-ideality of liquids might generate the liquid–liquid interface indeed and that biological systems might be generated from the non-ideal interface. The liquid–liquid interface has been, therefore, studied as a model of biological membrane.

From pairing two-phases of gas, liquid and solid, nine different pairs can be obtained, which include three homo-pairs of gas–gas, liquid–liquid and solid–solid pairs. The gas-gas interface, however, is practically no use under the ordinary conditions. Among the interfaces produced by the pairing, the liquid–liquid interface is most slippery and difficult to be studied experimentally in comparison with the gas–liquid and solid–liquid interfaces, as the liquid–liquid interface is flexible, thin and buried between bulk liquid phases. Therefore, in order to study the liquid–liquid interface, the invention of innovative measurement methods has a primary importance.

At the liquid–liquid interface, completely different properties of water and organic phases can be met in the two-dimensional boundary with a thickness of only 1 nm. In practical two-phase systems with highly miscible components, however, the formation of nano- and micro-droplets at the interfacial nano-region is suggested. The structural and dynamic properties of molecules at the interface are the most important subject in the study of physics and chemistry at the interface. The solution theory of the liquid–liquid interface has not been established yet, though the molecular dynamics simulations have been developed as a useful tool for depicting the molecular picture of the solvent and solute molecules in the interfacial region.

The adsorption of reactant molecules at the interface significantly affects the overall reaction rate in the two-phase system by the catalytic function of the interface. The liquid–liquid interface itself is a unique catalyst with such a flexible adsorbed area, which can be expanded or shrunk easily only by stirring or shaking. The increase of the adsorbed reactant molecules results in the promotion of reaction rate and the product will be extracted into the organic phase depending on its hydrophobicity.

The accumulation of solute molecule at the interface is ready to produce assemblies or aggregates at the interface with somewhat oriented structure. Molecular network and two-dimensionally stacked compound can be produced at the interface. These aggregates exhibit molecular recognizing ability very often. Studies of these functions are very important to understand the role of biological membrane and protein–interface interaction at the membrane.

This book is intended to make clear the front of the state-of-the art of the nanochemistry of the liquid–liquid interface. The plan to make this book had started from the discussion with Mr. Kenneth Howell of Kluwer Academic Publishers just after the Symposium on "Nano-Chemistry in Liquid–Liquid Interfaces" at the Pacifichem 2001 held in Hawaii. In the year of 2001, the Scientific Research on Priority Areas "Nano-Chemistry at the Liquid–Liquid Interfaces" (2001–2003) was approved by the Ministry of Education, Culture, Sports, Science and Technology of Japan. So, it will be timely to review some important studies accomplished in the project and to learn more about the liquid–liquid interfacial science by inviting outstanding researchers through the world as authors.

The title of this book is Interfacial Nanochemistry, but almost all the chapters are devoted to the research of the liquid–liquid interface and the unique chemistry at the interface. In spite of its being the most important interface for our biological world, we have the least knowledge about it. It might be our great pleasure if our readers could find any new concepts on the physical and chemical functions of the liquid–liquid interface in this book. I sincerely wish readers to improve their knowledge on the liquid–liquid interface and to produce any new ideas for the research or application of the liquid–liquid interface.

I would like to express my sincere thanks to the authors for submitting their worthy accomplishment and to the members of the Scientific Research of Priority Areas "Nano-Chemistry at the Liquid–Liquid Interfaces" for cooperating to build the new field of Interfacial Nanochemistry. I am deeply indebted to Dr. Hideo Akaiwa, a president of Gunma University, and Professor Fumiyuki Nakashio of Sojyo University for the success in our project. I also thank Mr. Kenneth Howell for his kind encouragement to produce this book, and Ms. Keiko Kaihatsu for her efforts on editing the manuscripts. This work was in part supported by the Ministry of Education, Culture, Sports, Science and Technology of Japan.

Hitoshi Watarai
Osaka, Japan

Contents

1

Second Harmonic Generation at Liquid/Liquid Interfaces

Jeremy G. Frey
School of Chemistry, University of Southampton, Southampton, UK

1.1. INTRODUCTION

Processes occurring at the interface between two immiscible liquids underlie many important phenomena in chemistry and biology, including liquid extraction, liquid chromatography, phase transfer catalysis, membrane processes and drug delivery. Understanding the role of structure and dynamics of these interfaces on adsorption to, solvation at and transfer across the interface is of direct relevance to these physicochemical processes. The study of such interfaces by macroscopic measurements such as surface tension while yielding significant information on the interfacial properties cannot yield microscopic or molecular detail. The non-linear optical techniques of second harmonic generation (SHG) and sum frequency generation (SFG) have been useful in probing the liquid/liquid interface.

SHG is a coherent process and in principle the experimental system needed to observe the response is very simple. The fundamental radiation from a laser source incident at an interface generates the harmonic beam via non-linear polarization of the medium. Typically, this beam is observed in reflection, but many studies have been undertaken in total internal reflection and transmission geometries. As the harmonic beam is well separated from the fundamental in frequency, it can be detected; the difficulties arise due to the inherent inefficiency of the harmonic generation and the low intensities that need to be detected. The sensitivity and selectivity of SHG to the interfacial species in the presence of the same species in the bulk phase provides the driving force to overcome these experimental difficulties.

There are several reviews of interfacial SHG which cover the theory and applications of SHG in general and describe some applications to the liquid/liquid interface [1–5]. In particular, the chapter by Brevet and Girault on Second Harmonic Generation at

Liquid/Liquid Interfaces [6] is an excellent discussion of the topic. In this review I will focus on a number of different examples of the application of SHG to liquid/liquid studies mainly from my own research group.

1.2. SHG THEORY

The SHG signals arise from the second-order polarization $\mathbf{P}^{(2)}$ induced in a non-centrosymmetric medium by the electric field $\mathbf{E}(\omega)$ of the incident fundamental radiation given by the tensor equation

$$\mathbf{P}^{(2)}(\omega) = \varepsilon_0 \chi^{(2)} \mathbf{E}(\omega) \mathbf{E}(\omega) \tag{1}$$

where $\chi^{(2)}_{ijk}$ is the third rank tensor expressing the second-order surface susceptibility of the material. In a centrosymmetric medium, no second-order polarization is possible in the dipole approximation. At an interface, the inversion symmetry is broken and a dipole contribution to $\mathbf{P}^{(2)}$ is allowed. The polarization at the interface is usually treated as a sheet of thickness much smaller than the wavelength of light. This polarized sheet gives rise to the harmonic wave generated in reflection or transmission, with the propagation directions being defined by conservation of momentum.

Equation (1) does hide some of the complexities as it emphasizes the local response, but the non-local terms can be significant and, for example, higher order quadroplar terms and terms involving the electric field gradient or magnetic terms lead to contributions to the SHG signal from the bulk. These terms all involve a derivative of the electric field vector. For most materials the magnetic terms are not significant and the electric quadrupole term provides the main contribution from the bulk. If SHG studies are extended to ferroelectric fluids, the magnetic term may need to be included.

The intensity $I(2\omega)$ of the SHG signal observed from an interface between two isotropic bulk phases illuminated with fundamental radiation of intensity $I(\omega)$ is given by [7]

$$I(2\omega) = \frac{32\pi^3\omega^2}{c^2} \frac{\sqrt{\varepsilon_1(2\omega)}}{\varepsilon_1(\omega)\left(\varepsilon(2\omega) - \varepsilon_1(2\omega)\sin^2\theta_1(2\omega)\right)} \left|\mathbf{e}(2\omega)\cdot\boldsymbol{\chi}^{(2)}\cdot\mathbf{e}(\omega)\mathbf{e}(\omega)\right|^2 I^2(\omega) \tag{2}$$

where $\mathbf{e}(\omega)$ and $\mathbf{e}(2\omega)$ are the polarization vectors for the fundamental and harmonic beams and include the appropriate combination of Fresnel factors. The refractive indices and permittivities ε_i, are defined for each layer of the three-layer model (Figure 1.1) and θ_1 is the angle of reflection of the harmonic beam in the upper layer. As written, Equation (2) applies when the permittivities are real and a more general expression is given by Brevet [7]. As explained by Brevet (chapter 7), for the three-layer model it is the real parts of the permittivity that are significant for the leading terms in Equation (2), though the full complex quantities are involved in the calculation of the Fresnel factors.

For an isotropic (in-plane) interface, only four of the tensor components, χ_{ZZZ}, χ_{ZXX}, χ_{ZXZ} and χ_{XYZ} where Z is the normal to the interface, contribute to the observed harmonic signal. The electric field of the S($E^{\mathrm{S}}_{2\omega}$) and P($E^{\mathrm{P}}_{2\omega}$) polarized components of the harmonic beam as a function of the linear polarization angle (γ) of the fundamental

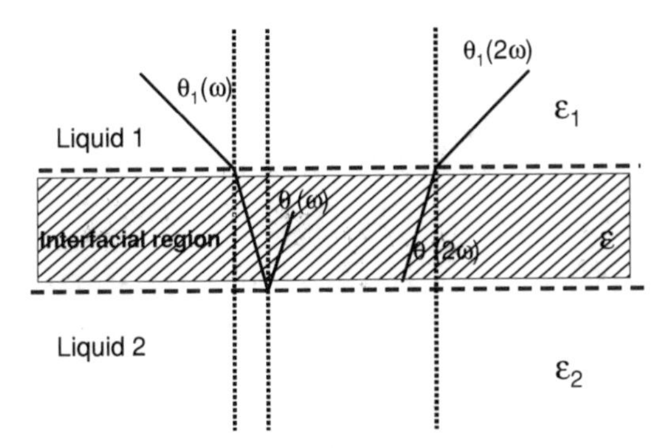

FIGURE 1.1. Layer model with the permittivity associated with each layer. The upper bulk phase is region 1 and the lower bulk phase surrounding the interface is region 2. The interface is considered as a microscopically thin region between the two bulk phases. In general, for a liquid/liquid or liquid/solid interface, dispersion may be significant and the reflection angle for the harmonic will differ slightly from the angle of incidence of the fundamental.

wave (assuming a pure linear polarization) are given by

$$E_{2\omega}^{S} = S\left(a_1 \chi_{XZX} \sin 2\gamma + a_6 \chi_{XYZ} \cos^2 \gamma\right) \tag{3}$$

$$E_{2\omega}^{P} = S\begin{pmatrix} (a_2 \chi_{XZX} + a_3 \chi_{ZXX} + a_4 \chi_{ZZZ}) \cos^2 \gamma + \\ a_5 \chi_{ZXX} \sin^2 \gamma - a_7 \chi_{XYZ} \sin 2\gamma \end{pmatrix} \tag{4}$$

The χ_{XYZ} component is only non-zero for chiral surfaces; the value of χ_{XYZ} for two enantiomers will be equal in magnitude but opposite in sign. As the tensor components can be complex quantities (especially near resonance), the harmonic wave can be elliptically polarized even for a linearly polarized fundamental, just as in conventional ellipsometry. This results in a variety of interesting effects of non-linear optical activity being observable in SHG [8–13]. The majority of observations of this type have been made on chiral films [14,15].

The a_i coefficients are combinations of Fresnel factors relating the electric fields in the interfacial region to the external field and they depend on the exact model for the interface chosen. For the second harmonic calculations, a simple three-layer model (Figure 1.1) is typically used. The non-linear region is the thin layer between bulk immersion medium layers. It is important to realize that this layer is assumed always to be vanishing thin, being on the order of the molecular dimension of the target, because thicker assemblies of molecules are usually centrosymmetric and do not generate even harmonics. The second harmonic calculations therefore assume that there is no significant light interference in this layer, although there is reflection at its upper and lower boundaries. This is in contrast to macroscopic layered structures often investigated by linear ellipsometry, where interference effects within the layer can contribute dominantly to the overall polarization changes.

In the absence of chiral effects it is often convenient to fit the polarization data, in the first instance, to the phenomenological equations (5) and (6) that describe the expected shape of the polarization behaviour [16]. For the P-polarized harmonic intensity, I_P is given by

$$I_{\mathrm{P}} = |A\cos^2\gamma + B\sin^2\gamma|^2 \tag{5}$$

and for the S-polarized harmonic intensity, I_{S} is given by

$$I_{\mathrm{S}} = |C\sin 2\gamma|^2 \tag{6}$$

where $\gamma = 0^\circ$ corresponds to P-polarized and $\gamma = 90^\circ$ to S-polarized fundamental fields. The parameters A, B and C, *which may be complex*, are linear combinations of the components of the second-order susceptibility, χ tensor, and the non-linear Fresnel coefficients, a_i, This allows the initial model to be fit without the concern of the often unknown interfacial refractive index that is required to evaluate the Fresnel coefficients, a_i. More details of this procedure and the model-dependent assumptions used in the data analysis are discussed later.

The additional terms that need to account for the main non-local effects include the effects of the field gradient at the interface. Following Brevet [7] we can write

$$\mathbf{P}^{(2)} = \chi^s\mathbf{EE} + \gamma\nabla[\mathbf{EE}] - \nabla[\chi_Q]\mathbf{EE} \tag{7}$$

Fortunately these terms can be cast into the same form as the surface dipole susceptibility giving effective surface tensor components which are given by

$$\begin{aligned}\chi_{XXZ}^{\text{interface}} &= \chi_{XXZ}^{\text{s}} + \chi_{XXZ}^{\text{eff}}\\ &= \chi_{XXZ}^{\text{s}} + \chi_{ZXXZ}^{\text{Q1}}\frac{\varepsilon(\omega)}{\varepsilon_1(\omega)} - \chi_{ZXXZ}^{\text{Q2}}\frac{\varepsilon(\omega)}{\varepsilon_2(\omega)}\end{aligned} \tag{8}$$

$$\begin{aligned}\chi_{XZX}^{\text{interface}} &= \chi_{XZX}^{\text{s}} + \chi_{XZX}^{\text{eff}}\\ &= \chi_{XZX}^{\text{s}} + \chi_{ZXZX}^{\text{Q1}}\frac{\varepsilon(\omega)}{\varepsilon_1(\omega)} - \chi_{ZXZX}^{\text{Q2}}\frac{\varepsilon(\omega)}{\varepsilon_2(\omega)}\end{aligned} \tag{9}$$

$$\begin{aligned}\chi_{ZXX}^{\text{interface}} &= \chi_{ZXX}^{\text{s}} + \chi_{ZXX}^{\text{eff}}\\ &= \chi_{ZXX}^{\text{s}} + \left(\gamma_1 + \chi_{ZZXX}^{\text{Q1}}\right)\frac{\varepsilon(\omega)}{\varepsilon_1(\omega)} - \left(\gamma_2 + \chi_{ZZXX}^{\text{Q2}}\right)\frac{\varepsilon(\omega)}{\varepsilon_2(\omega)}\end{aligned} \tag{10}$$

where the superscript s specifically indicates the dipole surface terms and eff the effective surface contribution of the bulk quadruple and gradient terms.

As pointed out by Brevet and Girault [17], this analysis shows that the discussion of SHG data from liquid/liquid interfaces must be cognisant of the possible contributions from the bulk and from field gradients at the interface. However, in the liquid/liquid case, the changes in optical constants from one bulk phase to the other will normally be less marked than that observed in air/liquid experiments. Of course in the limit of the same optical constants for the two phases, there would be no gradient effects but there would also be no reflection.

In order to extract information on the molecular orientation distribution, the relationship between the macroscopic surface susceptibilities and the molecular hyperpolarizabilities, β, needs to be considered. It is usual to consider the intrinsic non-linear response of each of the molecules as independent of the other molecules so that the interfacial response is an average over the orientational distribution and scales with the molecular density (squared). Even the modification of this response due to local field effects is usually considered in terms of mean field model and so does not alter the nature of this averaging (three additional diagonal terms need to be included).

$$\chi = \mathbf{T}\{N\langle\beta\rangle\} \tag{11}$$

where N is the surface number density and < > represents the average over the orientational distribution. The transformation between the molecular axis system *(ijk)* and the interfacial coordinate system (*IJK*) involves the various direction cosines, and the tensors involved are both third rank three rotation matrix terms, **R**, are present [18–20].

$$\chi_{IJK} = N\left\langle \sum\sum\sum R_{Ii}R_{Jj}R_{Kk}\beta_{ijk} \right\rangle \tag{12}$$

In many cases it is possible to simplify these equations because only relatively few components of β are significant, either because of the symmetry or the electronic structure of the molecules. When it is possible to reduce the number of distinct significant components to at most two, then it is often possible to extract the ratio of these components and geometric information directly from the observed values of the susceptibility. In these cases the values assumed for the interfacial refractive index and the role of the contributions from the bulk can make a dramatic difference to the derived geometric parameters [21–27].

While some qualitative inferences about the nature of the interface can be derived directly from the SHG observations, extracting detailed quantitative information from the SHG intensity and polarization data requires the construction of a model of the interface and frequently assumptions about some of the parameters for this model. Parameters such as the interfacial refractive index and roughness need to be determined separately, calculated or more frequently obtained by reasonable assumptions [20,21,23–27]. Some idea of the relationship between the model, assumptions and results is given in Figure 1.2.

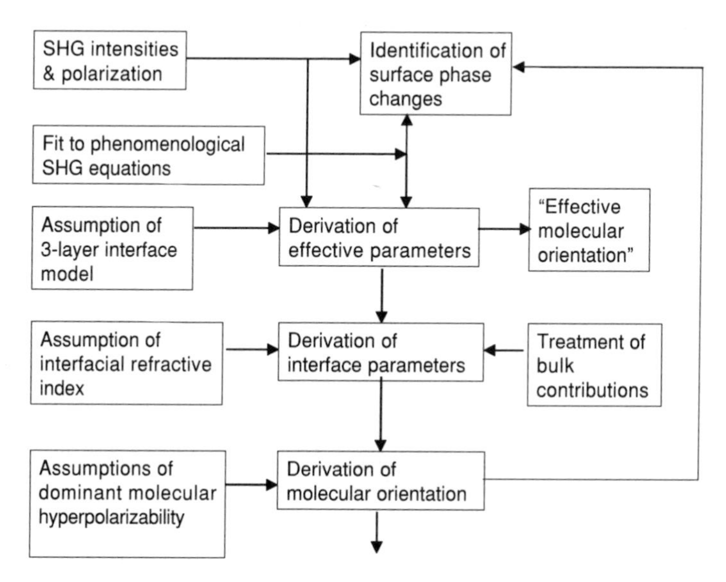

FIGURE 1.2. A guide to the modelling and assumptions needed to interpret the SHG intensity and polarization data.

The SHG results are at their most useful when they can be combined with other measurements and used to validate molecular dynamics simulations of the interface.

1.3. EXPERIMENTAL TECHNIQUES

The original surface SHG experiments were frequently performed with low repetition rate ns Nd:YAG lasers at 1064 or 532 nm (Figure 1.3). The low-intensity SHG signals were typically detected with a PMT and Box Car combination and averaged over many laser pulses. Subsequently, SHG spectra were obtained by using a dye laser. Higher repetition rates reduced signal-averaging times. The advantage of the higher intensities available with comparable overall energy (thus no more thermally disruptive of the interface) of ps Ti:Sapphire lasers was soon recognized. These also provide a degree of wavelength scanning. The use of ultra-short (fs) pulses to give a broad wavelength band (or indeed the use of continuum generation to provide an even broader coverage [28]) has enabled SHG spectra on some solid surface to be collected with a CCD detector so that single shot coverage of the whole spectral region is obtained. It is likely that similar techniques will soon be applied to air/liquid and liquid/liquid interfaces.

The interfacial SHG effect is inherently a weak process and there is not the opportunity to build up a phase-matched signal as used, for example, in a doubling crystal. It is therefore frequently necessary or at least useful to use electronically resonant systems to enhance the interfacial SHG signal. This introduces the added complication that $\chi_{ijk}^{(2)}$ and indeed the interfacial refractive index are complex quantities. Hence, in order to elucidate the real and imaginary components of $\chi_{ijk}^{(2)}$, knowledge of the polarization and phase of the surface SHG response is required. A sufficiently extensive set of SHG intensity measurements for a range of different input/output beam polarization combinations can provide the information needed. However, a more elegant approach is to introduce a rotating quarter-wave plate into the experiment to continuously modulate the polarization state of the fundamental beam incident at the interface.

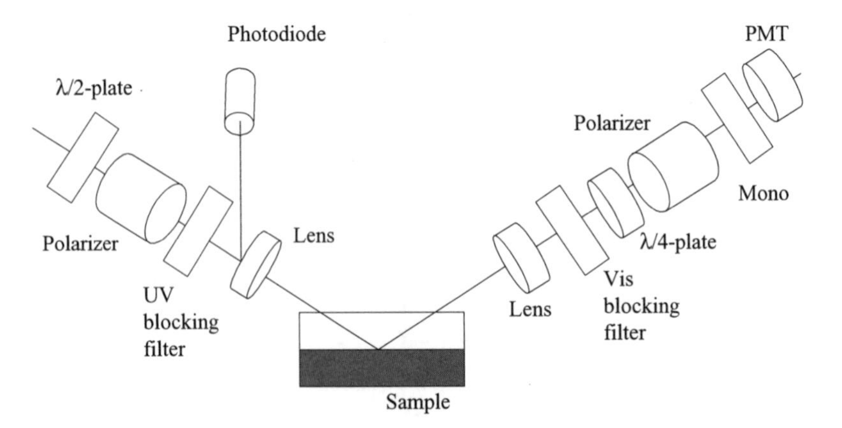

FIGURE 1.3. Diagram of the second harmonic ellipsometry apparatus used to investigate air/liquid and liquid/liquid interfaces.

The alternative is to employ a rotating optical compensator (quarter-wave plate) in the analysis of the SHG radiation generated at the surface to allow the full polarization characteristics of the reflected light to be determined by a Fourier analysis. Such a configuration we use here has been analyzed by Hauge [29,30] applied to the linear–optical ellipsometry case under the title of Generalized Rotating-Compensator Ellipsometry. A rotating compensator has the advantage over the more often employed rotating analyser system in that it enables the unambiguous determination of the polarization state of the light to be determined. An important practical point is that optical compensators usually have low beam deviation. Polarizers, by contrast, very often display beam deviation, and this is particularly true of high-power polarizers, which must be used in these experiments. This leads to the development of the Second Harmonic Ellipsometry technique [31], which will yield more efficient experiments in the future.

From the air/liquid interface, the SHG signal is typically observed in reflection, where the coherent harmonic beam propagates along the same direction as the reflected fundamental beam. The possibility of significant refractive index dispersion in the liquid means that for SHG experiments on the liquid/liquid interface, the harmonic beam path may deviate from the reflected fundamental.

1.4. THE BARE HYDROCARBON/WATER INTERFACE

SHG and SFG have been used to study the neat liquid/liquid interface despite the weak signals observed. At the alkane/water interface a more ordered environment is observed for even chain-length hydrocarbons compared to the odd chain lengths. The ordering is weak and decreases with increasing chain length [32–34]. The dodecane/water interface has been studied along with other hydrocarbon/water interfaces by molecular simulations (hexane/water [35], octane/water [36], nonane/water [37,38], decane/water [39]). From these and low-angle X-ray diffraction experiments, a general picture arises with an interface that is not molecularly sharp [40]. In a typical experimental apparatus the actual interfacial thickness will likely be dominated by the capillary wave motion and this will depend on the interfacial tension [21]. The interfacial tension is usually much less than, for example, the surface tension of water, so some differences can be expected between observations made at the hydrocarbon/water interface compared to the air/water interface. The simulations indicate that there is only a slight degree of interfacial ordering of the hydrocarbon chains, confirming the inferences made from the deviations from Kleinman symmetry observed in SHG experiments. The hydrocarbon chains have a preference for aligning parallel to the interface and there is a weak ordering of the water molecules at the interface in layers, with dipole pointing slightly towards the hydrocarbon and then a higher concentration of water molecules pointing on average away from the interface; it is similar at the water/*n*-alcohol interface [41].

1.5. ADSORPTION OF *PARA*-NITROPHENOL

A number of molecules have become favourites for study at liquid interfaces and there have been a number of investigations of the adsorption behaviour of *p*-nitrophenol

(PNP). PNP molecules exhibit a significant second-order non-linearity and have a UV absorption maximum that is moderately sensitive to the solvent environment though not as dramatic as ET(30) [42,43]. The approximate C_{2V} symmetry and the planar π glectron system suggest that the only significant components of β will be β_{ZZZ}, β_{ZXX} and β_{XZX}. While this has been conformed by several semi-empirical [44] and ab initio quantum calculations and certainly applies far from resonance, recently some doubt has been cast on these assumptions by DFT calculations [45]. As many of the experiments conducted with PNP exploit the UV resonance to obtain larger signals, care must be exercised with the results of the ab initio calulations as the current codes only evaluate the real part of β.

1.5.1. PNP Adsorption at the Hydrocarbon/Water Interface

The adsorption of PNP to a wide variety of interfaces has been studied by SHG. The adsorption isotherm at the dodecane/water interface is shown in Figure 1.4. The adsorption isotherms typically used to fit the surface tension and SHG data at the liquid/liquid interfaces are in terms of the interfacial coverage θ and the bulk phase concentration c, the Langmuir (13) and the Frumkin isotherms (14)

$$\frac{\theta}{1-\theta} = Kc \tag{13}$$

$$\frac{\theta}{1-\theta} e^{-2b\theta} = Kc \tag{14}$$

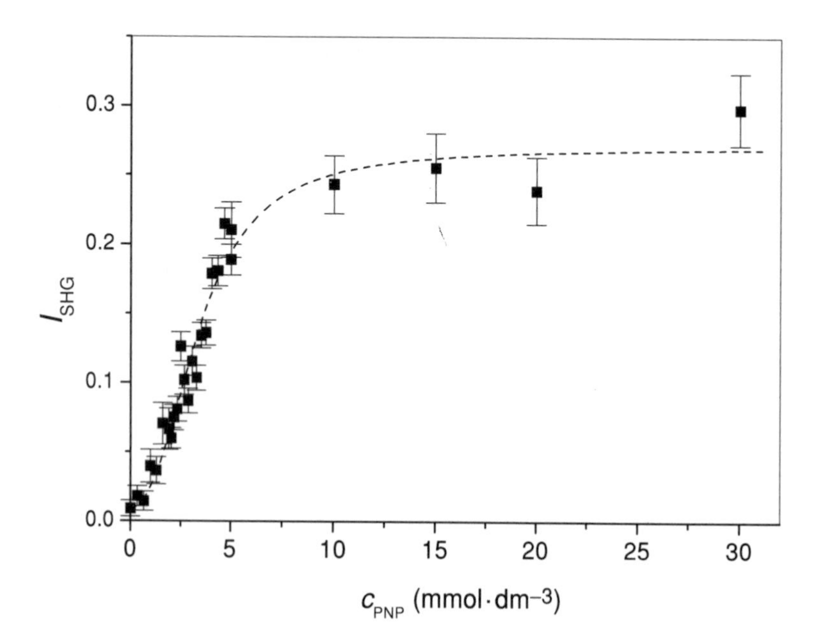

FIGURE 1.4. SHG signal from PNP at the dodecane/water interface. The fit to a Frumkin isotherm is shown as the dotted line. Adapted from ref. [48]; reproduced by permission of the PCCP Owner Societies.

TABLE 1.1. Comparison of the free energy of adsorption for PNP at the air/water and hydrocarbon/water interfaces.

Interface	$-\Delta_{\text{ads}}G^{\circ}$ (kJ · mol^{-1})	Method
Air/water	16.4 ± 0.2	Surface tension, Frumkin [46]
Hexane/water	17.2 ± 0.3	Surface tension, Frumkin [46]
Heptane/water	25.7	Langmuir [47]
Deodecane/water	25	Langmuir
Deodecane/water	22.0 ± 0.5	Frumkin [48]
Deodecane/water	20.0 ± 0.5	Frumkin and order terms [48]

where K is the adsorption equilibrium constant and b an interaction parameter that describes the interactions between adsorbate molecules. The isotherms are shown here in terms of the concentration of the adsorbate c in one of the liquid phases. If the isotherm is defined in terms of the mole fraction of the adsorbate in this phase then the equilibrium constants and associated free energies are related by a scaling factor (the molarity of the solvent; see Appendix).

Considerable care needs to be taken in extracting the interfacial concentration from the SHG intensities because of the interaction between surface density and surface order on the SHG process [49]. Table 1 shows a comparison of the values of $\Delta_{\text{ads}}G^{\circ}$ for PNP at the air/water and hydrocarbon/water interfaces determined by SHG methods. The different results obtained at the dodecane/water interface where different isotherms were used to fit the SHG data suggest that the determination of $-\Delta_{\text{ads}}G^{\circ}$ at the heptane/water interface using only a Langmuir isotherm gives a value that is too high and thus this value should be re-examined.

1.5.2. PNP Adsorption in the Presence of Tributyl Phosphate

The addition of tributyl phosphate (TBP) to the dodecane acts to reduce the SHG signal from an initial (partial) monolayer of PNP at the dodecane/water interface; the dependence shown in Figure 1.5 can be fitted to a Langmuir-like equation (15) from which an effective free energy of adsorption for TBP, $\Delta_{\text{ads}}G^{\circ}_{\text{TBP}}$, can be extracted [48].

$$I_{\text{SHG}} = I_0 \left(1 - \frac{kc}{1+kc}\right)^2 \tag{15}$$

The value of $\Delta_{\text{ads}}G^{\circ}_{\text{TBP}}$ determined in this manner is linearlydependent on the concentration of PNP. Extrapolating to zero PNP concentration gives $\Delta_{\text{ads}}G^{\circ}_{\text{TBP}} = -31.4 \pm 0.8\, kJ \cdot mol^{-1}$, which is consistent with the value determined by surface pressure and surface tension measurements at low TBP concentrations [50,51] (when corrected for the different interface standard states, see Appendix).

The possibility of absorption of the SHG signal by the upper medium complicates the interpretation of SHG from liquid/liquid studies compared to similar studies of the air/liquid interface. The same problem is of course faced by studies of the liquid/solid interface or total internal reflection studies at the air/liquid interface. In the case of the experiments on the dodecane/water interface, the possibility existed that the absorption

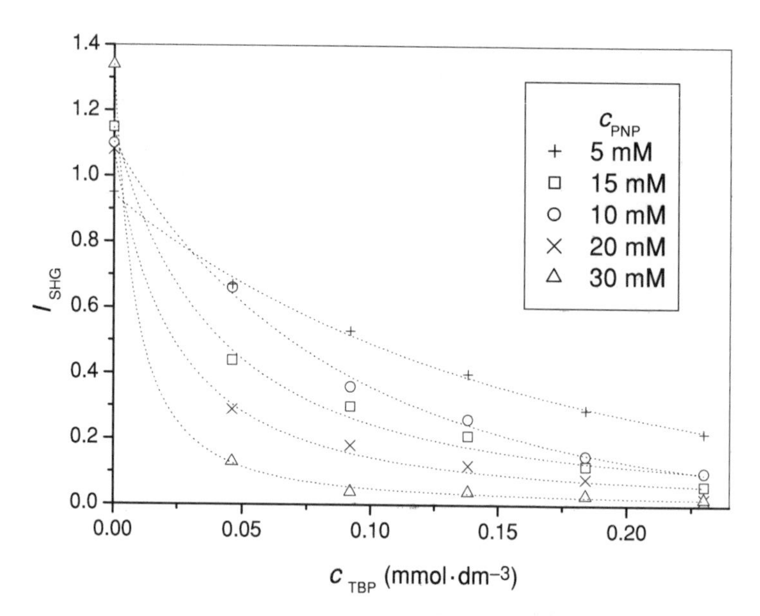

FIGURE 1.5. The SHG signal from PNP at the dodecane/water interface is reduced on addition of TBP to the dodecane. Adapted from ref. [48]; reprinted by permission of the PCCP Owner Societies.

of the harmonic wavelength (282 nm) by the TBP ($\lambda_{max} = 220$ nm) could account for the decrease in the observed SHG intensity as a function of increasing concentration of TBP. However, this decrease was observed for addition of TBP to an interface loaded with a high concentration of PNP, but an increase in SHG signal was observed when TBP was added at low PNP concentrations. It was thus possible to conclude that absorption by TBP was not the cause of the reduction in SHG intensity. Nonetheless, it is likely that some SHG intensity was lost by absorption by the dodecane solvent and the experiment's sensitivity could be improved by using a thinner layer of overlying solvent.

1.6. FLOW CELL EXPERIMENTS

The static experiments show that there is a complex between TBP and PNP at the dodecane/water interface. Even by simply mixing the two solutions it was clear that this interaction was time-dependent. However, the decay rate was sufficiently fast so that given the need to ensure uniform mixing in the bulk phases and the time required to accumulate a reasonable S/N, only the long time tail of the decay curve could be measured. No accurate estimate of the decay rate could be made in the Petri-dish. The solution was to construct a flow cell to measure the kinetics of the TBP and PNP interaction at the dodecane/water interface [48].

The design of the flow cell (Figure 1.6) ensured that a stable dodecane/water interface would form and then allow the two liquids to be separated. The flow rate was variable and the interface could be probed at various distances along the flow. The cell contained two horizontal flat glass plates located on the central plane of the cell that

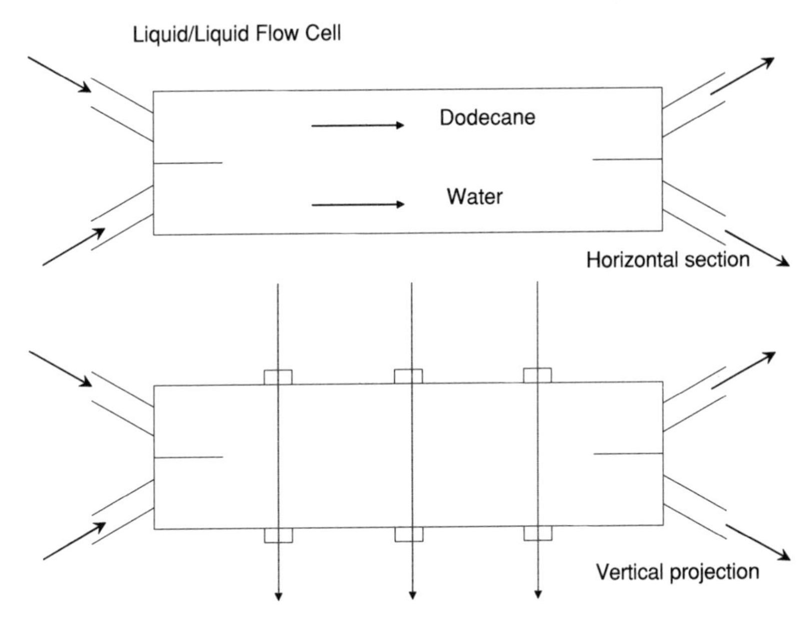

FIGURE 1.6. Liquid/liquid flow cell.

stabilize the incoming fluid flows and allow a stable interface to form. Initially the flow was driven by two peristaltic pumps each pumping one fluid from a main reservoir to the flow cell, from which a drain returned the fluid back to the main reservoir. The flow rates of the two fluids could be controlled independently to form an interface that was stable over a period of up to an hour. However, the peristaltic pumps produced ripples on the interface that were avoided by using gravity feed systems with the pumps used to refill small reservoirs; overflows in each reservoir ensured a constant pressure head.

CaF_2 windows along the cell were used as silica windows that developed significant SHG signals with exposure to the high laser beam energies. However, the tight focusing used for this experiment led to damage to the windows after prolonged use and were replaced as required. The thickness of the dodecane layer meant that higher laser intensities were required to achieve a good S/N ratio, than were needed for the static experiments.

The flow cell "translates" time into distance and the combination of the three and varying the flow rates gave a range of observations from 0 to 30 s. SHG measurements of the static aqueous/dodecane interface were made at each port before and after the flow experiment to calibrate the observations from each port. For a laminar (non-turbulent) flow, the two flow rates should be in the inverse ratio of the fluid viscosities; this ratio for dodecane on water is 0.65 at 25°C, very close to the observed flow rate ratio of 0.67. The bulk flow rates for each liquid were measured by collecting the volume of liquid flowing in a known time. Since the cell operates under non-turbulent conditions, the velocity of each layer at the interface must be the same, but the average velocities of the two layers are different. Ideally a model of the flow conditions inside the cell would be used to accurately determine the velocity of the interface. Since this was not

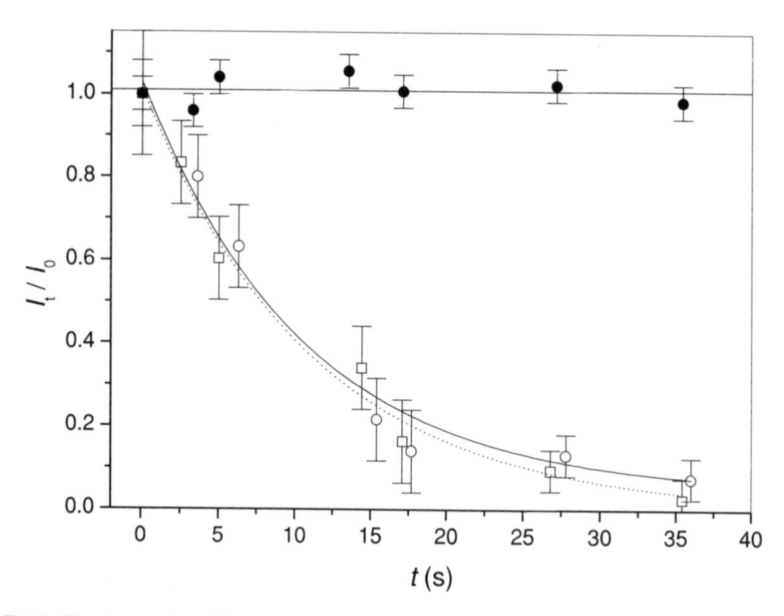

FIGURE 1.7. The decay of the SHG signal from PNP at the dodecane/water interface as a function of time from the formation of the interface. The full circles are the signals in the absence of TBP and the open circles and squares are two experiments in the presence of TBP. Adapted from ref. [48]; reproduced by permission of the PCCP Owner Societies.

available and since the flow rates used are very low, it is reasonable to make a first approximation that the velocity of the interface will be the average velocities of the two layers. The rapid equilibration of PNP at the newly formed dodecane/water interface is demonstrated in Figure 1.7. The PNP SHG signal, corrected for the water/dodecane background and the variation due to the windows, is constant over the time periods used for the later experiments. This is consistent with the rapid equilibration of the PNP signal at the air/water interface measured using a liquid jet. The SHG signals for the PNP/TBP system, similarly corrected and normalized to the signal at time zero are shown in Figure 1.7. As expected from the static experiments, for the concentrations of PNP and TBP used, the SHG signal decays monotonically to the background level. Assuming the intensity is proportional to the square of the surface coverage, the rate constant for reordering at the surface [48] is $k = 0.5 \pm 0.05\ s^{-1}$.

For comparison, the measured adsorption rate for PNP to the air/water interface gave a rate constant of $k = 4.4 \pm 0.2 \times 10^4\ s^{-1}$ but a desorption rate of $6 \pm 2\ s^{-1}$[52]. The rapid adsorption to the air/water interface is quite consistent with the very rapid establishment of the SHG signal from PNP at the dodecane/water interface observed in these flow cell experiments. However, the observed decay rate constant in the presence of TBP of ca. $0.5\ s^{-1}$ is much faster than the desorption rate constant that would be implied from the air/water experiments. This further implicates a reorganization process involving bonding between TBP and PNP as the cause of the loss of SHG intensity, which results in an overall loss of orientational order.

The addition of TBP in the dodecane increases the free energy of adsorption of PNP in a manner consistent with the formation of an interfacial complex between TBP and

PNP. We suggest that with the TBP present on the dodecane side of the interface, the preferential orientation of the TBP PNP complex will be opposite to that for uncomplexed PNP. For a mixture of complexed and uncomplexed PNP this will result in an overall reduction in the overall net orientation, that is, a reduction in the orientational order and a consequent reduction in the SHG intensity. This situation provides an explanation for a reduction in SHG intensity even when there is an increasing surface concentration of the PNP.

1.7. DYE MOLECULES AT THE DODECANE/WATER INTERFACE

Dye molecules have been popular adsorbates for optical studies at interfaces because of their accessible electronic resonances and the consequent large absorption and, most usefully, large fluorescence cross sections that enable even single molecule studies. Linear optical techniques have limitations when the molecules are present in the adjacent bulk phases as it is difficult to distinguish the interfacial species from the more prevalent bulk. SHG has proved useful in studying monolayers of dye molecules at the air/solid interface [53–60] and Langmuir–Blodgett films [61–66], situations in which the molecules are only present at the interfacial layer. The complications in interpretation of the non-linear technique compared to linear ellipsometric experiments become most worthwhile when the dye molecules at interface are in equilibrium with the bulk solvent. There have been a number of such studies at the air/water [67,68] and air/hydrocarbon interfaces, but relatively fewer at the corresponding liquid/liquid interface [69]. Some of the most popular dyes used in SHG studies are the rhodamine series, malachite green [70,71], oxazines [72] and eosin B [73,74]. The interactions between the dye molecules, their tendency to form dimers, clusters and large aggregates in a manner that depends on the environment also make these species ideal for probing the chemical physics of the interfacial region. Many studies have suggested that dimer species predominate at the air/solvent interface even at relatively low solution concentrations.

The adsorption isotherm of rhodamine 6G at the dodecane/water interface was recorded by SHG [75] at 564 nm and using the phase difference between the bare interface and the dye harmonic fields derived from the very low concentration limit of the isotherm, the second-order susceptibility from the dye can be derived from the SHG intensity, and the resulting data set is shown in Figure 1.8 together with a fit to a Frumkin isotherm at low concentrations and a separate Langmuir phase at higher concentrations. The phase transition is seen more clearly in the reciprocal plot where in addition the polarization changes (A and B from Equation 5) are also seen to occur at the same bulk dye concentration (ca. 6 μM). The coefficients obtained by least-squares fitting to both the Langmuir and Frumkin isotherms are summarized in Table 2.

The formation of dimers in solution at higher concentrations indicates an interaction between the dye molecules and so the b term in the Frumkin isotherm is not a surprise. The adsorption behaviour suggests that the dye is highly surface-active and formation of a monolayer is almost complete by the transition at ca. 6 μM. The standard free energy of adsorption derived from the isotherm is consistent with a monolayer coverage forming even at the micromolar aqueous concentrations. Once the first layer is substantially complete, subsequent adsorption takes place on this layer to form a second-ordered layer,

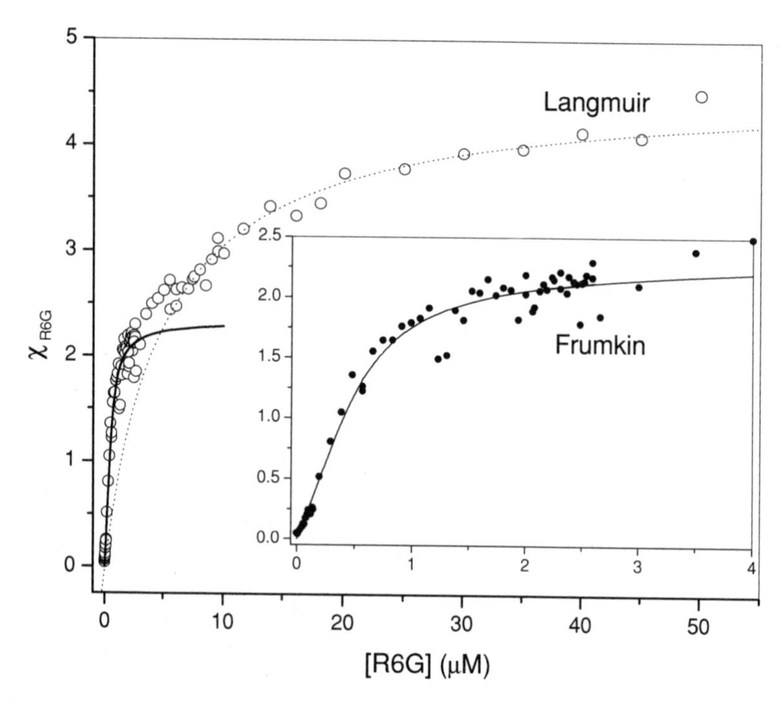

FIGURE 1.8. The "Frumkin" isotherm (full line) is a high quality fit to the low concentration region with a transition (see Figure 1.5) to a "Langmuir" isotherm (dotted line) for the higher concentration region.

though the polarization data suggests that the nature of the order in the "second" layer is different. The free energy of adsorption is lower for the second layer, but still substantial. The formation of second- and multiple-ordered dye layers has been previously observed at the air/solid interface [56] and air/liquid interface.

Extrapolating each isotherm to infinite concentration, gives the susceptibility of the full monolayer (which is the value of S on an arbitrary scale). The ratio $S_{\text{Lang}}/S_{\text{Frum}} \cong 2$ is very suggestive of a packing of the dye molecules that is twice as dense in the phase in equilibrium with the bulk above a concentration of 6 μM. This suggests that in keeping with previous observations of dimer formation at the interface, an alternative way to view such a pair of ordered layers may be as adsorbed dimers.

TABLE 1.2. The coefficients obtained by fitting the Frumkin below the 6-μM transition and Langmuir isotherms above the transition.

	Frumkin	Langmuir
S	2.34	4.57
K	0.993×10^6	0.2×10^6
B	−1.48	
ΔG° (kJ · mol^{-1})	−33.3	−29.4

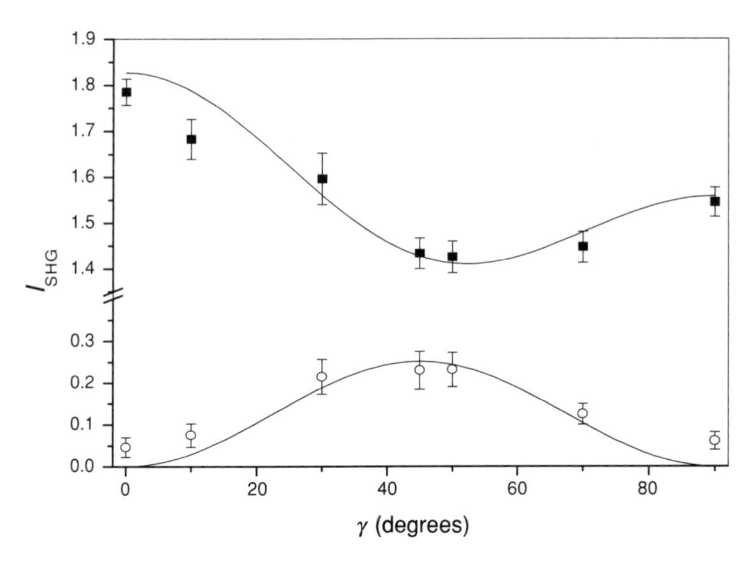

FIGURE 1.9. The dependence of the S- (circles) and P- (squares) polarized second harmonic as a function of the polarization of the fundamental for a dye concentration of 10 μM in the aqueous phase.

The polarization curves for concentrations higher than 6 μM can be fit by Equations (5) and (6) and Figure 1.9 shows the S- and P-polarized harmonic data for the 10-μM solution. Similar results have been reported for SHG experiments on the rhodamine dyes at various interfaces. The best fit is obtained by introducing a phase difference, η, between the parameters A and B consistent with the complex nature of the susceptibilities on resonance. A plot of the concentration dependence of the A/B, given in Figure 1.10, shows that a relatively sharp transition takes place in the structure of the interface at an aqueous dye concentration of ca. 6 μM.

The clear transition in the polarization behaviour that occurs at about 6 μM shows that the structure of the interface has changed. It is likely that this corresponds to reaching a critical packing density at which a change of phase takes place and the adsorbed layer corresponds to a collection of dye dimers at the interface. The formation of multiple layers has been observed with spin-coated films although multilayers of Rhodamine B have been deposited from solution without a change being observed in the layer structure. Similarly the predominance of dimers at interfaces has been inferred previously but in the current situation we are able to observe the transition between monomers and dimers at the interface.

1.8. ELECTROCHEMICAL LIQUID/LIQUID INTERFACES

While there have been many SHG studies at the solid electrode/liquid interface as both an in situ probe of the electrode interface and the influence of adsorption at the surface of the electrode, there have been far fewer studies of electrochemical processes occurring at the boundary between two immiscible electrolyte solutions.

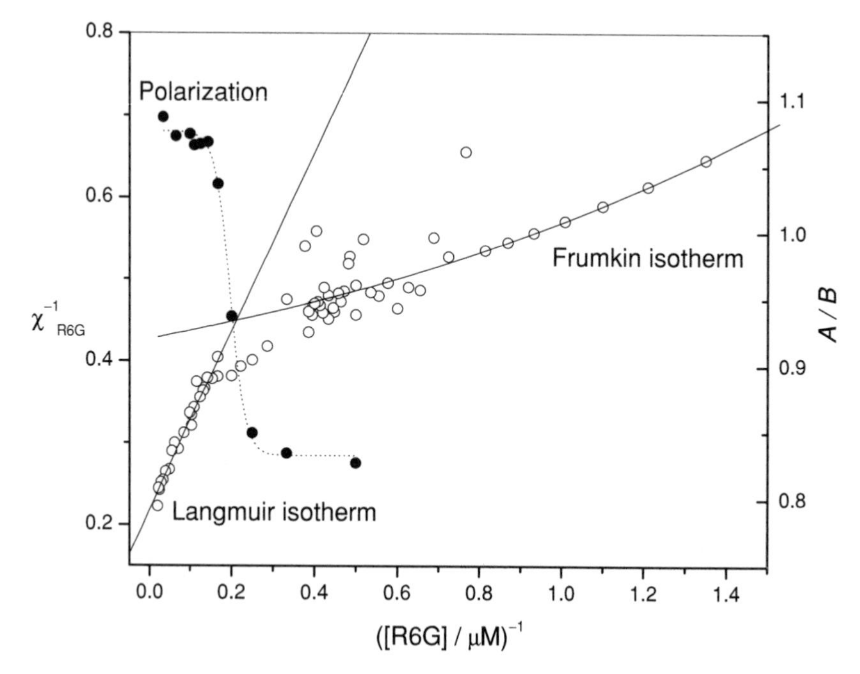

FIGURE 1.10. Plot of the reciprocal of the dye susceptibility vs. the reciprocal of the dye concentration in the aqueous phase. A relatively sharp transition is seen at around 6 μM and a similar transition is seen in the ratio A/B (see Equation 5) derived from the polarization curves.

Higgins and Corn [76,77] have studied the response of an adsorbed layer of 2-(*N*-octadecyl)aminonapthalene-6-sulfonate at the water/1,2-dichloroethane interface as a function of applied potential and showed how the ordering at the interface is influenced by the applied potential.

We have studied the interactions of the crown ether 4-nitrobenzo-15-crown-5 at the water/dichloroethane interface [78] (Figure 1.11). The variation of the SHG signal from this crown ether as a function of potential across the interface depends dramatically on the presence or absence of sodium ions. The neutral crown ether behaves quite differently from the charges cation-crown ether complex. At the hepatane/water interface, cation binding has been studied using dye-labelled crowns [79]. We are currently investigating the behaviour of other crown ethers at the air/water and solvent/water interfaces.

"Solid" models for these liquid/liquid interfaces in which one of the solutions is replaced by a polymer, often swollen with a significant proportion of solvent, have proved useful in SHG studies of analytically relevant studies of liquid/liquid junctions, for example, the study of crown ether ionophores imbedded in a PVC film.

1.9. CHIRAL MOLECULES AT LIQUID/LIQUID INTERFACES

The SHG provides a very useful method to study chiral molecules. The lack of inversion symmetry at the interface means that electric dipole terms can contribute to effects that have similar (but not identical) consequences to conventional optical activity.

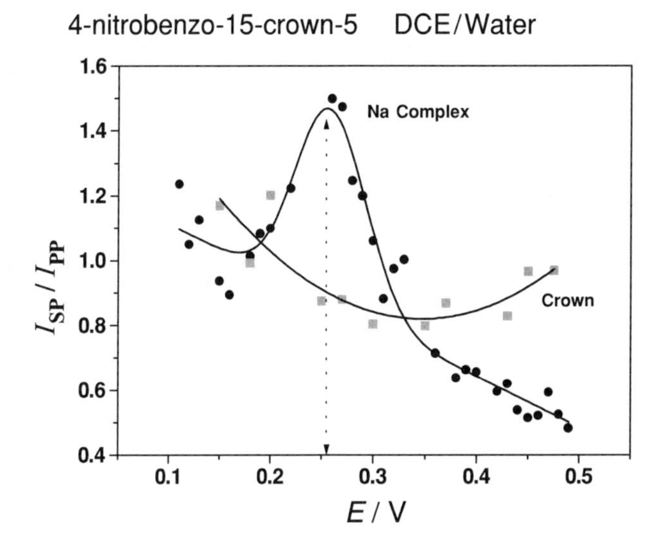

FIGURE 1.11. The variation of the crown ether SHG signal as a function of the applied potential across the dichloroethane/water interface.

The presence of the χ_{XYZ} term gives rise to a circular differential scattering (SHG-CD) and to an optical rotation (SHG-ORD) of the harmonic beam. For example, in the absence of this term, a p-polarized fundamental gives rise to a p-polarized harmonic, but in its presence the harmonic field is rotated; the rotation angle is wavelength-dependent and is opposite in sign for the two enantiomers. A number of systems have been studied at the solid and air/water interfaces and the experiments are now being extended to the liquid/liquid interface. Figure 1.12 shows a comparison of the SHG-ORD for the dipeptide Boc-Trp-Trp at the air/water and heptane/water interfaces.

1.10. SHG FROM MICELLES AND LIPOSOMES

SHG has more recently been shown to be a viable technique for the observation of even symmetrical (e.g. spherical) microparticle surfaces. Locally, the regions of the microparticle surface are non-centrosymmetric (inside vs. outside) and thus can generate the harmonic field. For large particles the field generated from sections of the interface on opposite sides of the particle would cancel as these interfaces point in opposite directions, that is, overall there is a centre of inversion. However, if the particle is comparable in size to the wavelength of the (incident) light, then the fields generated from opposite sides of the particle can add constructively. Taking an extreme view, for a particle of size λ, the phase of the fundamental field will be 180° different for the two interfaces of the microparticle, and thus the phase inversion of the radiation exactly counters the inversion of the molecules because of the opposite orientation of the interface [80–82].

Using this idea it has been possible to study the interfaces of polystyrene beads in aqueous suspensions [83], semiconductor and clay particles [84]; more relevant to

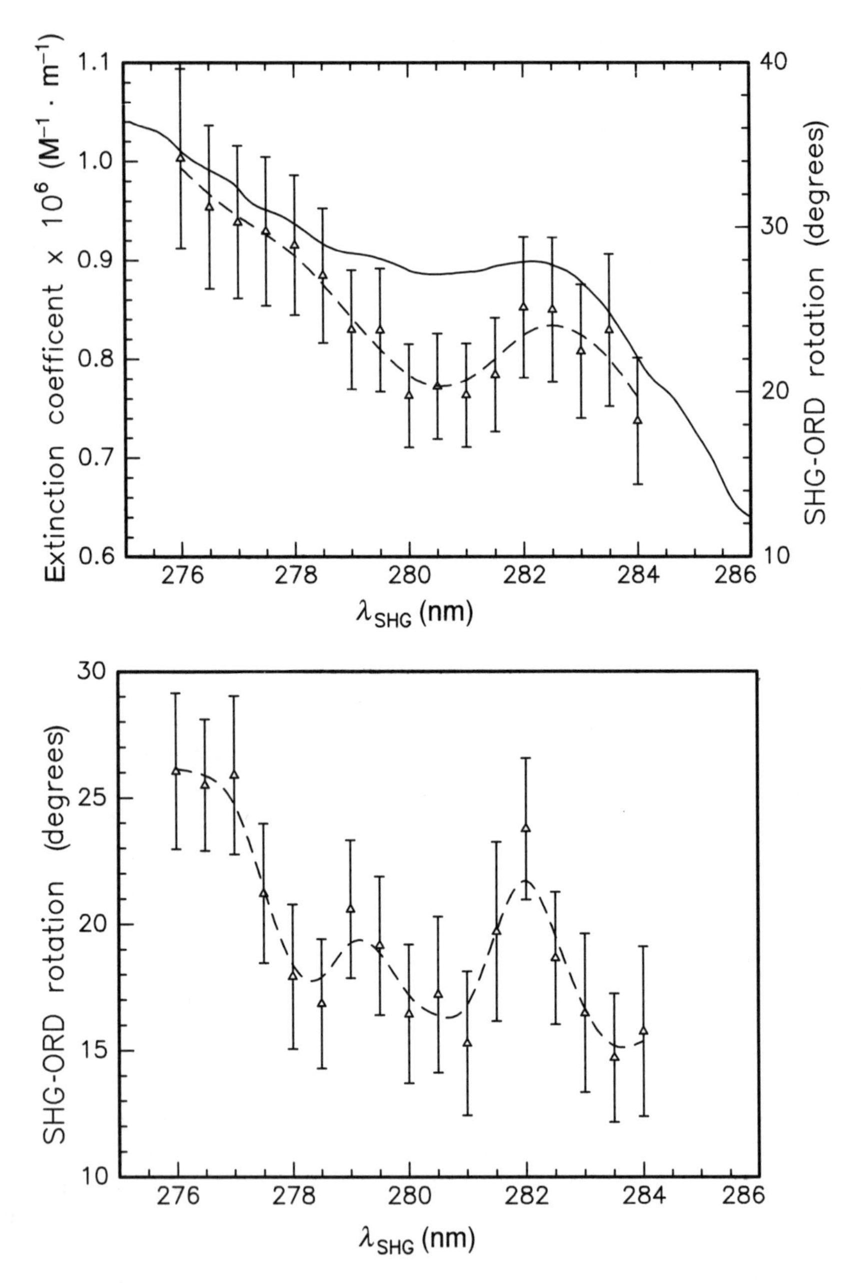

FIGURE 1.12. The SHG optical rotation from boc-Trp-Trp at the air/water interface (upper) and at the heptane/water interface (lower). The solid line is the UV absorption in water. Error bars are probably overestimated in this work.

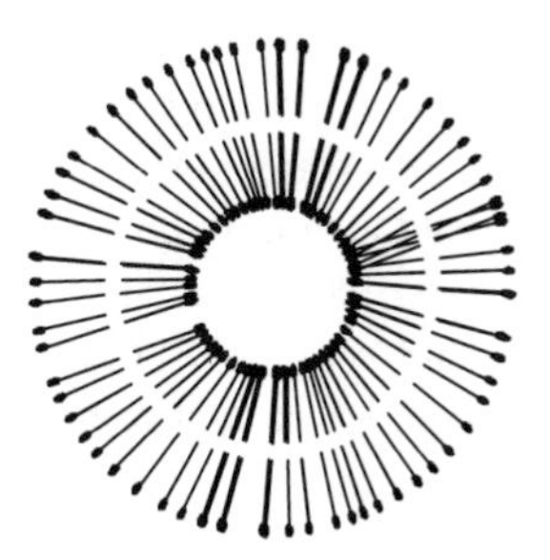

FIGURE 1.13. Diagram of a liposome showing external and internal aqueous regions and the symmetrical membrane.

this chapter are the experiments on oil/water emulsions [85] and the studies of transport across liposomes.

The case of the liposome is interesting because there are now four interfaces involved, the outside water/oil and inside oil/water interfaces on each side of the liposome (Figure 1.13). As the interface is narrow, typically of the order of a few nanometres, which is much less than the wavelength of the light, the opposing contributions to the SHG from the molecules on each side of the interface cancel (if identical). If molecules are added to the bulk solution, no SHG will be seen from the bulk but adsorption to the liposome interface will generate SHG as described above. As the resulting SHG signal depends on the difference between the outside and inside, then transport of the molecules through the liposome will result in a decrease in the observed SHG signal allowing the rate of transport to be determined [86,87].

$$I \propto |E_{2\omega}|^2 \propto |N_{\text{out}}(t) - N_{\text{in}}(t)|^2 \tag{16}$$

Using this technique Eisenthal's group has been able to monitor the influence of cholesterol on the transport of malachite green transport across the membrane of DOPG liposome. While the SHG technique applied to these microparticles and liposome does require that the molecules under study (either adsorbates or constituents of the membrane) give an SHG signal, it does not need special labelling to distinguish those adsorbed at the interface from those in the bulk. Nevertheless, obtaining structural information akin to that obtained at flat interfaces from the polarization dependence of the SHG is complicated by the geometry and in the case of a liposome by the external and internal interfaces, which due to stress effects need not be identical even if the membrane is a symmetrical bilayer.

1.11. CONCLUDING REMARKS

Second harmonic generation is a useful tool in probing the molecular behaviour at the liquid/liquid interface. While its limitations must be taken into account, particularly over the contributions of the observed signals from the bulk phases and the interrelated contributions of molecular density and orientation distribution, the ability to differentiate molecules at the interface from the bulk, is extremely useful. While the related technique of sum-frequency generation provides more specific molecular information when applied to vibrational resonances, the simplicity and flexibility of SHG, a single laser experiment, makes it suitable for a wider range of applications at present. The application of SHG to chiral molecules at the liquid/liquid interface shows particular promise for probing biological membrane systems. The studies on micellular systems shows that even the restrictions to non-centrosymmetric interfaces have a broad interpretation further extending the potential to biological applications [88].

ACKNOWLEDGEMENTS

Much of the work reported here was undertaken in my own research group and I would like to thank all the past and present members of my group who have worked on SHG projects, especially A.K. Alexander, A.J. Bell, M.J Crawford, S.G. Croucher, L. Danos, A.J.G. Fordyce, Y. Grudzkov, S. Haslam, C.G. Hickman, E. Rousay, A.J. Timson, and my colleagues, particularly T.J. VanderNoot and M.C. Grossel, for all their help in this research. Our work would not have been possible without all the technical support from our department in constructing and maintaining the equipment and the financial support of the UK research councils (Engineering and Physical Sciences Research Council & the Natural Environment Research Council) as well as contributions from the Royal Society and the University of Southampton and BNFL.

APPENDIX: CONCENTRATION SCALES, EQUILIBRIUM CONSTANTS AND STANDARD STATES

The use of the SHG technique to study a wide range of liquid interfaces has highlighted the need for care in the comparison of thermodynamics data derived from this and other sources because of the choice of different concentration scales and standard states. For all isotherms at very low bulk concentration of the adsorbate, c, the interfacial concentration, n, will be proportional to the bulk concentration

$$\frac{n}{n^\circ} = K_c \frac{c}{c^\circ} \tag{A1}$$

where K_c is the dimensionless equilibrium constant defined for standard states corresponding to an ideal solution with a concentration of $c^\circ = 1\ \text{mol·dm}^{-3}$ and a surface concentration of $n^\text{o} = 1\ \text{molecule} \cdot \text{m}^{-2}$. In the low concentration (or coverage) limit of the Langmuir isotherm,

$$\Theta = kc \tag{A2}$$

where Θ is the coverage and k is the bulk/interface equilibrium constant but with the surface concentrations measured in terms of a coverage; the interface standard state is the hypothetical full monolayer with unit activity. However, it is sometimes more useful to measure the solute concentrations on a mole fraction scale. The conversion to mole fraction X and the associated changes to the corresponding equilibrium constant depend on the solvent. For the adsorbate only in the water phase,

$$X = \frac{n}{n + n_{H_2O}} \approx \frac{c}{c_{H_2O}} = \frac{c \text{ mol} \cdot \text{dm}^{-3}}{55.5 \text{ mol} \cdot \text{dm}^{-3}} \tag{A3}$$

where X is the mole fraction of PNP in the aqueous solution. Thus at low concentrations,

$$\Theta = kc = KX \tag{A4}$$

K is defined in terms of standard states consisting of an ideal solution at unit mole fraction and for the surface an ideal two-dimensional gas with unit coverage. Therefore,

$$K = (55.5\text{mol} \cdot \text{dm}^{-3}) \times (k\text{mol}^{-1} \cdot \text{dm}^{3}) \tag{A5}$$

Similarly relating the equilibrium constants K_c and K gives

$$K_c = \frac{n^M}{c_{H_2O}} K \tag{A6}$$

In the example of the adsorption of TBP from the dodecane phase (density $\rho = 766 \text{ kg} \cdot \text{m}^{-3}$ and molar mass of $0.17 \text{ kg} \cdot \text{mol}^{-1}$) to the dodecane/water interface, the conversion to the mole fraction scale gives

$$K = c_{\text{Dodecane}} k = (4.506 \text{ mol} \cdot \text{dm}^{-3})(k\text{mol}^{-1} \cdot \text{dm}^{3}) \tag{A7}$$

Similarly, care needs to be taken when comparing the standard Gibbs energy of adsorption since the standard states may vary. A common way of obtaining $\Delta_{ads}G^o$ is from the concentration dependence of the surface or interfacial tension. For such surface pressure measurements at low concentrations, the standard state typically corresponds to a surface pressure of $1 \text{ mN} \cdot \text{m}^{-1}$, which differs from the standard state implied by the Langmuir and Frumkin isotherms. Assuming that a monolayer coverage of the adsorbate corresponds to approximately 2×10^{18} molecule $\cdot \text{m}^{-2}$ (2×10^{14} molecule $\cdot \text{cm}^{-2}$), the ideal two-dimensional gas equation gives a surface pressure for a full but ideal monolayer of $8 \text{ mN} \cdot \text{m}^{-1}$ at 298 K. The corresponding change in free energy due to the increase in pressure from 1 to $8 \text{ mN} \cdot \text{m}^{-1}$ is

$$\Delta\Delta_{ads}G = RT \ln \frac{P}{P^{\theta}} = 5.2 \text{ kJ} \cdot \text{mol}^{-1} \tag{A8}$$

REFERENCES

1. R.R. Naujok, D.A. Higgins, D.G. Hanken and R.M. Corn, J. Chem. Soc.-Faraday Trans. **91**, 2353 (1995).
2. R.M. Corn and D.A. Higgins, Chem. Rev. **94**, 107 (1994).
3. Y.R. Shen, Appl. Phys. A-Mater. Sci. Process. **59**, 541 (1994).
4. Y.R. Shen, Surf. Sci. **300**, 551 (1994).
5. K.B. Eisenthal, Chem. Rev. **96**, 1343 (1996).
6. P.F. Brevet and H.H. Girault, *Liquid/Liquid Interfaces, Theory and Methods,* Eds. A.G. Volkov and D.W. Deamer, CRC Press, Boca Raton, FL, 1996.
7. P.F. Brevet, *Surface Second Harmonic Generation,* Ed. C.D. Chimie, Presses Polytechniques et Universitaires Romandes, Lausanne, 1997.
8. J.M. Hicks and T. Petralli-Mallow, Appl. Phys. B-Lasers Opt. **68**, 589 (1999).

9. J.M. Hicks, T. Petralli-Mallow and J.D. Byers, Faraday Discuss. 341 (1994).
10. J.D. Byers, H.I. Yee, T. Petralli-Mallow and J.M. Hicks, Phys. Rev. B-Condens Matter **49**, 14643 (1994).
11. J.D. Byers, H.I. Yee and J.M. Hicks, J. Chem. Phys. **101**, 6233 (1994).
12. J.D. Byers and J.M. Hicks, Chem. Phys. Lett. **231**, 216 (1994).
13. L. Hecht and L.D. Barron, Mol. Phys. **89**, 61 (1996).
14. T. Verbiest, S. Van Elshocht, A. Persoons, C. Nuckolls, K.E. Phillips and T.J. Katz, Langmuir **17**, 4685 (2001).
15. M. Kauranen, S. Van Elshocht, T. Verbiest and A. Persoons, J. Chem. Phys. **112**, 1497 (2000).
16. V. Mizrahi and J.E. Sipe, J. Opt. Soc. Am. B-Opt. Phys. **5**, 660 (1988).
17. A.A.T. Luca, P. Hebert, P.F. Brevet and H.H. Girault, J. Chem. Soc.-Faraday Trans. **91**, 1763 (1995).
18. T.L. Mazely and W.M. Hetherington, J. Chem. Phys. **86**, 3640 (1987).
19. B. Dick, Chem. Phys. **96**, 199 (1985).
20. J.G. Frey, Chem. Phys. Lett. **323**, 454 (2000).
21. G.J. Simpson and K.L. Rowlen, Chem. Phys. Lett. **317**, 276 (2000).
22. G.J. Simpson and K.L. Rowlen, Acc. Chem. Res. **33**, 781 (2000).
23. G.J. Simpson and K.L. Rowlen, J. Am. Chem. Soc. **121**, 2635 (1999).
24. G.J. Simpson and K.L. Rowlen, Chem. Phys. Lett. **309**, 117 (1999).
25. G.J. Simpson and K.L. Rowlen, J. Phys. Chem. B **103**, 3800 (1999).
26. G.J. Simpson and K.L. Rowlen, J. Phys. Chem. B **103**, 1525 (1999).
27. J.A. Ekhoff and K.L. Rowlen, Anal. Chem. **74**, 5954 (2002).
28. D.M. Willard, K.Y. Kung, B.M. Luther and N.E. Levinger, Rev. Sci. Instrum. **68**, 3312 (1997).
29. P.S. Hauge and F.H. Dill, Opt. Commun. **14**, 431 (1975).
30. P.S. Hauge, Surf. Sci. **56**, 148 (1976).
31. A.J. Timson, R.D. Spencer-Smith, A.K. Alexander, R. Greef and J.G. Frey, Meas. Sci. Technol. **14**, 508 (2003).
32. J.C. Conboy, J.L. Daschbach and G.L. Richmond, J. Phys. Chem. **98**, 9688 (1994).
33. J.C. Conboy, J.L. Daschbach and G.L. Richmond, Appl. Phys. A-Mater. Sci. Process. **59**, 623 (1994).
34. A.A. Tamburello-Luca, P. Hebert, P. Brevet and H.H. Girault, J. Chem. Soc.-Faraday Trans. **91**, 1763 (1995).
35. I.L. Carpenter and W.J. Hehre, J. Phys. Chem. **94**, 531 (1990).
36. Y.H. Zhang, S.E. Feller, B.R. Brooks and R.W. Pastor, J. Chem. Phys. **103**, 10252 (1995).
37. D. Michael and I. Benjamin, J. Phys. Chem. **99**, 16810 (1995).
38. D. Michael and I. Benjamin, J. Chem. Phys. **114**, 2817 (2001).
39. A.R. Vanbuuren, S.J. Marrink and H.J.C. Berendsen, J. Phys. Chem. **97**, 9206 (1993).
40. D.M. Mitrinovic, Z.J. Zhang, S.M. Williams, Z.Q. Huang and M.L. Schlossman, J. Phys. Chem. B **103**, 1779 (1999).
41. R. Antoine, F. Bianchi, P.F. Brevet and H.H. Girault, J. Chem. Soc.-Faraday Trans. **93**, 3833 (1997).
42. H.F. Wang, E. Borguet and K.B. Eisenthal, J. Phys. Chem. B **102**, 4927 (1998).
43. H.F. Wang, E. Borguet and K.B. Eisenthal, J. Phys. Chem. A **101**, 713 (1997).
44. D.A. Higgins, M.B. Abrams, S.K. Byerly and R.M. Corn, Langmuir **8**, 1994 (1992).
45. H.H. Heinze, F. Della Sala and A. Gorling, J. Chem. Phys. **116**, 9624 (2002).
46. A.A. Tamburello-Luca, P. Hebert, P.F. Brevet and H.H. Girault, J. Chem. Soc.-Faraday Trans. **92**, 3079 (1996).
47. M.J. Crawford, *Second Harmonic Generation from Liquid Interfaces*, Ph.D. thesis, University of Southampton, Southampton, UK, 1995.
48. S. Haslam, S.G. Croucher, C.G. Hickman and J.G. Frey, Phys. Chem. Chem. Phys. **2**, 3235 (2000).
49. G.J. Simpson and K.L. Rowlen, Anal. Chem. **72**, 3407 (2000).
50. N.H. Sagert, Can. J. Chem. **57**, 1218 (1979).
51. N.H. Sagert and M.J. Quinn, J. Colloid Interface Sci. **99**, 297 (1984).
52. A. Castro, S.W. Ong and K.B. Eisenthal, Chem. Phys. Lett. **163**, 412 (1989).
53. G.J. Simpson and K.L. Rowlen, Acc. Chem. Res. **33**, 781 (2000).
54. J.A. Ekhoff, S.G. Westerbuhr and K.L. Rowlen, Langmuir **17**, 7079 (2001).
55. M.D. Elking, G. He and Z. Xu, J. Chem. Phys. **105**, 6565 (1996).
56. Y.A. Gruzdkov and V.N. Parmon, J. Chem. Soc.-Faraday Trans. **89**, 4017 (1993).
57. R.L. Hansen and J.M. Harris, Anal. Chem. **70**, 4247 (1998).

58. T. Kikteva, D. Star, Z.H. Zhao, T.L. Baisley and G.W. Leach, J. Phys. Chem. B **103**, 1124 (1999).
59. L. Werner, W. Hill, F. Marlow, A. Glismann and O. Hertz, Thin Solid Films **205**, 58 (1991).
60. A. Yamaguchi, T. Uchida, N. Teramae and H. Kaneta, Anal. Sci. **13**, 85 (1997).
61. V. Tsukanova, H. Lavoie, A. Harata, T. Ogawa and C. Salesse, J. Phys. Chem. B **106**, 4203 (2002).
62. K. Ishibashi, O. Sato, R. Baba, K. Hashimoto and A. Fujishima, J. Electroanal. Chem. **465**, 195 (1999).
63. A. Miura and N. Tamai, Chem. Phys. Lett. **328**, 23 (2000).
64. O.N. Slyadneva, M.N. Slyadnev, V.M. Tsukanova, T. Inoue, A. Harata and T. Ogawa, Langmuir **15**, 8651 (1999).
65. V. Tsukanova, O. Slyadneva, T. Inoue, A. Harata and T. Ogawa, Chem. Phys. **250**, 207 (1999).
66. X.Y. Zheng, A. Harata and T. Ogawa, Chem. Phys. Lett. **316**, 6 (2000).
67. A. Castro, E.V. Sitzmann, D. Zhang and K.B. Eisenthal, J. Phys. Chem. **95**, 6752 (1991).
68. D. Zimdars, J.I. Dadap, K.B. Eisenthal and T.F. Heinz, J. Phys. Chem. B **103**, 3425 (1999).
69. S. Ishizaka, K. Nakatani, S. Habuchi and N. Kitamura, Anal. Chem. **71**, 419 (1999).
70. T. Kikteva, D. Star and G.W. Leach, J. Phys. Chem. B **104**, 2860 (2000).
71. M.J.E. Morgenthaler and S.R. Meech, Chem. Phys. Lett. **202**, 57 (1993).
72. D.A. Steinhurst and J.C. Owrutsky, J. Phys. Chem. B **105**, 3062 (2001).
73. R. Antoine, A.A. Tamburello-Luca, P. Hebert, P.F. Brevet and H.H. Girault, Chem. Phys. Lett. **288**, 138 (1998).
74. A.A. Tamburello-Luca, P. Hebert, R. Antoine, P.F. Brevet and H.H. Girault, Langmuir **13**, 4428 (1997).
75. S. Haslam, *Second Harmonic Generation from Liquid/Liquid Interfaces*, Ph.D., University of Southampton, Southampton, UK, 1998.
76. D.A. Higgins and R.M. Corn, J. Phys. Chem. **97**, 489 (1993).
77. D.A. Higgins, R.R. Naujok and R.M. Corn, Chem. Phys. Lett. **213**, 485 (1993).
78. M.J. Crawford, J.G. Frey, T.J. VanderNoot and Y.G. Zhao, J. Chem. Soc.-Faraday Trans. **92**, 1369 (1996).
79. K. Nochi, A. Yamaguchi, T. Hayashita, T. Uchida and N. Teramae, J. Phys. Chem. B **106**, 9906 (2002).
80. H. Wang, E.C.Y. Yan, E. Borguet and K.B. Eisenthal, Chem. Phys. Lett. **259**, 15 (1996).
81. P. Allcock, D.L. Andrews, S.R. Meech and A.J. Wigman, Phys. Rev. A **53**, 2788 (1996).
82. S. Yamada and I.Y.S. Lee, Anal. Sci. **14**, 1045 (1998).
83. E.C.Y. Yan, Y. Liu and K.B. Eisenthal, J. Phys. Chem. B **102**, 6331 (1998).
84. E.C.Y. Yan and K.B. Eisenthal, J. Phys. Chem. B **103**, 6056 (1999).
85. H.F. Wang, E.C.Y. Yan, Y. Liu and K.B. Eisenthal, J. Phys. Chem. B **102**, 4446 (1998).
86. E.C.Y. Yan and K.B. Eisenthal, Biophys. J. **79**, 898 (2000).
87. E.C.Y. Yan, Y. Liu and K.B. Eisenthal, J. Phys. Chem. B **105**, 8531 (2001).
88. J.S. Salafsky and K.B. Eisenthal, Chem. Phys. Lett. **319**, 435 (2000).

2

Vibrational Sum-Frequency Spectroscopic Investigations of Molecular Interactions at Liquid/Liquid Interfaces

Mark R. Watry and Geraldine L. Richmond
Rocky Mountain College, Billings, MT 59102; University of Oregon, Eugene, OR 97403

2.1. INTRODUCTION

Liquid surfaces play a key role in many processes affecting our everyday lives. Some of the most interesting chemical, industrial, biological and environmental reactions are facilitated by these interfaces. Separations, including liquid chromatographies and solvent extractions, are possible because of the hydrophobic/hydrophilic properties of liquid/solid and liquid/liquid interfaces. Emulsification takes advantage of the interfacial properties of liquids and makes a myriad of products and processes possible, including paints, detergents, soaps, cosmetics and processed foods. Biological chemistry is highly dependent on the properties of interfacial water considering that a cell is essentially a small sack of aqueous solution that is connected to the outside world through chemical and physical interactions that are mediated by the cell membrane/water interface. In atmospheric chemistry, many important reactions occur on the surface or in the interior of water droplets. Unfortunately, few details about molecular interactions at these fluid interfaces have been uncovered, largely because of the inability of most experimental methods to access these surfaces and to distinguish the molecular properties of the thin surface region from the overwhelming properties of the adjacent bulk phases. The difficulties have been even greater for the study of liquid/liquid surfaces, which cannot be readily accessed by the scattering techniques (neutron scattering, for example) that have been very valuable for studying the air/water interface. Vibrational sum-frequency spectroscopy (VSFS) is one of a select few molecular techniques that can both access buried liquid interfaces and is inherently surface-specific. As a relatively new method for measuring vibrational spectra at surfaces, it has been applied to a variety of solid/liquid [1–4], liquid/liquid [5–7] and air/liquid [8–12] interfaces.

This chapter examines the orientation and conformation of adsorbate and solvent molecules at liquid/liquid interfaces, and from this information, it further examines the interactions between the molecules present at these interfaces. The summary begins with a description of the theoretical underpinnings of the technique and describes the experimental apparatus employed in our studies of these interfaces. In each case, one of the liquids is water. The remainder of the chapter describes studies of molecular adsorbates at liquid/liquid interfaces and studies of neat liquid/liquid interfaces.

2.2. THEORETICAL CONSIDERATIONS OF VSFS

2.2.1. *Linear and Non-Linear Polarization of a Medium*

When light interacts with matter, and the photons are not absorbed, it does so by inducing a polarization in the medium. Since the interaction energy between the electric field of the incident radiation and the molecules making up the medium is small compared to the total energy of the molecules, the incident radiation can be treated as a perturbation to the total energy of the medium. (This is true for pulsed laser beams as well as ambient light [13].) Therefore, the polarization of the medium, P, can be expanded as a power series in the electric field [13,14].

$$\vec{P} = \alpha_0 + \overleftrightarrow{\chi}^{(1)} \cdot \vec{E} + \overleftrightarrow{\chi}^{(2)} : \vec{E}\vec{E} + \overleftrightarrow{\chi}^{(3)} \vdots \vec{E}\vec{E}\vec{E} + \cdots \tag{1}$$

Here, α_0 is the static (natural) polarization of the medium, the $\overleftrightarrow{\chi}^{(n)}$ are the susceptibilities of order n containing the frequency dependence of the polarization and the Fresnel factors associated with the geometry of the system and $\vec{E}$ is the electric field vector. Since the electric field is oscillating in time, the polarization induced in the medium is an oscillating polarization. This oscillating polarization is primarily due to the movement of electrons in response to the electric field, and these moving (accelerating) electrons can be the sources of electromagnetic radiation. The terms in the expansion rapidly become small so that unless the electric field is very large, the second term dominates the interaction. This term is responsible for all linear optical effects including reflection, refraction and propagation through a medium (transparency), and the generated light has the same frequency as that of the incident light.

If the electric field(s) are large (e.g. laser light), subsequent terms in the expansion become important. The third term is the one that is of importance in the present work. It is the second-order non-linear polarization, $\vec{P}^{(2)}$.

$$\vec{P}^{(2)} = \overleftrightarrow{\chi}^{(2)} : \vec{E}\vec{E} \tag{2}$$

If there are two distinct electric fields present, as in sum-frequency generation, $\vec{P}^{(2)}$ is composed of four terms.

$$\vec{P}^{(2)} = \overleftrightarrow{\chi}^{(2)}\vec{E}_1\vec{E}_1 + \overleftrightarrow{\chi}^{(2)}\vec{E}_1\vec{E}_2 + \overleftrightarrow{\chi}^{(2)}\vec{E}_2\vec{E}_1 + \overleftrightarrow{\chi}^{(2)}\vec{E}_2\vec{E}_2 \tag{3}$$

The first and last terms correspond to oscillating polarizations that emit light at twice the frequency of E_1 and E_2 respectively, and the other two terms correspond to sum- and difference-frequency mixing of the two frequencies. In general, experiments can be conducted in such a way as to strongly favour one non-linear process above the others. However, if this is not possible, steps can be taken to detect the output of a single process through the use of appropriate filters, spatial separation of the emitted beams and choice of detector.

Since the experiments to be described later utilize sum-frequency generation, only the sum-frequency contribution to the polarization is developed here. The non-linear polarization induced in the medium at $\omega_{\text{sfg}} = \omega_1 + \omega_2$ by the oscillating electric fields $E(\omega_1)$ and $E(\omega_2)$ can be expressed as

$$\vec{P}^{(2)}(\omega_{\text{sfg}}) = \overleftrightarrow{\chi}^{(2)}_{\text{sfg}} : \vec{E}(\omega_1)\vec{E}(\omega_2). \tag{4}$$

$\overleftrightarrow{\chi}^{(2)}_{\text{sfg}}$ is a third rank tensor composed of 27 elements. Equation (4) is often written as

$$\vec{P}^{(2)}_i(\omega_{\text{sfg}}) = \overleftrightarrow{\chi}^{(2)}_{ijk} : \vec{E}_j(\omega_1)\vec{E}_k(\omega_2) \tag{5}$$

to make explicit the connection between the components of the fields and the components of the polarization of the medium.

2.2.2. *Vibrational Sum-Frequency Spectroscopy*

Vibrational sum-frequency spectroscopy (VSFS) is a second-order non-linear optical technique that can directly measure the vibrational spectrum of molecules at an interface. Under the dipole approximation, this second-order non-linear optical technique is uniquely suited to the study of surfaces because it is forbidden in media possessing inversion symmetry. At the interface between two centrosymmetric media there is no inversion centre and sum-frequency generation is allowed. Thus the asymmetric nature of the interface allows a selectivity for interfacial properties at a molecular level that is not inherent in other, linear, surface vibrational spectroscopies such as infrared or Raman spectroscopy. VSFS is related to the more common but optically simpler second harmonic generation process in which both beams are of the same fixed frequency and is also surface-specific.

In a VSFS experiment, the pulses from a visible laser beam and a tunable IR laser beam are coincident in time and space at the interface. The top of Figure 2.1 gives an example of an experimental geometry for studying liquid/liquid interfaces where the organic liquid is more dense than water, and the bottom of Figure 2.1 gives an example of a configuration for organic liquids less dense than water. The polarizations indicated by P and S in the figure are for light polarized in the plane of incidence and perpendicular to the plane of incidence, respectively. The high-intensity electric fields of the laser pulses induce a coherent non-linear polarization in the molecules at the interface, and this oscillating non-linear polarization is the source of light radiating from the surface at frequencies in addition to the frequencies of the incident light. Among these are the harmonics of the input frequencies and the sum- and difference-frequencies of the two input beams. Generally one of the above processes will be the most efficient given the design of the experiment. For the measurements described herein, the light reflecting

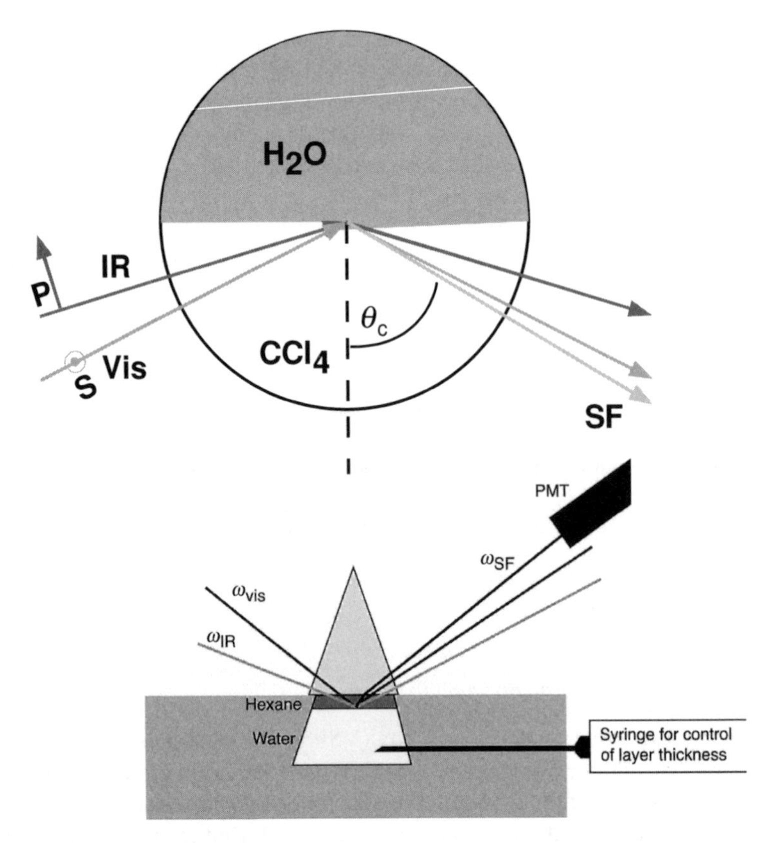

FIGURE 2.1. Experimental geometries. Top: CCl_4/water interfaces showing the polarization scheme ssp and the critical angle θ_c. Bottom: alkane/water interfaces with adjustable thickness of alkane layer to minimize IR absorption by the alkane.

from the surface at a frequency that is the sum of the two incident IR and visible fields is collected and recorded to generate the VSF spectrum. Unlike most other spectroscopic techniques, VSFS does not result in a change in the internal energy of the molecules under study. The interfacial molecules merely act as a medium in which the two incoming laser beams are coupled to produce a third coherent beam. In this regard it is similar to elastic scattering.

As mentioned above, the intensity of the VSFS signal is dependent on the non-linear polarization induced in the medium. The polarization that gives rise to sum-frequency generation, $P^{(2)}_{\text{sfg}}$, is in turn dependent on the surface non-linear susceptibility $\chi^{(2)}_{\text{sfg}}(\omega_{\text{sfg}} = \omega_{\text{vis}} + \omega_{\text{ir}})$ as

$$I_{\text{sfg}} \propto \left|P^{(2)}_{\text{sfg}}\right|^2 \propto \left|\chi^{(2)}_{\text{NR}} e^{i\phi_{\text{NR}}} + \sum_{v} \left|\chi^{(2)}_{\text{R}_v}\right| e^{i\gamma_v}\right|^2 I_{\text{vis}} I_{\text{ir}} \tag{6}$$

with $\chi^{(2)}_{\text{NR}}$ and $\chi^{(2)}_{\text{R}_v}$ being the non-resonant and resonant parts of $\chi^{(2)}_{\text{sfg}}$ respectively. ϕ_{R} is the relative phase of the non-resonant susceptibility, and γ_v is the relative phase of the vth vibrational mode. The non-resonant part of $\chi^{(2)}_{\text{sfg}}$ depends primarily on the polarizability of the molecules at the interface and must be included in the full spectral analysis. (It cannot merely be subtracted as if it were a background.) The resonant term arises from a coincidence in frequency between the tunable infrared light and a vibrational mode in the molecule of interest. Because of the coherent nature of VSFS, and as Equation (6) indicates, spectral contributions are summed prior to squaring the terms, in contrast to linear spectroscopy where the contributions are squared prior to summation.

$$I_{\text{sfg}} \propto \left|\chi^2_{\text{NR}} e^{i\phi_{\text{NR}}} + \chi^2_{1\text{R}} e^{i\gamma_1} + \chi^2_{2\text{R}} e^{i\gamma_2} + \chi^2_{3\text{R}} e^{i\gamma_3} + \cdots\right|^2 \tag{7}$$

Since the non-linear susceptibility is generally complex, each resonant term in the summation is associated with a relative phase, γ_v, which describes the interference between overlapping vibrational modes. The resonant macroscopic susceptibility associated with a particular vibrational mode v, $\chi^{(2)}_{\text{R}_v}$, is related to the microscopic susceptibility also called the molecular hyperpolarizability, β_v, in the following way

$$\chi^{(2)}_{\text{R}_v} = \frac{N}{\varepsilon_0} \langle \beta_v \rangle \tag{8}$$

where N is the number of molecules contributing to the sum-frequency signal and $\langle \beta_v \rangle$ is the orientationally averaged molecular hyperpolarizability. From Equations (6) and (8) it is seen that the square root of the sum-frequency intensity depends on the number of molecules giving rise to the response and their average orientation through the molecular hyperpolarizability. These dependencies can be exploited to determine changes in the orientation of interfacial molecules induced by changes in system parameters such as interfacial concentration or liquid-phase composition. It is the molecular hyperpolarizability, β_v, that is responsible for the enhancement of the sum-frequency intensity when the frequency of IR radiation is resonant with a sum-frequency active vibration.

The interference between different vibrations (including those of different molecules) resulting from the coherent nature of the experiment makes the analysis of VSFS spectra considerably more complicated than that of spectra recorded with linear spectroscopic techniques. However, this complexity can be exploited to provide orientational information if a complete analysis of the VSF spectrum is employed taking into account the phase relationships of the contributing vibrational modes to the sum-frequency response [15,16]. In the analysis it is possible to constrain the average orientation of the molecules at the surface by relating the macroscopic second-order susceptibility, $\chi^{(2)}_{\text{sfg}}$, of the system to the molecular hyperpolarizabilities, β_v, of the individual molecules at the interface [17,18].

The vibrationally resonant hyperpolarizability of a molecule can be described with the following expression obtained utilizing perturbation theory [17] (assuming that the only interactions between the electric fields and the media are dipolar interactions).

$$\beta_{lmn,v} = \frac{\langle g \left|\alpha_{lm}\right| v\rangle \langle v \left|\mu_n\right| g\rangle}{\omega_{\text{IR}} - \omega_v + i\Gamma_v} \tag{9}$$

The subscripts l, m and n represent the molecular inertial axes (a, b and c); α_{lm} and μ_n represent the Raman and dipole vibrational transition elements, respectively

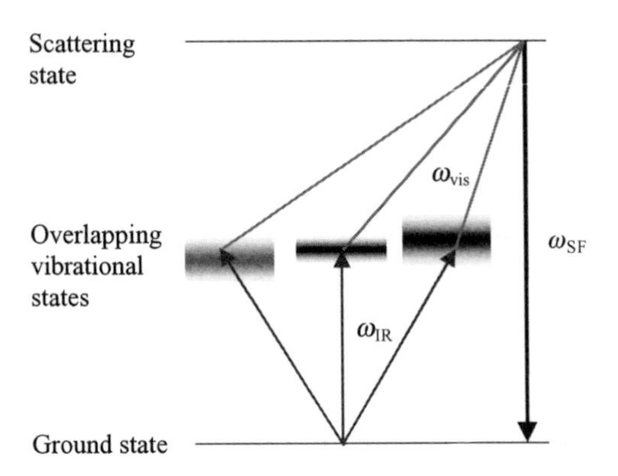

FIGURE 2.2. Energy diagram showing interference between vibrational modes. Interference arises because VSFS is a coherent process. The multiple paths are analogous to those in the double slit experiment.

for a particular vibrational mode; and a Lorentzian distribution of resonant transition energies is assumed. The absolute square of the numerator of β_v can be considered a sum-frequency "transition probability". An energy diagram illustrating the interference between different vibrational modes is shown in Figure 2.2. It is helpful to compare the interference of different vibrations in VSFS to those in the classic double slit experiment. In both cases, particles passing through *indistinguishable* intermediate states give rise to distinct interference patterns.

The macroscopic property observed in sum-frequency experiments, $\chi^{(2)}_{sfg}$, is a sum of the molecular hyperpolarizabilities, β_v, over all vibrational modes and all of the molecules at the interface, which takes into account the orientation of each molecule. Orientational information is obtained from the experimental spectra through consideration of the relationship between the observed Cartesian components of the macroscopic second-order susceptibility $\chi^{(2)}_{IJK,\,v}$ and the corresponding spectroscopically active components of the molecular hyperpolarizability, β_{lmn}. This is accomplished through an Euler angle rotation of the molecular axis system into the laboratory axis system as defined through the use of the rotational matrix $\mu_{IJK:lmn}$. The general expression for the transformation from a molecular-fixed axis system to a laboratory-fixed system is

$$\chi_{IJK,v} = \sum_{lmn} \mu_{IJK:lmn} \cdot \beta_{lmn,v}. \tag{10}$$

The indices I, J and K represent the lab frame coordinates X, Y or Z appropriate to a specific experiment. The indices l, m and n run through the molecular coordinates a, b and c. The orientation of the molecular axis system in the lab frame is defined by the transformation tensor, $\mu_{IJK:lmn}$, through the Euler angles θ, φ and χ. The Euler angles are defined in Figure 2.3. If the signs of $\beta_{lmn,v}$ and $\chi^{(2)}_{IJK,v}$ are known, then the average orientation of the molecules can be determined by analyzing how the sign of the transformation tensor changes with respect to the angles θ, φ and χ. A complete table of the appropriate transformation equations is given by Hirose *et al.* in their exhaustive general treatment of CH_2 and CH_3 vibrations in sum-frequency spectra [17,18]. In the case of the liquid/liquid interfaces discussed here, the signs of the $\chi^{(2)}_{IJK,v}$ terms can be determined through a comprehensive fit of the observed sum-frequency spectra to Equations (6) and

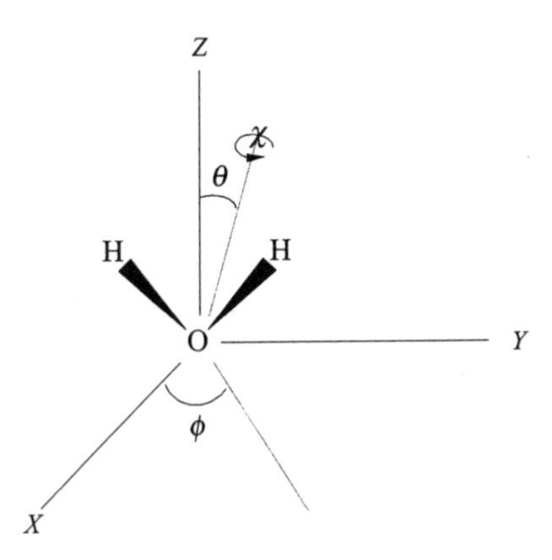

FIGURE 2.3. Definition of the Euler angles.

(9) (see below), and the signs of the $\beta_{lmn,v}$ components can be determined through ab initio calculations [19,20].

Combining Equations (8)–(10) yields the following expression for the resonant macroscopic susceptibility

$$\chi^{(2)}_{\mathrm{R}_v} \propto \frac{A_K M_{IJ}}{\omega_v - \omega_{ir} - i\Gamma_v} \tag{11}$$

still assuming that a Lorentzian distribution of vibrational energies and the dipole approximation are employed. In this expression A_K is the IR transition moment, M_{IJ} is the Raman transition probability, ω_v is the resonant mode frequency and Γ_v is the natural line width of the transition. Since sum-frequency active modes must be both IR- and Raman-active, any vibrational mode that has an inversion centre cannot be sum-frequency-active. This result coupled with the coherent nature of sum-frequency generation precludes any sum-frequency response from bulk isotropic media.

The surface susceptibility $\chi^{(2)}_{\mathrm{sfg}}$ is a 27-element tensor that can generally be reduced to a handful of non-vanishing elements after consideration of the symmetry of the system. In particular, the interface between two isotropic bulk media is isotropic in the plane of the interface (it has $C_{\infty v}$ symmetry), and $\chi^{(2)}_{\mathrm{sfg}}$ reduces to the following four independent non-zero elements

$$\chi^{(2)}_{zzz}; \quad \chi^{(2)}_{xxz} = \chi^{(2)}_{yyz}; \quad \chi^{(2)}_{xzx} = \chi^{(2)}_{yzy}; \quad \chi^{(2)}_{zxx} = \chi^{(2)}_{zyy} \tag{12}$$

where z is the direction normal to the interface. The four independent values can in principle be determined after the acquisition of sum-frequency under four different polarization combinations (ssp, sps, pss, ppp) with the polarizations listed in order of decreasing frequency (sum-frequency, visible, IR). p-polarized light is defined as having its electric field vector parallel to the plane of incidence, whereas s-polarized light is polarized perpendicular to the incident plane (see Figure 2.1). The ssp polarization combination probes modes with IR transition moment components perpendicular to the interfacial plane, sps and pss polarization combinations probe modes that have IR transition moment components in the plane of the interface, and the ppp polarization combination probes all components of the allowed vibrations.

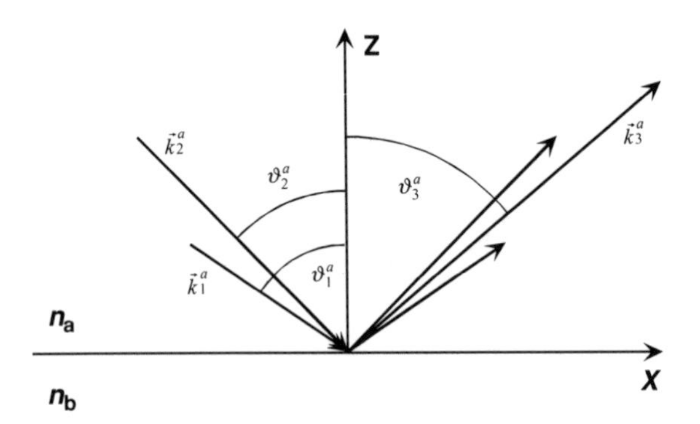

FIGURE 2.4. Notation for indices of refraction and angles of incidence and reflection.

In addition to the tensor element dependence of the sum-frequency intensity, there is also a dependence on the geometry of the experiment that manifests itself in the linear and non-linear Fresnel factors that describe the behaviour of the three light beams at the interface. Fresnel factors are the reflection and transmission coefficients for electromagnetic radiation at a boundary and depend on the frequency, polarization and incident angle of the electromagnetic waves and the indices of refraction for the media at the boundary [16,21].

For sum-frequency light collected in reflection (Figure 2.4), these dependencies are

$$I_{\mathrm{ppp}} \propto \left|\tilde{f}_z f_z f_z \chi^{(2)}_{zzz} + \tilde{f}_z f_x f_x \chi^{(2)}_{zxx} + \tilde{f}_x f_z f_x \chi^{(2)}_{xzx} + \tilde{f}_x f_x f_z \chi^{(2)}_{xxz}\right|^2,$$
$$I_{\mathrm{ssp}} \propto \left|\tilde{f}_y f_y f_z \chi^{(2)}_{yyz}\right|^2, \quad I_{\mathrm{sps}} \propto \left|\tilde{f}_y f_z f_y \chi^{(2)}_{yzy}\right|^2, \quad I_{\mathrm{pss}} \propto \left|\tilde{f}_z f_y f_y \chi^{(2)}_{zyy}\right|^2, \tag{13}$$

where the subscript i stands for x or y, and f and $\tilde{f}$ are the linear and non-linear Fresnel factors, respectively.

The linear and non-linear Fresnel factors are [16]

$$f_{i,x} = \frac{2n_i^a \cos\vartheta_i^b}{n_i^b \cos\vartheta_i^a + n_i^a \cos\vartheta_i^b}, \qquad \tilde{f}_x = \frac{\cos\vartheta_3^b}{n_3^a \cos\vartheta_3^b + n_3^a n_3^b \cos\vartheta_3^a}$$
$$f_{i,y} = \frac{2n_i^a \cos\vartheta_i^a}{n_i^b \cos\vartheta_i^b + n_i^a \cos\vartheta_i^a}, \qquad \tilde{f}_y = \frac{1}{n_3^a \cos\vartheta_3^a + n_3^b \cos\vartheta_3^b}$$
$$f_{i,z} = \frac{2(n_i^a/n_i^b)^2 n_i^b \cos\vartheta_i^a}{n_i^b \cos\vartheta_i^a + n_i^a \cos\vartheta_i^b}, \qquad \tilde{f}_z = \frac{\sin\vartheta_3^a}{n_3^a n_3^b \cos\vartheta_3^b + (n_3^b)^2 \cos\vartheta_3^a}. \tag{14}$$

The subscripts i denote the incident beam, the superscripts a and b denote the medium the beam is propagating in (see Figure 2.4), n denotes the frequency-dependent index of refraction, and ϑ_i^b denote the angles of incidence. The angles of the transmitted waves, ϑ_i^b, are given by Snell's law

$$n_i^a \sin\vartheta_i^a = n_i^b \sin\vartheta_i^b \tag{15}$$

and the angle of the SF wave reflected into medium a is given by [22]

$$\sin\vartheta_3^a = \frac{\omega_2}{\omega_3}\frac{n_2^a}{n_3^a}\sin\vartheta_2^a + \frac{\omega_1}{\omega_3}\frac{n_1^a}{n_3^a}\sin\vartheta_1^a. \tag{16}$$

2.2.3. *Total Internal Reflection Geometry*

Some interfacial systems have a weak sum-frequency response whereas others suffer from significant and undesirable heating from high-intensity laser beams. Utilization of a total internal reflection (TIR) geometry can minimize these problems with the added bonus of generating all of the sum-frequency light in reflection. If the index of refraction of medium a is greater than that of medium b, there exist incident angles for which the corresponding angle of transmission is purely imaginary, $\sin\vartheta_i^b > 1$, and the incident beam will be totally reflected. The angle at which this phenomenon first occurs is called the critical angle. In a TIR geometry, one or more Fresnel factors take on an imaginary value. As a result, the local field intensity can exceed that of the incident wave because of the formation of an evanescent wave at the interface [23,24]. Near this angle, the sum-frequency response increases dramatically as a function of incident angle [25]. The sum-frequency signal is enhanced the most at the critical angle, and enhancement is obtained for any and all beams at their critical angle including the generated sum-frequency. A TIR geometry has been shown to increase the sum-frequency signal by more than two orders of magnitude over external reflection geometries [22–24,26–29].

2.2.4. *VSFS Line Shape and Sum-Frequency Intensities*

Equation (6) shows that the sum-frequency intensity is dependent on the second-order polarization $P_{\text{sfg}}^{(2)}$. However, that polarization is a vector quantity $\vec{P}_{\text{sfg}}^{(2)}$. Within the electric dipole approximation the local fields $\vec{E}_1^b$ and $\vec{E}_2^b$ are coupled by the second-order non-linear susceptibility, $\overleftrightarrow{\chi}^{(2)}$, [16], inducing the following non-linear polarization at $\omega_3 = \omega_1 + \omega_2$:

$$\vec{P}_{\text{sfg}}^{(2)} = \overleftrightarrow{\chi}^{(2)}(\omega_1) : \vec{E}_2^b(\omega_2)\vec{E}_1^b(\omega_1). \tag{17}$$

The subscripts and superscripts are consistent with those used in Figure 2.4, and the notation for $\overleftrightarrow{\chi}^{(2)}$ makes it clear that only the IR is near resonance. The local fields $\vec{E}_i^b$ can be described in terms of the incident fields $\vec{E}_i^a$.

$$\vec{E}_i^b(\omega_i) = \overleftrightarrow{L}_i^b(\omega_i)\overleftrightarrow{f}_i^{ba}(\omega_i)\vec{E}_i^a(\omega_i). \tag{18}$$

Within the polarization scheme used here where p-polarized light is polarized in the plane of incidence, and s-polarized light is polarized perpendicular to the plane, the Cartesian components of $\vec{E}_i^a(\omega_i)$ are given by the projection of the s and p components of the incident wave

$$\begin{aligned} E_x^a(\omega_i) &= E_p^a(\omega_i)\cos\vartheta_i^a,\ E_y^a(\omega_i) = E_s^a(\omega_i), \\ E_z^a(\omega_i) &= E_p^a(\omega_i)\sin\vartheta_i^a \end{aligned} \tag{19}$$

$\overleftrightarrow{f}$ and $\overleftrightarrow{L}$ are second-rank tensors that represent the macroscopic and microscopic field corrections. The components of the macroscopic field corrections were presented in Equation (14). The microscopic field corrections $\overleftrightarrow{L}$ account for dipole–dipole interactions between molecules. In the work presented here, this term is assumed to be unity [30].

As mentioned earlier, there are only four independent non-zero elements of $\overleftrightarrow{\chi}^{(2)}$ for a system with $C_{\infty v}$ symmetry. Four experimental polarization combinations can be employed to obtain all four (three individually and the fourth using the other three).

$$\begin{aligned}
P_y^{\rm ssp}(\omega_3) &= \chi_{yyz}^{(2)} E_y^b(\omega_2) E_z^b(\omega_1), \quad P_y^{\rm sps}(\omega_3) = \chi_{yzy}^{(2)} E_z^b(\omega_2) E_y^b(\omega_1),\\
P_z^{\rm pss}(\omega_3) &= \chi_{zyy}^{(2)} E_y^b(\omega_2) E_y^b(\omega_1),\\
P_x^{\rm ppp}(\omega_3) &= \chi_{xxz}^{(2)} E_x^b(\omega_2) E_z^b(\omega_1) + \chi_{xzx}^{(2)} E_z^b(\omega_2) E_x^b(\omega_1),\\
P_z^{\rm ppp}(\omega_3) &= \chi_{zzz}^{(2)} E_z^b(\omega_2) E_z^b(\omega_1) + \chi_{zxx}^{(2)} E_x^b(\omega_2) E_x^b(\omega_1)
\end{aligned} \tag{20}$$

The intensities of electromagnetic radiation emitted at the frequency ω_3 from these oscillating polarizations are

$$\begin{aligned}
I^{\rm ssp}(\omega_3) &= \frac{2\pi}{c} n_3^a \omega_3^2 \left| \tilde{f}_y L_{yy}^b P_y^{\rm ssp}(\omega_3) \right|^2,\\
I^{\rm sps}(\omega_3) &= \frac{2\pi}{c} n_3^a \omega_3^2 \left| \tilde{f}_y L_{yy}^b P_y^{\rm sps}(\omega_3) \right|^2,\\
I^{\rm pss}(\omega_3) &= \frac{2\pi}{c} n_3^a \omega_3^2 \left| \tilde{f}_z L_{zz}^b P_z^{\rm pss}(\omega_3) \right|^2,\\
I^{\rm ppp}(\omega_3) &= \frac{2\pi}{c} n_3^a \omega_3^2 \left| \tilde{f}_x L_{xx}^b P_x^{\rm ppp}(\omega_3) + \tilde{f}_z L_{zz}^b P_z^{\rm ppp}(\omega_3) \right|^2.
\end{aligned} \tag{21}$$

Again, the L_{ii}^b are microscopic field corrections and are taken to be unity in the work that follows, and the $\tilde{f}_i$ have been given above. After combining Equations (18)–(21), the full expressions become

$$\begin{aligned}
I^{\rm ssp}(\omega_3) &= \tfrac{2\pi}{c} n_3^a \omega_3^2 \left| \tilde{f}_y \chi_{yyz}^{(2)} f_{2,y} E_s^a(\omega_2) f_{1,z} E_p^a(\omega_1) \sin\vartheta_1^a \right|^2\\
I^{\rm sps}(\omega_3) &= \tfrac{2\pi}{c} n_3^a \omega_3^2 \left| \tilde{f}_y \chi_{yzy}^{(2)} f_{2,z} E_p^a(\omega_2) (\sin\vartheta_2^a) f_{1,y} E_s^a(\omega_1) \right|^2\\
I^{\rm pss}(\omega_3) &= \tfrac{2\pi}{c} n_3^a \omega_3^2 \left| \tilde{f}_z \chi_{zyy}^{(2)} f_{2,y} E_s^a(\omega_2) f_{1,y} E_s^a(\omega_1) \right|^2\\
I^{\rm ppp}(\omega_3) &= \tfrac{2\pi}{c} n_3^a \omega_3^2 \left| \begin{array}{l} \tilde{f}_x \left[\begin{array}{l} \chi_{xxz}^{(2)} f_{2,x} E_p^a(\omega_2)(\cos\vartheta_2^a) f_{1,z} E_p^a(\omega_1)(\sin\vartheta_1^a) \\ + \chi_{xzx}^{(2)} f_{2,z} E_p^a(\omega_2)(\sin\vartheta_2^a) f_{1,x} E_p^a(\omega_1)(\cos\vartheta_1^a) \end{array} \right] \\ + \tilde{f}_z \left[\begin{array}{l} \chi_{zzz}^{(2)} f_{2,z} E_p^a(\omega_2)(\sin\vartheta_2^a) f_{1,z} E_p^a(\omega_1)(\sin\vartheta_1^a) \\ + \chi_{zxx}^{(2)} f_{2,x} E_p^a(\omega_2)(\cos\vartheta_2^a) f_{1,x} E_p^a(\omega_1)(\cos\vartheta_1^a) \end{array} \right] \end{array} \right|^2
\end{aligned} \tag{22}$$

If the incident angles and the indices of refraction (as a function of wavelength) are known, the intensity can be corrected for Fresnel factors and angles of incidence. If the IR intensity is known as a function of wavelength, that too can be factored out; and, since the visible frequency is constant, it can be factored out into the constant. Since the sum-frequency wavelength is far from a transition, n_3^a is nearly constant. ω_3 can be corrected for; however, this correction is insignificant over the small wavelength range in the studies presented later. After these manipulations, all that is left is the dependence of the intensity on the non-linear susceptibility of the molecules under study:

$$I = \text{const} \times \left| \chi_{ijk}^{(2)} \right|^2 \tag{23}$$

Using Equations (6) and (11) we arrive at

$$I_{\text{sfg}} = \text{const} \times \left| \chi_{\text{NR}}^{(2)} e^{i\phi_{\text{NR}}} + \sum_{v} \frac{A_K M_{IJ}}{\omega_v - \omega_{\text{ir}} - i\Gamma_v} e^{i\gamma_v} \right|^2 . \tag{24}$$

This line shape assumes a Lorentzian distribution of energies.

Lorentzian line shapes are frequently used to take into account the interference between different vibrations, although other more complex line shapes are used. Unfortunately, the broad "wings" characteristic of a Lorentzian distribution often overestimate the amount of overlap, and therefore the amount of interference, between widely separated peaks. Particularly useful, although computationally more intensive, is the line shape profile described by Bain *et al.* [4,12].

$$I_{\text{sfg}} = \text{const.} \times \left| \chi_{\text{NR}}^{(2)} e^{i\phi} + \sum_{v} \int_{-\infty}^{\infty} \frac{A_v e^{i\varphi_v} e^{-[\omega_L - \omega_v / \Gamma_v]^2}}{\omega_{\text{IR}} - \omega_L - i\Gamma_L} \mathrm{d}\omega_L \right|^2 \tag{25}$$

Similar to a Voigt profile, this line shape expression is a convolution of Equation (11) and a Gaussian distribution to account for inhomogeneous broadening [12,31]. This profile is the best profile currently employed for the analysis of spectra exhibiting significant overlap of modes of different phases.

Figure 2.5 shows a few of the Lorentzian line shapes in the Gaussian distribution of the real part (top) and imaginary part (bottom) of the resonant non-linear susceptibility

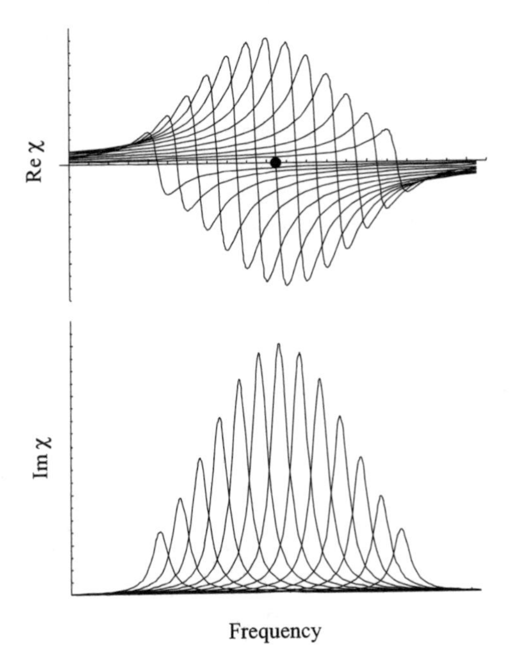

FIGURE 2.5. Selected components of the real and imaginary parts of χ with a phase of zero and centered on the band origin.

with the Gaussian centered at the band centre and a relative phase of zero radians for the vibration. Each of the resonant susceptibilities is composed in this way, and then added together along with the non-resonant susceptibility. The absolute square of this result yields the fit line to the entire spectrum.

Since VSFS is a coherent technique, and as Equation (25) implies, the oscillating electric field from each vibrational state involved in the generation of sum-frequency light can interfere with that from every other state and with that from the non-resonant response. This is quite different from IR spectroscopy where spectra are simple superpositions of intensity from individual vibrational modes. As such, VSFS leads to interesting line shapes that can be interpreted incorrectly if spectral intensities are compared visually without fitting the spectrum.

2.3. EXPERIMENTAL CONSIDERATIONS

2.3.1. Spectroscopy

The sum-frequency response at aqueous interfaces is very weak because of the small number of adsorbed molecules present and the poor polarizability of most liquids. To compensate for the low sum-frequency efficiency, pulsed lasers are used. Since the sum-frequency intensity increases with the peak intensity of the incident beams, picosecond and femtosecond pulses are optimal, although these shorter pulses result in larger IR bandwidths. Nanosecond systems are generally much simpler to operate and have narrower IR bandwidths, but can contribute to significant heating of the interface unless an optical coupling scheme such as total internal reflection (TIR) [6] or other mechanisms such as sample rotation [32] are employed.

Most VSFS studies have focused on the vibrational region around 3 μm because nanosecond and picosecond systems generate the highest IR power densities there. Tunable IR light has been produced by a number of optical parametric generation (OPG), oscillation (OPO) and amplification (OPA) systems as well as difference-frequency mixing and stimulated Raman scattering. In our laboratory we use nanosecond and picosecond systems. The nanosecond systems are primarily used to study liquid/liquid interfaces and employ a TIR geometry to couple the light to the interface (see Figure 2.1). This arrangement results in a sensitivity that is comparable to picosecond systems operating in an external reflection geometry. In the TIR geometry, the incident laser beams pass through the higher index medium and strike the interface near their respective critical angles to generate sum-frequency in reflection at its critical angle. Since the IR must pass through the higher index medium, it must be transparent to these IR frequencies, or the path through the high index phase must be minimized s that absorption is not significant (see Figure 2.1).

In our laboratory we use the 1064-nm output of an injection-seeded Nd:YAG laser that pumps a potassium titanyl phosphate (KTP) OPO/OPA assembly. This system produces 3.5nanosecond pulses and tunable IR from 2.5 μm (4000 cm^{-1}) to 5 μm (1975 cm^{-1}) with energies available at the interface ranging from 4 mJ to 1 mJ at the two respective limits. It operates with $\sim$1 cm^{-1} resolution and has a variable repetition rate (1–100 Hz).

Since the surface of a liquid provides an energetically attractive site for molecules that have both polar and apolar parts, obtaining a clean interface is a serious challenge in any VSF experiment. For example, our experiments have shown remarkable sensitivity in the VSF water spectrum to trace amounts of impurities that migrate to the interface [7]. If these impurities are highly surface-active (as is generally the case), the spectrum of the impurity can dominate the interfacial spectrum. Since typical surface densities of surface-active species are $\sim 10^{14}$ molecules $\cdot$ cm^{-2}, nanomolar concentrations of highly surface-active impurity are a significant problem. Therefore, samples and solvents must be highly purified, and all equipment that comes in contact with the samples must be extraordinarily clean.

2.3.2. *Surface Tension*

The sum-frequency intensity generated from a sample is directly proportional to the square of the number of molecules probed in the experiment (Equations (6) and (8)). A determination of the number of molecules adsorbed to a liquid interface can be made by measuring the surface tension isotherm. Equilibrium surface tensions are recorded for a series of solute concentrations, and surface excess concentrations are extracted from the isotherm. The Wilhelmy plate method [33,34] was used to make the surface tension measurements in these studies. In the Wilhelmy method, a thin rectangular hydrophilic plate is suspended from a sensitive balance or tensiometer into an aqueous solution and a meniscus forms on the plate as it is wetted drawing it into the solution. A hydrophilic plate is used for measurements of aqueous solutions so that the contact angle is zero, and the only component of the force on the plate is in the vertical direction. If the contact angle is not zero, the angle must be determined. Often the uncertainty on such an angle measurement is unacceptably large. The apparatus measures a line tension along the line of contact between the water surface and the plate. If the balance has been calibrated, the reading on the balance is the surface tension (software packages will also report the surface pressure).

The balance used in the measurements presented below is a KSV 5000 with a dynamic range of 0–100 mN/m. Samples were prepared in crystallizing dishes 60 mm in diameter and 35 mm deep. The measurements were made by immersing about 1/3 of the plate in the lower phase. Measurements were made using a roughened platinum plate that had been cleaned with NoChromix solution and flamed until red hot to make it hydrophilic.

2.4. APPLICATIONS

2.4.1. *Simple Surfactants*

The class of compounds most extensively studied at liquid surfaces by VSF is the alkyl ionic surfactants. The simplest type of these surfactants consists of a charged polar headgroup and a long hydrocarbon chain and represents typical surfactants used in commercial products and industrial processes. In a typical soap or detergent solution, if the concentration of surfactant is high enough, the surfactant molecules form micelles in aqueous solution. The surfactant molecules in the micelles are oriented with their hydrocarbon chains on the inside away from the water, with their hydrophilic headgroups on the outside solvated by the water. Dirt and oil are then solubilized inside the micelle, and the water carries the micelle away. The key to understanding surfactant behaviour is in understanding the molecular interactions within the micelle and between the micelle and the water and the oil. With conventional techniques it is very difficult to examine these interactions separately from the rest of the system. However, if the surfactant is allowed to accumulate at an interface, VSFS can be used to investigate the monolayer itself where the molecular interactions are similar to those in the micelle. In addition, many surfactant systems are designed with the primary function of reducing the surface tension of water to increase emulsification of a poorly (or non) soluble component or to increase mass transport across the interface. VSFS studies can shed light on these systems as well.

2.4.1.1. Hydrocarbon Chain Order. The initial VSFS studies of surfactants at a liquid interface were conducted with a focus on the conformation of the alkyl chains as a function of the number density of molecules at the interface and as a function of alkyl chain length for a complete monolayer [1,6,35,36]. The C-H stretch modes of these chains lie in the 3-μm region, coincident with the region most readily available with tunable IR lasers used in VSF experiments. In these experiments, as in other vibrational spectroscopic studies, the ratio of the integrated methyl symmetric stretch intensity to the integrated methylene symmetric stretch intensity has been used as a parameter for examining the relative conformation (ordering) of alkyl chains at an interface. In VSFS using p-polarized IR, an all-trans chain has the transition moment for the methyl symmetric stretch aligned along the direction of the electric field and produces a maximum SF signal. However, since there are local inversion centers at the midpoint of each C-C bond in an all-trans alkyl chain, intensity in the methylene symmetric stretch band is forbidden under the dipole approximation. (The Raman and IR moments of the methylene units are 180° out of phase, resulting in cancellation of the SF response.) Conversely, under this same polarization combination (ssp), highly disordered chains exhibit a more isotropic orientation of the methyl groups producing less SF intensity in the methyl symmetric stretch, while the increased number of gauche defects in the alkyl chains results in an increase in intensity in the methylene symmetric stretch. Therefore, the intensity ratio CH_3SS/CH_2SS approaches infinity for perfect order with a strong methyl symmetric stretch and the absence of methylene symmetric stretch intensity, and a smaller ratio is expected for increasing disorder as the methylene mode grows in intensity and the methyl mode loses intensity.

From studies of Langmuir films (insoluble surfactants) and films adsorbed to solid substrates, alkyl chains are known to be well ordered. For soluble surfactants at oil/water, however, the picture is much different. Several studies from this laboratory have demonstrated these differences [6,37–39]. Figure 2.6 shows the ssp spectra for sodium dodecyl sulfate (SDS) at the CCl_4/D_2O interface at monolayer coverage (squares) and at extremely low surface coverage (circles) with the spectra normalized to the methyl symmetric stretch [6].

There are three clear features in these spectra. First, the signal to noise is high even for a very low surface concentration, exhibiting the sensitivity of VSFS. Second,

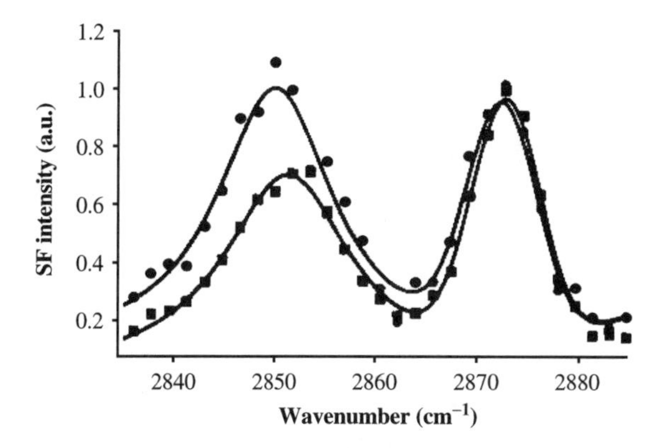

FIGURE 2.6. VSF spectra of the symmetric stretch region for alkyl chains in SDS at the CCl_4/D_2O interface. The spectra are for 0.1 mM (•) and 10 mM (■) aqueous SDS concentration, corresponding to low and high (monolayer) surface coverage. The spectra were acquired under ssp polarization. From Ref. [6].

there is a strong methylene symmetric stretch intensity at monolayer coverage suggesting that there are still a significant number of gauche defects in the chains. Third, the CH_3SS/CH_2SS ratio is larger for the monolayer than for the fraction of a monolayer, indicating more order in the alkyl chains in the full monolayer. The order parameter was examined as a function of bulk concentration, and it was seen that the order parameter increased up to a concentration of 2 mM (monolayer coverage) and then levelled off. As the surfactant molecules become more congested at the interface, the interactions between the hydrophobic chains would be expected to increase, resulting in a decrease in gauche defects in the chains. This is exactly what is observed. However, even at monolayer coverage at the liquid/liquid interface, the appreciable signal from the methylene symmetric stretch indicates that these chains have significant disorder relative to ordered chains observed for some surfactants at solid/air and air/water interfaces where this mode disappears at high coverage. The significant differences between this interface and the solid interface are that the molecules are not bound to adsorption sites and that the chains are solvated by CCl_4, which reduces their interactions and ability to order.

2.4.1.2. Surfactant Headgroup. Just as VSFS allows the determination of molecular conformation at the liquid interface as a function of surface coverage, it also allows for the determination of surfactant headgroup effects on the molecular conformation. The left side of Figure 2.7 shows spectra acquired under the ssp polarization configuration for sodium dodecyl sulfate (SDS), sodium dodecyl sulfonate (DDS), dodecyltrimethylammonium chloride (DTAC) and dodecylammonium chloride (DAC) at approximately equal surface coverage, and the right side shows spectra acquired under the sps polarization configuration [37]. The two cationic surfactants exhibit greater order than the anionic surfactants, and after an analysis of ordering as a function of surface concentration, the anionic surfactants were determined to have the same degree of ordering while DAC orders better than DTAC. The ordering relationship between surfactants of like charge is understandable based on headgroup size. SDS and DDS have very similar headgroups while DAC has a much smaller headgroup than DTAC. It has been concluded in these

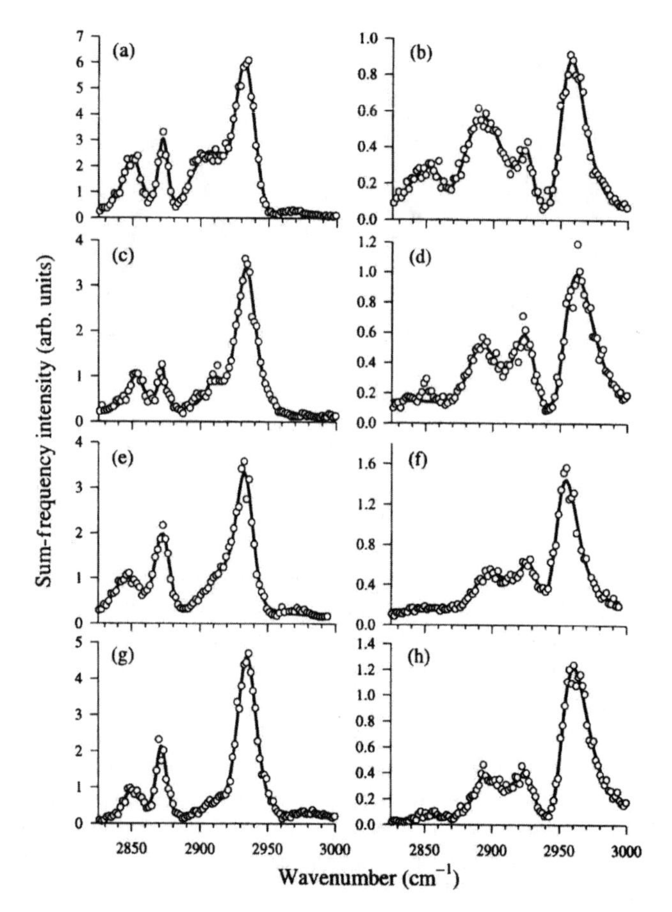

FIGURE 2.7. VSFS spectra of four surfactants, each with an alkyl chain of 12 carbons, but with different headgroups. Each spectrum was obtained for a bulk concentration of 5.0 mM. Spectra acquired under ssp polarization: (a) SDS, (c) DDS, (e) DTAC and (g) DAC. Spectra acquired under sps polarization: (b) SDS, (d) DDS, (f) DTAC and (h) DAC. From Ref. [37].

studies that the differences in alkyl order between the cationic and anionic surfactants are probably due to solvation effects and the depth of penetration of the headgroup into the aqueous phase.

The headgroup of a surfactant can also be probed with VSFS, provided there are vibrational modes that fall within the tuning range of the IR beam. Sodium dodecylbenzenesulfonate (DBS) is an important industrial and commercial surfactant used in cleansers and detergents. It is highly soluble and extremely surface-active. To investigate the properties that make this compound such a quality surfactant, VSFS studies were conducted at the CCl_4/D_2O and air/water interfaces examining both the alkyl chains and the phenyl moiety of the headgroup [40].

Alkyl chain conformation was examined at the CCl_4/D_2O and air/water interfaces and compared to sodium dodecylsulfonate (DDS), which has the same chain length, at the same interfaces. DDS exhibits typical simple surfactant behaviour at both interfaces.

The chains become more ordered as monolayer coverage is approached, but the chains do not achieve a high degree of order as indicated by the significant methylene symmetric stretch intensity at monolayer coverage. DBS, on the other hand, does not undergo any detectable ordering of the chains as monolayer coverage is approached although the particular locations of the gauche conformations may be changing. The chains are highly disordered at all surface concentrations at both interfaces. Even in the presence of excess salt, which screens the charged headgroups and allows the DBS molecules to pack more tightly, there is no significant change in the order of the alkyl chains.

A comparison of the DBS and DDS data indicate that the phenyl headgroup has a significant influence on chain order for DBS relative to DDS. If the phenyl rings were present in a thin horizontal plane at the interface analogous to what might be expected for their adsorption onto a solid surface, then one might expect that the much bulkier phenyl headgroup could reduce the van der Waals interactions between the first few methylene units on adjacent chains, leading to the disruption of chain ordering in DBS. However, surface tension experiments showed that DDS and DBS have very similar surface concentrations at monolayer coverage. This suggests that the DBS molecules are in a staggered arrangement at monolayer coverage. In this picture, the headgroups penetrate to different degrees into the interfacial region, which molecular dynamics calculations have indicated as ~7 Å (for CCl_4/H_2O) [41,42]. The polarizable nature of the phenyl ring should at some level facilitate this staggered distribution in an interface of this thickness relative to the less polarizable DDS. Such staggering of adjacent headgroups would reduce the interaction between the first few methylene groups on adjacent chains leading to disruption of chain ordering.

Phenyl orientation in DBS was examined by following the strongest phenyl mode (analogous to υ_2 in benzene), which has its IR transition moment pointing from the alkyl chain to the sulfonate. Spectra from the CCl_4/D_2O and air/D_2O interfaces showed an obvious change in SF intensity as the bulk concentration was increased in each case. There are two possible contributions to these increases. The first is the increased number of molecules present at the interface, and the second is a changing orientation of the transition moment as a result of crowding and intermolecular interactions. Figure 2.8 shows the SF intensity dependence on surface concentration, and it was found that intensity at the air/water interface was dependent on the number of molecules at the interface and a change in orientation, while at the CCl_4/D_2O interface, intensity was dependent only on the number of molecules (as indicated by the linear relationship). In addition, it is quite clear that most of the orientational ordering at the air/water interface occurs very close to surface saturation. The reorientation of the phenyl groups at the air/D_2O interface as monolayer is approached is attributed to an increased interaction between neighbouring phenyl groups and/or congestion at the interface because of the bulky headgroups, while the lack of reorientation of the phenyl groups at the CCl_4/D_2O interface along with spectral evidence for the phenyl rings to orient along the normal to the interface is attributed to solvation of the polarizable phenyl group in the interfacial region. It is interesting to note that the changes in ring orientation, or lack thereof, appear to have no effect on the order of the alkyl chains.

From these studies, a picture of the DBS micelle emerges that is consistent with good detergency. The compound is highly surface-active, and the micelles are likely to

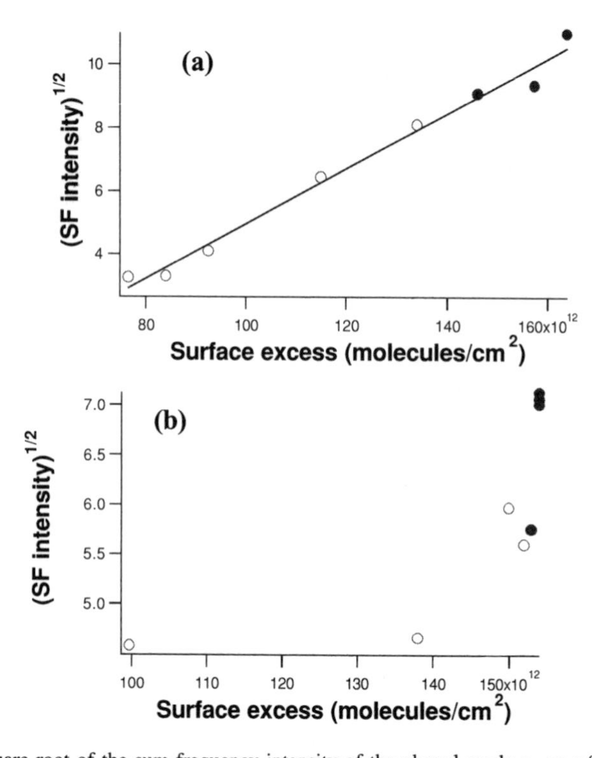

FIGURE 2.8. Square root of the sum-frequency intensity of the phenyl mode υ_2 as a function of surface concentration for (a) DBS at the CCl_4/water interface with 0.1 M NaCl. (The line is a fit to the data.) (b) DBS at the air/water interface. The solid data points correspond to surface concentrations at and near full monolayer coverage. Adapted from Ref. [40].

have a liquid-like interior (i.e. the capability to dissolve organic matter) as evidenced by the high disorder in the alkyl chains at all interfacial packing densities.

2.4.2. *Lipid Monolayers and Anesthetic Action*

2.4.2.1. Lipid Conformation in Phospholipid Monolayers. Phospholipid monolayers have been the subject of intense interest for over three decades [43–45] because they are the major component of most cell membranes and consequently are important as models for these membranes. Phospholipids consist of a charged headgroup connected to a pair of long acyl chains by means of a three-carbon glycerol backbone. They form Langmuir monolayers at liquid interfaces, exhibiting a host of different phases and morphologies. Understanding the rich thermodynamic behaviour found in phospholipid monolayers has helped to elucidate the nature of the more complex phase behaviour that takes place in bilayer systems. Since properties such as permeability, compressibility and phase transition temperatures depend on the identity of the phospholipid headgroup and acyl chain lengths and saturation, considerable effort has focused on correlating monolayer structure with function. Given that under many conditions a phospholipid monolayer closely approximates half of a lipid bilayer, a molecular level picture of the interactions in the monolayer will lead to a better understanding of the interactions in the full bilayer.

Numerous techniques have been employed to examine the monolayer structure of phospholipids at the air/water interface including surface tension, fluorescence, neutron and X-ray reflection, and IR and Raman spectroscopy. In contrast, very few techniques are suitable to examine monolayers at the oil/water interface. Surface tension and fluorescence microscopy [46–48] have shed some light on these buried monolayers, but most other surface techniques are hampered because of effects from the bulk liquids. Since VSFS is insensitive to the bulk, it is an excellent technique for probing these monolayers.

Since the solidity or fluidity of the bilayer membrane is likely to depend on the alkyl chain interactions and consequently their length, an understanding of the relationship between chain order and chain length for tightly packed monolayers of phospholipids are important. As an example of how VSFS can be employed to study phospholipids at a liquid surface, a series of saturated symmetric chain phosphatidylcholines (PCs) were examined at the air/water and CCl_4/water interfaces [49]. At the air/D_2O interface, chain order within the monolayer was found to increase as the length of the chains increased (Figure 2.9a) under conditions of constant phospholipid head group area.

This increase in order is attributed to the ability of the chains to interact through van der Waals forces and longer chains having more interaction sites. When these phospholipids were examined at the CCl_4/water interface, the trend was reversed. Monolayers composed of longer chain phosphatidylcholines show greater chain disorder (Figure 2.9b). This increase in disorder with chain length is attributed to solvent penetration into the monolayer that disrupts the van der Waals interactions between neighbouring chains.

Another set of studies from this laboratory examined the assembly of symmetric and asymmetric phosphatidylcholines at the CCl_4/water interface [50]. In these studies, a series of saturated symmetric and asymmetric chain PCs were examined. Symmetric PCs with 16 or fewer carbons per acyl chain and highly asymmetric PCs were found to produce relatively disordered films at the CCl_4/water interface as measured by VSFS. However, the longest chain PCs studied, 1,2-distearoyl-sn-PC (C18:C18), 1-stearoyl-2-palmitoyl-sn-PC (C18:C16) and 1-palmitoyl-2-stearoyl-sn-PC (C16:C18), formed well-ordered crystalline phase monolayers at room temperature. These results have been explained in terms of enhanced chain–chain interactions among the longer, nearly symmetric hydrocarbon chains that reduce the intercalation of solvent.

Solvent structure can also be monitored with VSFS. Another study by Walker *et al.* [51,52] showed that, at least for DLPC (C12:C12) at the CCl_4/water interface, the hydrogen bonding structure in the plane of the interface is significantly different from that perpendicular to the interface. The energy of the symmetric stretch of symmetrically hydrogen-bonded water molecules shifts to higher energy for the out-of-plane response relative to the in-plane response. The symmetric stretch of symmetrically hydrogen-bonded water was monitored as a function of time while the monolayer was forming, and it was found that the intensity oscillated in a complicated but systematic and reproducible fashion, suggesting that the water is sensitive to two-dimensional phase transitions in the monolayer as it forms.

2.4.2.2. Halothane's Effect on Lipid Monolayers. For more than 100 years the effects of inhaled anesthetics have been studied with the ultimate goal of understanding the

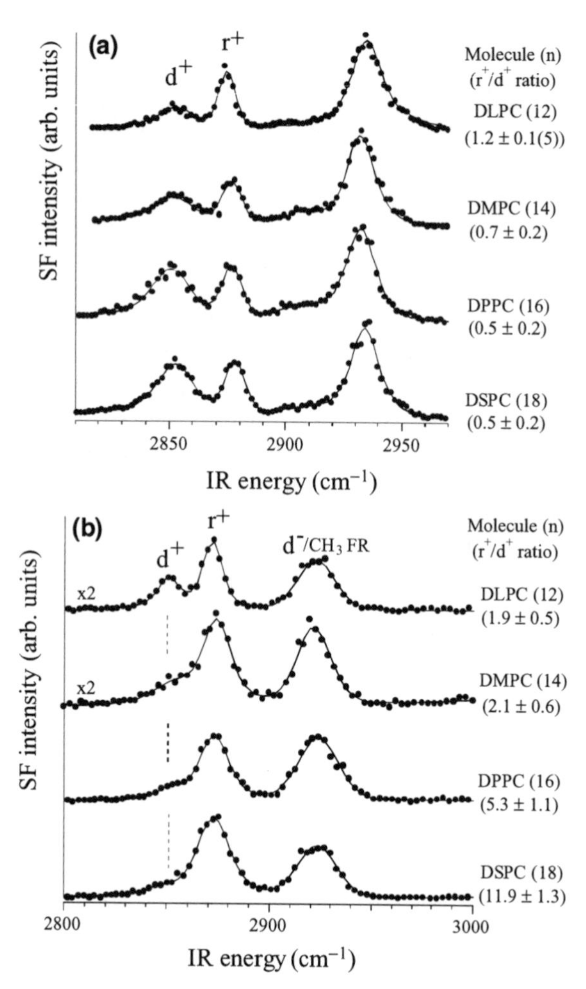

FIGURE 2.9. VSF spectra of tightly packed monolayers composed of DLPC ($n = 12$), DMPC ($n = 14$), DPPC ($n = 16$) and DSPC ($n = 18$) where n is the number of carbon atoms in the acyl chains. (a) The CCl_4/water interface. (b) The air/water interface. Order parameters appear beneath the phosphocholine acronyms. Spectral features were fit with Voigt profiles, although the solid lines in the figures serve only as guides to the eye. From Ref. [49].

mechanism, or action site, of general anesthesia [53,54]. Although tremendous effort has been expended in this endeavour, and an overwhelming number of experiments have been completed, the controversy remains whether the primary site of action for inhaled anesthetics is the membrane lipid or the proteins embedded in the membrane.

Despite the continuing debate, there is much that is known about anesthetic action. It is clear that local anesthetics significantly change ion currents across neural membranes, that the ion channels through which they pass are composed of protein, and that these proteins are embedded in lipid [53]. However, there is little knowledge of the mechanism by which lipid soluble anesthetics of such widely different structures affect the ion channels. Very recently, interest in the role of membrane lipids was renewed when Cantor [55,56] hypothesized that perturbations in membrane structure induced by inhaled anesthetics induces changes in protein function. However, recent experiments examining this hypothesis utilizing relatively simple model lipid systems do not agree on the active sites occupied by inhaled anesthetics in the membrane. Experimentally suggested sites include the acyl chain domain [57], various locations in the headgroup region [58,59] and the lipid/water interface [60,61].

In the study described here, VSFS spectra of a series of phospholipid monolayers at the CCl_4/water interface were acquired and the effect of a common inhaled anesthetic, halothane ($CF_3CHClBr$), on their conformation was examined [62]. Phospholipid monolayers at an organic/aqueous interface are a useful model system for the study of cell membranes because they provide a realistic model of the hydrophilic and hydrophobic environments commonly found in vivo. In addition, the understanding and thermodynamics of these monolayer systems has been extensively described and rigorous theoretical analyses have been reported [63]. In these studies, the CCl_4/D_2O interface was used for experimental convenience, and although this interface is non-biological, it does mimic the hydrophobic/aqueous interface found in many biological systems [64].

The phospholipid monolayers examined in this study were all saturated, symmetric, 1,2-diacyl-*sn*-glycero-3-phosphate-based lipids. Four different lipid headgroups attached to the phosphate were examined: choline, ethanolamine, glycerol and serine. Each lipid features a glycerol backbone, two saturated fatty acid chains and a phosphatidyl headgroup.

The spectra from each of the four monolayer systems exhibited similar relative peak intensities and similar peak positions throughout the C-H stretching region. The spectra show considerable intensity in both the CH_2SS and the CH_3SS modes for each of the phospholipids examined, indicating that the monolayers are all relatively disordered. The spectra also indicate that the nature of the headgroup does not significantly impact the ordering of the acyl chains. Information extracted from spectral fits indicate that DPPC, DPPE and DPPG are all similarly ordered with parameters of 0.62, 0.67 and 0.65, respectively, and that DPPS is the most disordered with an order parameter of 0.55. (Uncertainties are on the order of 5%.) The degree of chain order for these monolayers agrees well with previous results from this laboratory for phosphatidylcholines [49]. The large degree of disorder of these monolayers compared to monolayers at the air/water interface can be attributed to the intercalation of CCl_4 into the chains, which reduces van der Waals interactions between the chains. As a comparison, the order parameters for a series of diacyl phosphatidylcholines at the air/water interface (determined in this laboratory) were found to be: 1.9 for 12 carbon chains, 2.1 for 14 carbon chains, 5.3 for 16 carbon chains (DPPC) and 11.9 for 18 carbon chains [49].

An attempt was made to study changes in interfacial water structure in the presence of halothane in the monolayer; however, total absorption of the IR just above 3000 cm^{-1} by halothane in the CCl_4 prevents the acquisition of a complete, continuous spectrum.

Figure 2.10 shows the methylene and methyl symmetric stretch region of the spectrum for each headgroup before (solid squares) and after (solid triangles) the exposure of the monolayer to halothane. For each spectrum shown, there is a small increase in the overall intensity when halothane is present. This increase occurs across the entire spectrum for each of the phospholipids studied. We believe that the small change in intensity

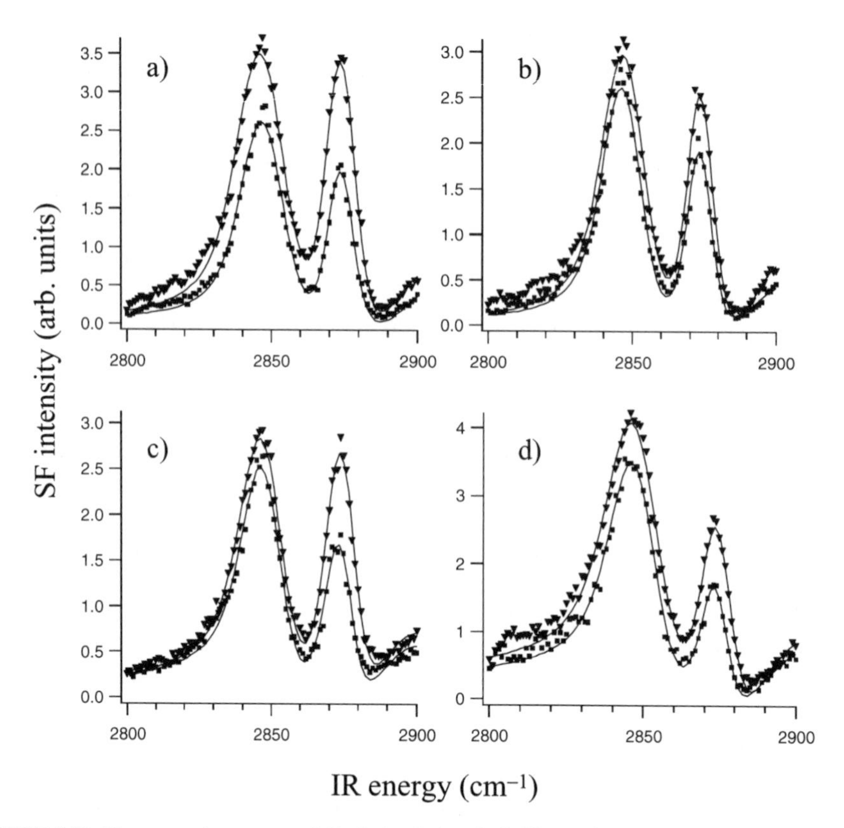

FIGURE 2.10. Representative spectra of dipalmitoyl phospholipid monolayers at the CCl_4/D_2O interface under the ssp polarization scheme showing the methyl and methylene symmetric stretch: (a) DPPC, (b) DPPE, (c) DPPG, (d) DPPS. Spectra of the monolayers are shown in solid squares. Spectra of the monolayers with halothane are shown with solid triangles. The lines are fits to the data. From Ref. [62].

is due to the alteration in the index of refraction in the interfacial region because of the presence of halothane. In the TIR geometry used, the SF intensity is highly sensitive to the incident angles of the incoming beams that are in turn dependent on the index of refraction. The spectra are corrected for the indices of refraction of the two input beams and the outgoing SF beam based on the indices for the bulk liquids and cannot be corrected for the change in index of the interface because of halothane, as that change is unknown.

For each phospholipid shown in Figure 2.10, the CH_3SS and the CH_2SS peaks are affected by the introduction of halothane. Visual inspection of the spectra suggests that the order parameter increases in the presence of halothane with the CH_3SS intensity increasing more than the CH_2SS. However, verification of this increase in the order parameter requires fitting the data as previously described. When this is done it is found that the order parameter does increase for each of the monolayers studied. For DPPC, the order parameter increases from 0.62 to 0.74; DPPE shows a change from 0.67 to 0.71, DPPG from 0.65 to 0.76 and DPPS from 0.55 to 0.60. The results from monolayers composed of lipids with dimyristoyl chains instead of dipalmitoyl chains are qualitatively similar, also showing a slight increase in chain order. These results demonstrate that halothane has an effect on the interactions between phospholipid molecules in the monolayer, leading to changes in the average molecular conformation. The effect is small suggesting that halothane is not strongly bound to the monolayer. The interaction could be local, a particular binding site for example, or non-local, i.e. the halothane just takes up space forcing the chains to order due to steric effects.

Since chain order could be affected indirectly through halothane interactions with the phospholipid headgroup, the second objective of this study was to determine whether the headgroup has a significant effect on the chain ordering induced by halothane. The distribution of localized charges in the headgroups of the lipids studied here are quite varied, suggesting that any interactions between halothane and the phospholipid headgroup would be quite varied as well. The ethanolamine and choline are zwitterions, with the ethanolamine being less hydrophilic and having a more localized positive charge than the choline. The glycerol has a negative charge localized on the phosphate, and the serine exhibits the zwitterion of the ethanolamine in addition to a negative charge on the carboxyl group adjacent to the amine. Considering the significant differences in the polarities of these headgroups, the fact that each monolayer was affected in a qualitatively similar manner suggests that the mechanism of halothane action is not strongly associated with the headgroup or the headgroup/water interface. However, the present study does not preclude the glycerol backbone as the site of interaction, which has been suggested as the probable site in the molecular dynamics study of the DPPC bilayer referred to above.

The subtle changes in monolayer conformation noted here suggest that the hypothesis presented by Cantor that small perturbations to membrane structure could be responsible for anesthetic action is a reasonable one [55,56]. In that work, it was shown that perturbations in the membrane would result in large changes in the lateral stresses in different parts of the membrane because the stresses themselves vary greatly in magnitude as a function of position along the bilayer normal. Although the changes in the stresses are small relative to the stresses themselves, the absolute magnitude of the change in the force felt by a transmembrane protein may be large enough to affect the function of the protein.

2.4.3. The Neat Organic/Aqueous Interface

Processes such as protein folding, membrane formation, ion transport and micellar assembly are all mediated by the interaction between water and a hydrophobic fluid. The driving force behind these processes is an anomalously large entropy loss usually explained as an enhancement in the hydrogen bond structure of water in the vicinity of apolar molecules, biological molecules and hydrophobic surfaces. Experimental measurements providing a molecular level view of these interfaces has been generally lacking because of the difficulties in selectively examining the interfacial molecules. Therefore, most of the understanding of intermolecular water structure at these interfaces has come from theory. Here we will discuss some recent experiments examining water structure at non-polar organic/aqueous interfaces [7,65,66].

2.4.3.1. Experiment. The anomalously high surface tension at a vapour/water interface is generally recognized as the result of strong hydrogen bonding between surface water molecules. At organic/water interfaces, the interfacial tension is known to decrease with the polarity of the organic phase until the phases are miscible. Spectral interpretation in the 2800–3200 cm^{-1} region of the alkane/H_2O systems is complicated by the fact that CH stretching modes absorb incident IR radiation in this spectral region and can complicate the interpretation. The CCl_4/H_2O system is void of this complication. Hence this system allows us to take a more detailed analytical look at the interfacial water environments that contribute to the VSF spectrum of the neat CCl_4/H_2O interface and how this interfacial water structure compares with water at other surfaces and with theoretical calculations. Figure 2.11 shows the vibrational spectrum of the OH stretching modes of water at the CCl_4/H_2O interface. The frequencies of these modes are highly sensitive to intermolecular hydrogen bonding. As the hydrogen bonding interaction decreases, the OH oscillator is strengthened, and the spectral peaks sharpen and shift to the blue. The intensity between 3100 and 3400 cm^{-1} is characteristic of tetrahedrally coordinated water molecules participating in strong hydrogen bonding interactions with their neighbours. This is the region in which one would expect a significant amount of intensity for an interface that is highly structured.

In an effort to separate the broad VSF CCl_4/H_2O spectral envelope into discrete H_2O spectral bands, corresponding to water molecules in distinct interfacial H-bonding environments, FTIR and VSF isotopic exchange studies have been performed. The results of these studies and FTIR spectra of water monomers in CCl_4 were used to make the assignments and generate the peaks and fits displayed in Figure 2.11.

Over the course of extensive experiments in our laboratory involving the CCl_4/H_2O interface, it has become clear that the structure and H-bonding of interfacial water molecules are *highly* sensitive to trace amounts of impurities that tend to concentrate at the interface. As the impurities are progressively removed, spectral features in the 3400–3600 cm^{-1} region increase and spectral intensity in the 3200 cm^{-1} region diminishes. Similar changes have been observed at several alkane/H_2O interfaces [7]. These findings have therefore modified previous interpretations, and can readily account for the differences between our observations and previous CCl_4/H_2O [67] and hexane/H_2O [68] VSF studies.

This study presents insight into the nature of the hydration of hydrophobic planar liquids. In particular, the studies show that in contrast to models describing the hydrophobic effect where a small non-polar solute in water tends to enhance water hydrogen bonding in its vicinity, water at a planar hydrophobic surface has weaker hydrogen bonding interactions with other water molecules than that occurs in the bulk phase. This weakening is manifested in a dominance of spectral intensity in the region where water molecules show minimal interaction with other water molecules. Accompanying the weakening of the H_2O–H_2O interactions is the increased importance of CCl_4–H_2O and hydrocarbon–H_2O interactions. Such interactions are strong enough to result in significant orientation of water molecules that either straddle the interface or orient into the organic phase. The existence of these organic–water interactions, although not explicitly addressed in most theoretical models of water/hydrophobic interfaces, is manifested in our everyday experience. A small drop of insoluble oil on a water surface will spread to form a monolayer on that surface. Additional evidence is provided by the stability of thin films of water

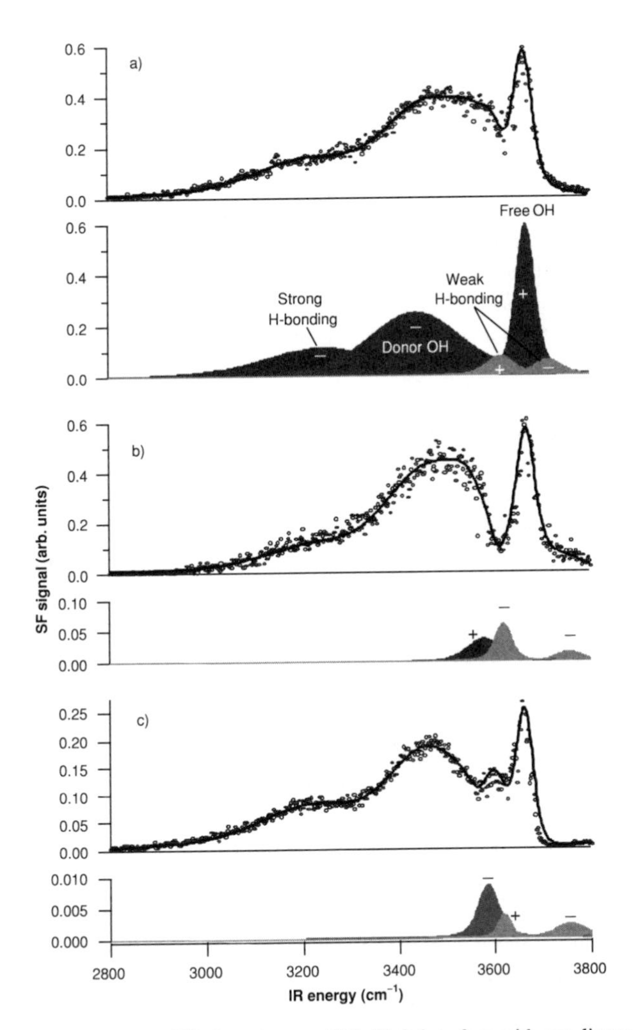

FIGURE 2.11. (a) VSF spectrum of H_2O at the neat CCl_4/H_2O interface with non-linear least-squares fit to the data superimposed as a solid line. Shown directly below are peaks contributing to the spectrum for the neat CCl_4/H_2O interface. (b) VSF spectrum of the CCl_4/H_2O interface at 10 nM SDS bulk aqueous phase concentration. The solid line is the fit to the data upon inclusion of contributing peaks (shown directly below) for OH modes of water molecules solvating the cationic headgroup of SDS at the CCl_4/H_2O interface. (c) VSF spectrum of the CCl_4/H_2O interface at 47 nM DTAC bulk aqueous phase concentration. The solid line is the fit to the data upon inclusion of contributing peaks (shown directly below) for OH modes of water molecules solvating the cationic headgroup of DTAC at the CCl_4/H_2O interface. All spectra are taken with ssp polarization. The shading of the solvating peaks corresponds to the water molecules displayed in Figure 2.15 with similar shading. The positive/negative sign associated with each solvating H_2O peak corresponds to the phase (amplitude sign) of the vibrational mode.

on hydrophobic surfaces, the basis behind mineral flotation. The results presented here are also consistent with the difference in the thermodynamics of the hydration of a small non-polar solute and the hydration of a large hydrophobic surface. From a free energy perspective, both cases are unfavourable [69]. However, the entropy and enthalpy terms act in very different ways. As the molecular dimension of the hydrated species increases, the system can no longer be in a regime where water can continue to hydrogen-bond to neighbouring water molecules as it did in the presence of a small solute. Hence, the entropy term for the solute is negative whereas for the large hydrophobic surface, it is positive. For a hydrocarbon surface, the interfacial energy, or free energy needed to create the interface, decreases with temperature because of the increasing importance of the positive entropy term. For the non-polar solute, the free energy of transfer decreases with temperature because of the negative entropy term. From a fundamental perspective, the outstanding question is one of dimension—what molecular scale is appropriate for these two different effects and what happens at intermediate length scales [70–72]?

2.4.3.2. Simulation. In this study, VSFS and molecular dynamics calculations were employed to examine the structure and dynamics of the hydrogen bonding network of water at the hexane/water, heptane/water and octane/water interfaces in detail [66]. The complementary nature of the approaches has allowed a more detailed understanding of the interface. The calculations provide information not available in the spectroscopic studies, namely the interactions between interfacial water molecules that are isotropically oriented. The direct and iterative comparison of experiment with theory allows for the improvement of the models used to describe water–water and water–solute interactions.

The results are compared to those above for the CCl_4/H_2O interface. Several properties of alkane/water and CCl_4/water interfaces suggest that their interfacial characteristics should be similar. The measured interfacial tensions are 49.7 mN/m for hexane/water and 45 mN/m for CCl_4/water [73,74], with molecular dipole polarizabilities of 11.9 and 11.2×10^{-24} cm^3 respectively [75]. However, IR experiments by Conrad and Strauss [76,77] show that water molecules dissolved in an alkane solvent are free to rotate while water dissolved in CCl_4 is relatively constrained. It is the details of these molecular interactions that dominate interfacial structure and dynamics.

VSF spectra of three alkane/water interfaces are shown in Figure 2.12 along with CCl_4/water for comparison. These spectra (acquired under ssp polarization) are very similar, suggesting that the hydrogen bonding environments of the water molecules present at these interfaces are also quite similar. Although these spectra show similar general features, there are notable differences in intensities around the free OH resonance. The free OH vibrational frequency, which is a sensitive probe of oil–water interactions, is shifted 5 cm^{-1} to the blue relative to the CCl_4/water interface. In addition, the integrated intensity of the free OH relative to the donor OH is smaller for hexane/water than for CCl_4/water. Since the free OH and donor OH populations must be equal, this difference must be due to differences in orientation or differences in the frequency distribution of the vibrations. The results of MD simulations shed some light on this issue. A subtle difference in the shape of the spectra in the free OH region is due to the absence of very weakly hydrogen-bonded species found at the CCl_4/water interface. These are water molecules with both hydrogen atoms not hydrogen-bonded to other water molecules. There are two types, unbonded water monomers and water molecules that are hydrogen bond acceptors. These environments are not seen in spectra of the alkane/water interfaces and are attributed to a stronger CCl_4–H_2O interaction that orients these molecules, resulting in their observation in VSFS spectra.

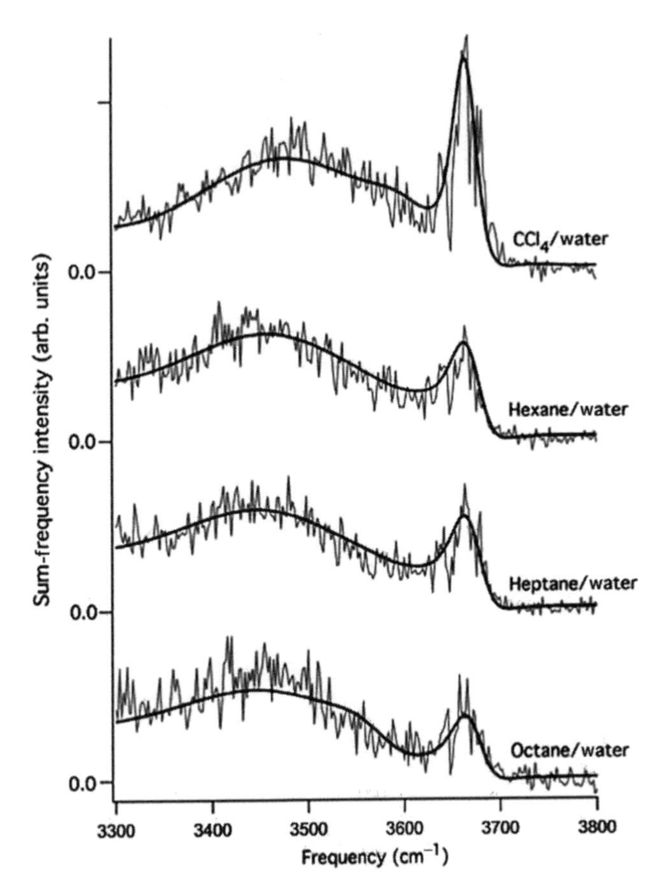

FIGURE 2.12. VSF spectra of the CCl_4/water, hexane/water, heptane/water and octane/water interfaces. The solid lines are fits to the data. From Ref. [66].

From the results of MD simulations, the non-linear susceptibility, $\chi^{(2)}_{\text{ssp}}$, can be calculated for each interfacial species of water molecule as a function of distance along the simulation cell (see Figure 2.13) to determine how each species contributes to the SF signal and to the depth that SF intensity is generated. Although this representation is only a first approximation of the SF probe depth, it is the most relevant measure of interfacial thickness for SF experiments because it indicates the depth to which water molecules are affected by the presence of the interface. To make a direct comparison to experiment, the contribution from each OH oscillator to the total $\chi^{(2)}_{\text{ssp}}$ is multiplied by a factor, linear in frequency, that accounts for the IR vibrational response dependency on frequency. For example, an OH vibration at 3400 cm^{-1} is approximately 12 times stronger in SF intensity than the free OH.

MD calculations of these interfaces give rise to simulated spectra that reproduce the experimental results (Figure 2.14) including the differences in intensity in the free OH region.

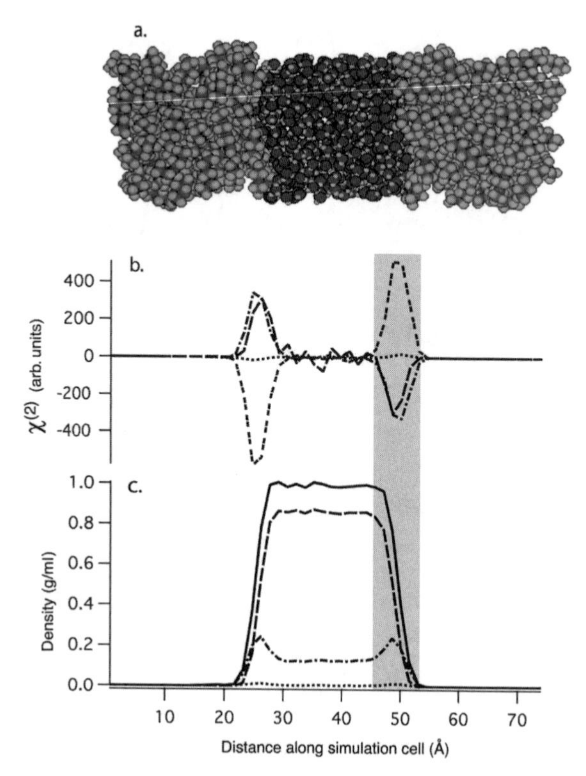

FIGURE 2.13. (a) Snapshot of the simulation cell for hexane/water (water is in the centre of the box). (b) Non-linear susceptibility obtained from MD simulation. The contributing species are the free OH (-- --), donor OH (-·-), water molecules with both hydrogens bonded (– –), and those with neither hydrogen bonded (··). (c) Density histogram for water and the contributing species from (b). The solid line is the total density. Donor OH and free OH contributions are equal. From Ref. [66].

Although no intermolecular cooperativity was built into the model calculations, the agreement between experiment and simulation supports the observation that the vibrations probed in the VSF spectrum arise primarily from water molecules with little cooperative motion. The simulations show that there is a relatively large density of water molecules that are hydrogen-bonded to other water molecules through both hydrogens, but their orientation tends to be in plane and isotropic, resulting in very little contribution to the intensity of the VSF spectrum. The agreement between theory and experiment suggests that the understanding of the spectroscopy of non-polar interfaces is converging.

2.4.4. Charges at the Organic/Aqueous Interface

Ion transport across an aqueous/non-aqueous boundary layer is one of the most important and prevalent physical processes that occur in living systems. The mechanistic role that solvating water molecules play in this transport process continues to be poorly understood on a molecular level. The same is true of our understanding of the role

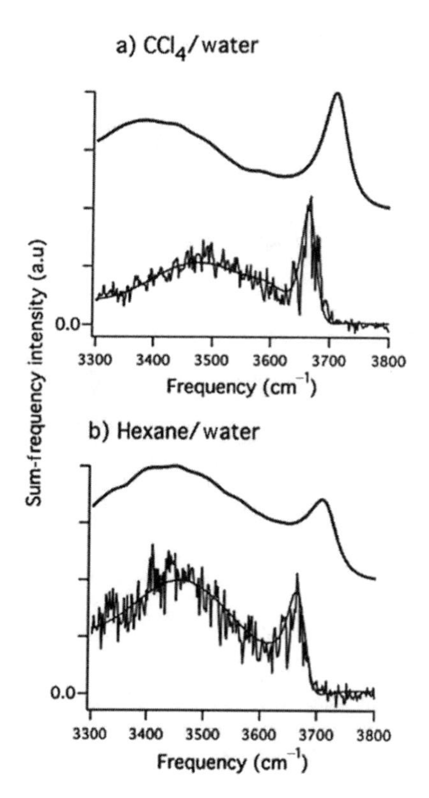

FIGURE 2.14. Calculated (offset) and experimental SF spectra of the CCl_4/water and hexane/water interfaces. From Ref. [66].

that solvating water molecules surrounding bound charge on macromolecules play in three-dimensional structure and assembly. Obtaining a microscopic understanding of processes such as protein folding, macromolecular assembly and ion transport at these surfaces requires a molecularly sophisticated understanding of how water solvates surface charges, an understanding that to date has largely been derived from theory [45,78–88].

These studies examined two charged surfactants, sodium dodecyl sulfate (SDS) and dodecyltrimethylammonium chloride (DTAC), adsorbed at the CCl_4/H_2O interface. The experiments were configured to probe vibrational modes in the water molecules that have their dipole transition components perpendicular to the interfacial plane. Figures 2.11a–c show the spectra of the OH stretch region of interfacial water molecules at the neat CCl_4/H_2O interface and the same interface with trace amounts of anionic (SDS) and cationic (DTAC) surfactant adsorbed at the interface from nanomolar aqueous phase surfactant concentrations. Relative to the neat interface (Figure 2.11a), the presence of trace amounts of SDS at the interface (Figure 2.11b) produces a large dip in VSF intensity near 3600 cm^{-1} and slightly increased intensity near 3700 cm^{-1}. For DTAC (Figure 2.11c), adsorption produces increased intensity near 3600 cm^{-1} and a decrease near 3700 cm^{-1}. As the concentration of these two surfactants increase up to bulk

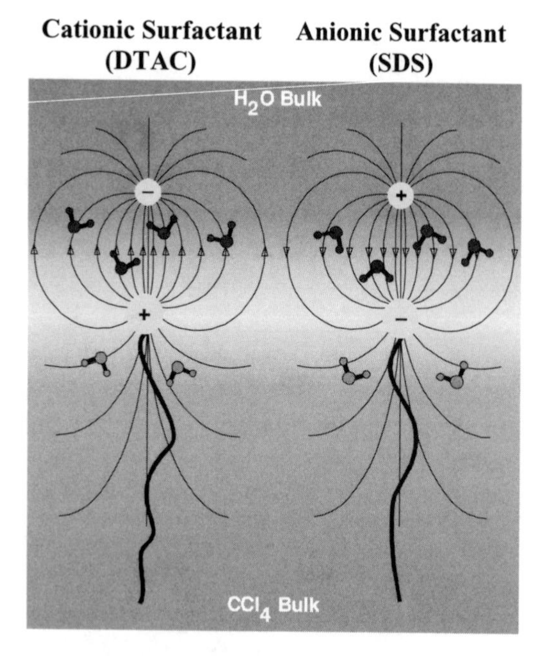

FIGURE 2.15. Schematic of solvating H_2O molecules displaying charge-dependent orientation due to charge–dipole interactions with anionic/cationic surfactants adsorbed at the CCl_4/H_2O interface. Solvating H_2O molecules in a CCl_4-rich environment are shown to orient in opposing directions corresponding to the charge on the surfactant headgroup. H_2O molecules in an aqueous rich environment are shown to orient similarly.

concentrations of ~200 nM, the observations become more pronounced but the spectral regions affected remain the same.

The spectra clearly show that the adsorption of charged surfactants at the CCl_4/H_2O interface at nanomolar aqueous phase surfactant concentrations results in a significant modification of the interfacial water behaviour. Wilhelmy balance surface pressure measurements [89] show that at these concentrations (headgroup areas of >4000 Å$^2 \cdot$ molecule^{-1}), the water molecules responsible for the observed spectral changes solvate *isolated* interfacial surfactant headgroups. The inverse spectral changes observed in the 3600 cm^{-1} and 3700 cm^{-1} regions for these two differently charged surfactants represent the opposite orientation of the solvating water molecules as influenced by the different electrostatic field orientation.

Figure 2.15 shows a schematic representation of the orientation and environment of the solvating water molecules assigned to the peaks shown in Figure 2.11b and c. For SDS, the peaks at 3619 ± 1 cm^{-1} and 3756 ± 10 cm^{-1} are attributed to solvating water molecules that lie in the CCl_4-rich portion of the interface. The assignment is facilitated by the neat interfacial studies described above. The negative amplitude of the 3619 cm^{-1} peak for SDS indicates that these water molecules are oriented by the electrostatic field to have their H atoms directed towards the aqueous phase. For DTAC, the OH modes of solvating water molecules that lie in the CCl_4-rich portion occur at the same frequencies but, as indicated by the sign derived for these peaks, these

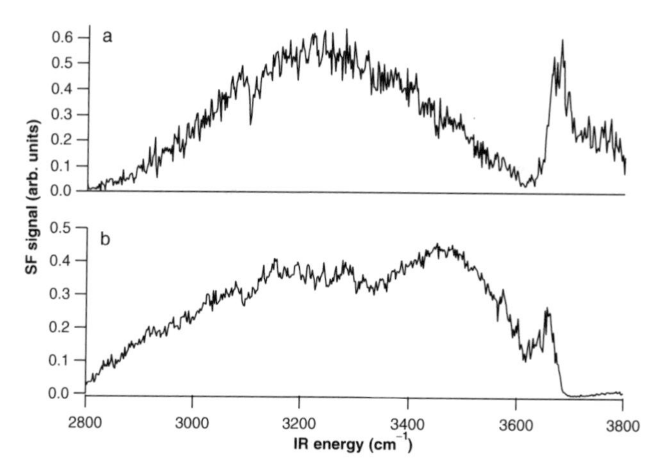

FIGURE 2.16. (a) VSF spectra of the CCl_4/H_2O interface at 292 nM aqueous phase concentration SDS. (b) VSF spectra of the CCl_4/H_2O interface with 399 nM aqueous phase concentration DTAC.

solvating water molecules are oriented in the opposite direction, with their H atoms pointed in the direction of the CCl_4 phase. The lower energy and broader peaks at $3577 \pm 2\ cm^{-1}$ (SDS) and $3585 \pm 2\ cm^{-1}$ (DTAC) are due to solvating H_2O molecules in an aqueous rich environment. These measured frequencies are coincident with those observed in previous spectroscopic studies of solvating H_2O molecules in aqueous salt solutions and are in general agreement with frequency shift trends observed for halides and oxyanions in aqueous solutions [90–94]. These lower energy peaks are of opposite phase (orientation), consistent with the different charge on SDS and DTAC, which orients these water molecules in opposite direction in the aqueous phase. Furthermore, these solvating water molecules in the aqueous environment are of opposite orientation to their counterparts residing in the organic rich phase. To emphasize, all of these water molecules show minimal bonding interactions with other water molecules at these low concentrations.

Interestingly, the intensities of the free OH peak, the H-donor OH peak and the peak representing tetrahedral bonding among water molecules are not significantly affected by the presence of the surfactant for the concentrations depicted in Figure 2.11b and c, or for bulk concentrations up to ~200 nM (L.F. Scatena and G.L. Richmond, in preparation). The CH stretch modes near 2800–3000 cm^{-1} are also not apparent at these low concentrations. Figure 2.16 shows that this behaviour changes with higher surfactant concentrations. Near 300 nM (~500 $Å^2 \cdot molecule^{-1}$), the intensity of the free OH mode (3669 cm^{-1}) begins to decline, indicating that at this point a measurable portion of free OH bonds are either bonding to the surfactant or are reoriented by the surfactant charge. Accompanying this decrease in intensity of the free OH mode is a progressively large increase in intensity near 3200 and 3400 cm^{-1}, indicative of stronger hydrogen bonding interactions between water molecules and the onset of an interfacial double layer.

The SF response at this point begins to sample and increase volume of oriented interfacial water, as shown in previous higher concentration studies [95,96].

2.5. SUMMARY AND FUTURE DIRECTIONS

The studies presented here illuminate just a few of the exciting possibilities for the use of VSFS to study chemistry at liquid/liquid surfaces. Solvents and adsorbates can be probed and orientations and conformations obtained. Molecular dynamics has recently been employed to gain additional information using the constraints provided by the spectroscopy. The future of this technique lies in expanding the spectrum to longer wavelengths so that more vibrations can be probed in each molecule and more complicated molecules can be studied. The study of interfacial dynamics will also offer exciting opportunities for the future.

ACKNOWLEDGEMENTS

The authors gratefully acknowledge the National Science Foundation (CHE-9725751), the Office of Naval Research, and the Department of Energy, Basic Energy Sciences for funding the studies described from this laboratory.

REFERENCES

1. P. Guyot-Sionnest, J.H. Hunt and Y.R. Shen, Phys. Rev. Lett. **59**, 1597–1600 (1987).
2. Y.R. Shen, Nature **337**, 519–525 (1989).
3. Q. Du, R. Freysz, Y.R. Shen, Phys. Rev. Lett. **72**, 238–241 (1994).
4. C.D. Bain, P.B. Davies, T.H. Ong, R.N. Ward and M.A. Brown, Langmuir **7**, 1563–1566 (1991).
5. D.E. Gragson and G.L. Richmond, J. Phys. Chem. (Feature Article) **102**, 3847–3861 (1998).
6. J.C. Conboy, M.C. Messmerand G.L. Richmond, J. Phys. Chem. **100**, 7617–7622 (1996).
7. L.F. Scatena and G.L. Richmond, Science **292**, 908–911 (2001).
8. H.C. Allen, E.A. Raymond and G.L. Richmond, Curr. Opin. Colloids Surf. **5**, 74–80 (2000).
9. K.B. Eisenthal, Chem. Rev. **96**, 1343–1360 (1996).
10. P.B. Miranda, Q. Du and Y.R. Shen, Chem. Phys. Lett. **286**, 1–8 (1998).
11. C. Schnitzer, S. Baldelli, D.J. Campbell and M.J. Schultz, J. Phys. Chem. A **103**, 6383–6386 (1999).
12. S.R. Goates, D.A. Schofield and C.D. Bain, Langmuir **15**, 1400–1409 (1999).
13. J. Zyss and D.S. Chemla, *Quadratic Nonlinear Optics and Optimization of Second-Order Nonlinear Optical Response of Molecular Crystals*, Eds. D.S. Chemla and J. Zyss, Academic Press, Orlando, FL, 1987, Vol. 1, pp. 23–191.
14. P.N. Butcher and D. Cotter, *The Elements of Nonlinear Optics*, Cambridge University Press, Cambridge, UK, 1990, Vol. 9.
15. K. Wolfrum and A. Laubereau, Chem. Phys. Lett. **228**, 83–88 (1994).
16. J. Lobau and K. Wolfrum, J. Opt. Soc. Am. B **14**, 2505–2512 (1997).
17. C. Hirose, N. Akamatsu and K. Domen, Appl. Spectrosc. **46**, 1051–1072 (1992).
18. C. Hirose, N. Akamatus and K. Domen, J. Chem. Phys. **96**, 997–1004 (1992).
19. D.R. Fredkin, A. Komornicki, S.R. White and K.R. Wilson, J. Chem. Phys. **78**, 7077–7092 (1983).
20. C.D. Bain, J. Chem. Soc. Faraday Trans. **91**, 1281–1296 (1995).
21. Y.J. Yang, R.L. Pizzolatto and M.C. Messmer, J. Opt. Soc. Am. B **17**, 638–645 (2000).
22. N. Bloembergen and P.S. Pershan, Phy. Rev. **128**, 606–622 (1962).
23. N. Bloembergen, H.J. Simmon and C.H. Lee, Phys. Rev. **181**, 1261–1271 (1969).
24. B. Dick, A. Gierulsky and G. Marowsky, Appl. Phys. B **38**, 107–116 (1985).
25. J.C. Conboy, *Investigation of Immiscible Liquid/Liquid Interfaces with Second Harmonic Generation and Sum-Frequency Vibrational Spectroscopy*, University of Oregon, Eugene, 1996.

26. J.C. Conboy, J.L. Daschbach and G.L. Richmond, J. Phys. Chem. **98**, 9688–9692 (1994).
27. P. Guyot-Sionnest, Y.R. Shen and T.F. Heinz, Appl. Phys. B **42**, 237–238 (1987).
28. S.R. Hatch, R.S. Polizzotti, S. Dougal and P. Rabinowitz, J. Vac. Sci. Technol. **11**, 2232–2238 (1993).
29. N. Bloembergen, Opt. Acta. **13**, 311–322 (1966).
30. P.B. Miranda and Y.R. Shen, J. Phys. Chem. B **103**, 3292–3307 (1999).
31. F. Schreier, Quant. Spectrosc. Radiat. Transfer **48**, 743–762 (1992).
32. G.R. Bell, Z.X. Li, C.D. Bain, P. Fischer and D.C. Duffy, J. Phys. Chem. B **102**, 9461–9472 (1998).
33. J.T. Davies and E.K. Rideal, *Interfacial Phenomena*, 2nd ed., Academic Press, New York, 1963.
34. E. Rame, J. Colloid Interface Sci. **185**, 245–251 (1997).
35. G.R. Bell, C.D. Bain and R.N. Ward, J. Chem. Soc. Faraday Trans. **92**, 515–523 (1996).
36. D.E. Gragson, B.M. McCarty and G.L. Richmond, J. Phys. Chem. **100**, 14272–14275 (1996).
37. J.C. Conboy, M.C. Messmer and G.L. Richmond, J. Phys. Chem. **101**, 6724–6733 (1997).
38. J.C. Conboy, M.C. Messmer, R. Walker and G.L. Richmond, *Prog. Colloid Polym. Sci., AmphiphilesIn-terfaces* **103**, 10–20 (1997).
39. M.C. Messmer, J.C. Conboy and G.L. Richmond, SPIE Proc. **2547**, 135–341 (1995).
40. M. Watry and G.L. Richmond, J. Am. Chem. Soc. **122**, 875–883 (2000).
41. D. Michael and I. Benjamin, J. Chem. Phys. **107**, 5684 (1997).
42. T.-M. Chang and L.X. Dang, J. Chem. Phys. **104**, 6772–6783 (1996).
43. G. Cevk and D. Marsh, *Phospholipid Bilayers*, John Wiley and Sons, New York, 1987, Vol. 5.
44. M.C. Phillips and D. Chapman, Biochim. Biophys. Acta **163**, 301–313 (1968).
45. M.N. Jones and D. Chapman, *Micelles, Monolayers and Biomembranes*, Wiley-Liss, New York, 1994.
46. X.F. Wang, K. Florine-Casteel, J.J. Lemasters and B. Herman, J. Fluoresc. **5**, 71–84 (1995).
47. K.J. Stein and C.M. Nobler, Ultramicroscopy **47**, 23–24 (1992).
48. A. Periasamy, X.F. Wang, G.G. Wodnick, S. Kwon, P.A. Diliberto and B. Herman, J. Microsc. Soc. Am. **1**, 13–23 (1995).
49. R.A. Walker, J.A. Gruetzmacher and G.L. Richmond, J. Am. Chem. Soc. **120**, 6991–7003 (1998).
50. B. Smiley and G.L. Richmond, J. Phys. Chem. B **103**, 653–659 (1999).
51. R.A. Walker and G.L. Richmond, Colloids Surf. A **154**, 175–185 (1999).
52. R.A. Walker, B.L. Smiley and G.L. Richmond, Spectroscopy **14**, 18–27 (1999).
53. I.C.P. Smith, M. Auger and H.C. Jarrell, *Molecular Details of Anesthetic-Lipid Interaction*, Eds. E. Rubin, K.W. Miller, and S.H. Roth, New York Academy of Sciences, New York, 1991, pp. 668–684.
54. L. Koubi, M. Tarek, M.L. Klein and D. Scharf, Biophys. J. **78**, 800–811 (2000).
55. R.S. Cantor, Biochemistry **36**, 2339–2344 (1997).
56. R.S. Cantor, Biophys. J. **76**, 2625–2639 (1999).
57. R.G. Eckenhoff, Proc. Natl. Acad. Sci. U.S.A. **93**, 2807–2810 (1996).
58. J. Baber, J.F. Ellena and D.S. Cafiso, Biochemistry **34**, 6533–6539 (1995).
59. C. North and C.S. Dafiso, Biophys. J. **72**, 1754–1761 (1997).
60. P. Tang, B. Yan and Y. Xu, Biophys. J. **72**, 1676–1682 (1997).
61. Y. Xu and P. Tang, Biochim. Biophys. Acta-Biomembranes **1323**, 154–162 (1997).
62. M.R. Watry and G.L. Richmond, Langmuir **18**, 8881–8887 (2002).
63. K.S. Birdi, *Lipid and Biopolymer Monolayers at Liquid Interfaces*, Plenum Press, New York, 1989.
64. C. Tanford, *The Hydrophobic Effect: Formation of Micelles and Biological Membranes*, Wiley-Interscience Publications, New York, 1973.
65. L.F. Scatena and G.L. Richmond, J. Phys. Chem. B **105**, 11240–11250 (2001).
66. M.G. Brown, D.S. Walker, E.A. Raymond and G.L. Richmond, J. Phys. Chem. B **107**, 237–244 (2003).
67. D.E. Gragson and G.L. Richmond, Langmuir **13**, 4804–4806 (1997).
68. Q. Du, E. Freysz and Y.R. Shen, Science **264**, 826–828 (1994).
69. M.E. Schrader and G.I. Loeb, *Modern Approaches to Wettability*, Plenum Press, New York, 1992.
70. C.Y. Lee, J.A. McCammon and P.J. Rossky, J. Chem. Phys. **80**, 4448–4455 (1984).
71. G. Hummer, S. Garde, A.E. García, M.E. Paulaitis, L.R. Pratt, J. Phys. Chem. B **102**, 10469–10482 (1998).
72. K. Lum, D. Chandler and J.D. Weeks, J. Phys. Chem. B **103**, 4570–4577 (1999).
73. D. Bartel, J. Phys. Chem. **52**, 480–484 (1952).
74. L.A. Girifalco and R.J. Good, J. Phys. Chem. **61**, 904–909 (1957).
75. *CRC Handbook of Chemistry and Physics*, 73rd ed., Boca Raton, FL, 1993.

76. M.P. Conrad and H.L. Strauss, Biophys. J. **48**, 117–124 (1985).
77. M.P. Conrad and H.L. Strauss, J. Phys. Chem. **91**, 1668–1673 (1987).
78. P.M. Wiggins, Microbiol. Rev. **54**, 432–449 (1990).
79. P.M. Wiggins, Physica A **238**, 113–128 (1997).
80. G. Wilse Robinson and C.H. Cho, Biophys. J. **77**, 3311–3318 (1999).
81. B.H. Honig, W.L. Hubbell and R.F. Flewelling, Ann. Rev. Biophys. Chem. **15**, 163–193 (1986).
82. K.J. Schweighofer, U. Essmann and M. Berkowitz, J. Phys. Chem. B **101**, 10775–10780 (1997).
83. K.J. Schweighofer, U. Essmann and M. Berkowitz, J. Phys. Chem. **101**, 3793–3799S (1997).
84. S.H. Lee and P.J. Rossky, J. Chem. Phys. **100**, 3334–3345 (1994).
85. I. Benjamin, Science **261**, 1558–1560 (1993).
86. I. Benjamin, Chem. Rev. **96**, 1449–1476 (1996).
87. P.A. Fernandes, M.N.D.S. Cordeiro and J.A.N.F. Gomes, J. Phys. Chem. B **103**, 6290–6299 (1999).
88. Y. Marcus, *Ion Solvation*, John Wiley & Sons, New York, 1986.
89. A.W. Adamson and A.P. Gast, *Physical Chemistry of Surfaces*, 6th ed., John Wiley & Sons, New York, 1997.
90. T.T. Wall and D.F. Hornig, J. Chem. Phys. **47**, 784–792 (1967).
91. G.E. Walrafen, J. Chem. Phys. **52**, 4176–4198 (1970).
92. G.E. Walrafen, J. Chem. Phys. **55**, 768–792 (1971).
93. I.M. Strauss and M.C.R. Symons, J. Chem. Soc., Faraday Trans I **74**, 2518–2529 (1978).
94. J.W. Schultz and D.F. Hornig, J. Phys. Chem. **65**, 2131–2138 (1961).
95. D.E. Gragson and D.E. Richmond, J. Am. Chem. Soc. **120**, 366–375 (1998).
96. D.E. Gragson and G.L. Richmond, J. Phys. Chem. B **102**, 569–576 (1998).

3

Observation of Dynamic Molecular Behaviour at Liquid/Liquid Interfaces by Using the Time-Resolved Quasi-Elastic Laser Scattering Method

Hiroharu Yui, Yasuhiro Ikezoe and Tsuguo Sawada
The University of Tokyo, 5-1-5-403, Kashiwanoha, Kashiwa, Chiba, 277-8561, Japan

3.1. INTRODUCTION

Physical and chemical properties of a liquid/liquid interface change with adsorption/desorption or chemical reactions of molecules. Especially for non-equilibrium systems, it is essential to make use of experimental probes with time-resolved measurement capability. The time-resolved quasi-elastic laser scattering (TR-QELS) method has advantages as a tool for *in situ*, non-contact time-resolved measurements of dynamic behaviour of molecules at liquid/liquid interfaces [1–11]. The method monitors the frequencies of capillary waves, which are spontaneously generated by a thermal fluctuation at liquid/liquid interfaces. Since the capillary wave frequency is a function of interfacial tension, and the change in the interfacial tension reflects the change in the number density of surfactant molecules at the interface, the TR-QELS method allows observation of dynamic changes of liquid/liquid interfaces such as the change in number density of surfactant molecules and the formation of a lipid monolayer. Owing to its improved time resolution, each power spectrum can be obtained in 1 ms to 1 s, so the method can be used to monitor dynamic changes at liquid/liquid interfaces in a real environment. Furthermore, this method has good interface selectivity for overcoming the problem of interference by bulk phases, because the capillary wave is a characteristic phenomenon at the interfaces and its frequency can be detected by an optical heterodyne technique.

In these past 10 years, it has been demonstrated that the TR-QELS method is a versatile technique that can provide much information on interfacial molecular dynamics [1–11]. In this chapter, we intend to show interfacial behaviour of molecules elucidated by the TR-QELS method. In Section 3.2, we present the principle, the historical background and the experimental apparatus for TR-QELS. The dynamic collective behaviour of molecules at liquid/liquid interfaces was first obtained by improving the time resolution of the TR-QELS method. In Section 3.3, we present an application of the TR-QELS method to a phase transfer catalyst system and describe results on the scheme of the catalytic reactions. This is the first application of the TR-QELS method to a practical liquid/liquid interface system. In Section 3.4, we show chemical oscillations of interfacial tension and interfacial electric potential. In this way, the TR-QELS method allows us to analyze non-linear adsorption/desorption behaviour of surfactant molecules in the system.

3.2. TIME-RESOLVED QUASI-ELASTIC LASER SCATTERING METHOD

3.2.1. *Capillary Waves*

Capillary waves occur spontaneously at liquid surfaces or liquid/liquid interfaces as a result of thermal fluctuations of the bulk phases. These waves have been known as surface tension waves, ripples or ripplons since the last century, and Lamb described their properties in his book *Hydrodynamics* in 1932 [12]. Before that, William Thomson (Lord Kelvin) mentioned these waves in some of his many writings.

Here we briefly present the relevant theory of capillary waves. The thermally excited displacement $\xi(\boldsymbol{r}, t)$ of the free surface of a liquid from the equilibrium position normal to the surface can be Fourier-decomposed into a complete set of surface modes as

$$\xi(\boldsymbol{r}, t) = \xi_0 \sum_k \exp[i\boldsymbol{kr} + \Omega t] \tag{1}$$

The complex wave frequency $\Omega(= i\omega_0 - \Gamma)$ is related to $\boldsymbol{k}$ via a dispersion relation. For an inviscid liquid, Lamb's equation is well-known as a classical approximation for the dispersion relation [12]

$$f = \frac{1}{2\pi}\left(\frac{\gamma}{\rho_U + \rho_L}\right)^{1/2} k^{3/2} \tag{2}$$

where f is the capillary wave frequency, γ the interfacial tension, ρ_U the density of the upper phase, ρ_L the density of the lower phase, and k the wavenumber of the capillary wave. This equation is valid for both liquid surfaces and liquid/liquid interfaces. For capillary waves at a liquid surface, ρ_U is regarded as zero in Equation (2). In this equation, only the normal component of the interfacial tension is considered to act as a restoring force on the thermally excited displacement $\xi(\boldsymbol{r}, t)$. Although this equation neglects the effect of surface viscosity, it gives a good fit in the relation between frequency and wavenumber of capillary waves at a liquid surface such as ethanol, anisol or water [13].

For the surface of a low-viscosity liquid, Levich[14] has derived the following dispersion relation:

$$D(S) = (S + 1)^2 + y - (2S + 1)^{1/2} = 0 \tag{3}$$

where $S = \Omega/2\nu k^2$ and $y = \sigma/4\rho\nu^2 k$. Here ν is the kinematic viscosity (η/ρ:η = the viscosity of the liquid). In this equation, both the normal components of the surface tension and the stress by the viscosity are considered to act as restoring forces on the displacement $\xi(\boldsymbol{r}, t)$. On the basis of this equation, a heterodyne power spectrum in the frequency domain becomes a Lorentzian profile centered at

$$\omega_0 = 2\pi f = |\mathrm{Im}\,(\Omega)| \tag{4}$$

having the full width at half-maximum (FWHM)

$$\pi\Delta f = |\mathrm{Re}\,(\Omega)| \tag{5}$$

Equation (3) is simplified to

$$f = \frac{1}{2\pi}\left(\frac{\gamma}{\rho}\right)^{1/2} \boldsymbol{k}^{3/2} \tag{6}$$

and

$$\Delta f = 2\nu k^2/\pi. \tag{7}$$

Equation (6) is the same as Lamb's classical approximation (2).

3.2.2. *Historical Overview*

Capillary waves at liquid surfaces were observed some years before the invention of lasers. Goodrich [15] reported on using a cathetometer to observe the damping of water waves by monomolecular films. Strictly speaking, this wave was a forced one and not a spontaneously generated capillary wave. The observation of spontaneously generated capillary waves was first reported by Katyl and Ingard [16,17]. They measured spectra from a methanol surface and an isopropanol surface using a He–Ne laser. Lucassen [18,19] theoretically derived the transverse mode of capillary waves and experimentally verified its existence. Bouchiat and Meunier [20] measured the liquid/gas interface of carbon dioxide near the critical point and reported agreement with theoretical predictions on the surface tension and viscosity. Mann *et al.* [21] investigated the dispersion relation at air/water interface using the QELS method and reported conformity with Lamb's equation. They also mentioned the possibility of application to interfacial tension measurements. Hård *et al.* [22] measured surface tensions using the QELS method and reported that the experimental values agreed with the theoretical ones within deviations of 2–10%. Sano *et al.* [13] measured the surface tensions of water and anisole and reported they were within deviations of 5% of the theoretical values. They also reported that the first-order approximation, i.e. Lamb's equation, was sufficient to calculate the surface tensions from capillary wave frequencies. Earnshaw and co-workers [23–25] have investigated liquid surfaces of surfactant solutions and reported on such fundamental properties as surface viscosity. Tsuyumoto and Uchikawa [26] proposed a homodyne detection technique with a simple optical alignment. They obtained the power spectra of a ripplon with 50–100 times higher S/N ratios. Furthermore, very recently some QELS instruments combined with another probe, for example an electrochemical cell [10,11,27] or a surface reflectometer [28], have been developed. Hosoda *et al.* recently reported an interesting application of the QELS method as a surface flow meter [29]. The QELS method is suitable for many kinds of liquid with low viscosity; on the other hand, the surface deformation method by laser picking-up developed by Sakai *et al.* [30] is noticeable because it can be used for highly viscous liquids like syrup.

As reviewed above, there have been many QELS studies on liquid surfaces. However, until several years ago, reports were scarce on molecular dynamics at liquid/liquid interfaces that used time courses of capillary wave frequency. Molecular collective behaviour at liquid/liquid interfaces from a QELS study was first reported by Zhang *et al.* in 1997 [3], and after that, other relevant experiments were reported [4–11].

3.2.3. Principle

The incident beam normal to the interface is quasi-elastically scattered by the capillary wave with a Doppler shift at an angle determined by the following equation (Figure 3.1):

$$K \tan\theta = k \tag{8}$$

where K and k are the wavenumbers of the incident beam and the capillary wave, respectively. Thus, the wavenumber k of the capillary wave is obtained by giving θ. A transmission diffraction grating is arranged in front of the cell to adjust the angle θ [20]. The angle θ is determined by the following equation using the spacing d and the order

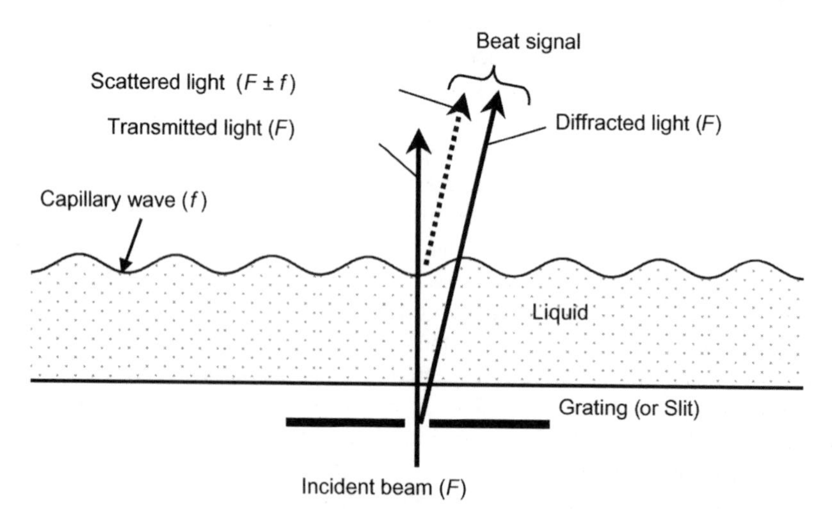

FIGURE 3.1. Principle of the quasi-elastic laser scattering method.

n of the diffraction grating,

$$d \sin\theta = n\lambda \tag{9}$$

where λ is the wavelength of the laser beam. (When we use a slit with spacing d, "$n + 1/2$" should be substituted for "n" in this equation.)

From Equations (8) and (9), we obtain the wavenumber k and the wavelength Λ of the observed capillary wave.

$$k = 2\pi n/d \tag{10}$$

$$\Lambda = d/n \tag{11}$$

The capillary wave frequency is detected by an optical heterodyne technique. The laser beam, quasi-elastically scattered by the capillary wave at the liquid/liquid interface, is accompanied by a Doppler shift. The scattered beam is optically mixed with the diffracted beam from the diffraction grating to generate an optical beat in the mixed light. The beat frequency obtained here is the same as the Doppler shift, i.e., the capillary wave frequency. By selecting the order of the mixed diffracted beam, we can change the wavelength of the observed capillary wave according to Equation (11).

3.2.4. *Experimental Apparatus*

A schematic diagram of the experimental set-up is shown in Figure 3.2. Figure 3.2a is an enlarged sketch of the sample space in Figure 3.2b. The beam from a diode-pumped YAG laser is incident on the transmitting diffraction grating or slit and passes through the bottom of the sample cell. The cell is made of quartz glass and has an optically flat bottom, which is indispensable to maintaining good reproducibility of the experimental results. After passing through the sample, the diffracted beams are mixed with scattered light from the capillary wave, and one of them is selected by an aperture positioned in front of

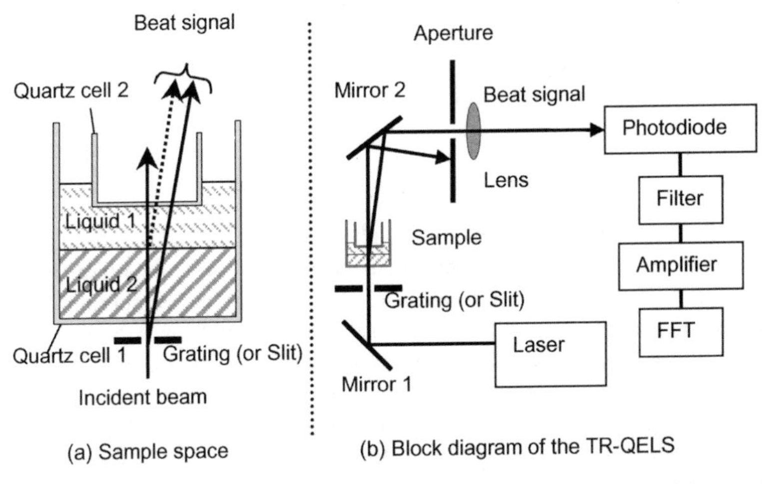

FIGURE 3.2. Schematic diagram of the experimental set-up. (a) An enlarged sketch of the sample space, and (b) a block diagram of the TR-QELS method for liquid/liquid interface investigation.

a photodiode. The optical beat of the mixed light is measured by the photodiode. Quartz cell 2 is set to prevent a light scattered by the capillary waves at the air/liquid interface from entering the photodiode. The signals are Fourier-transformed and saved by a digital spectrum analyzer. As needed, a thermostat (Section 3.3) or electrodes (Section 3.4, for a measurement of interfacial electrical potential) or syringe pump (Section 3.4, for continuous supply of an aqueous surfactant solution) may be combined with this set-up.

3.3. A PHASE TRANSFER CATALYTIC REACTION AT A LIQUID/LIQUID INTERFACE

3.3.1. Phase Transfer Catalysis

Phase transfer catalysis has attracted much attention because the catalyst activates the reaction by transferring across the liquid/liquid interface and brings about high product yield and high selectivity by a recurred cycle [31–35]. Tetrabutylammonium bromide (TBAB, $(n\text{-}C_4H_9)_4N^+Br^-$) is a commonly used phase transfer catalyst, which behaves as a surfactant and activates organic synthesis reactions [7,36]. It is empirically known that TBAB, whose carbon chain number is 4, brings about higher product yield than other quaternary ammonium bromides. Longer carbon chains of the quaternary ammonium bromide cause more difficult separation of product, and thus shorter carbon chains have an advantage as a phase transfer catalyst [37,38]. However, the reason TBAB shows a higher product yield than tetrapropylammonium bromide (TPRAB, $(C_3H_7)_4N^+Br^-$) and tetraethylammonium bromide (TEAB, $(C_2H_5)_4N^+Br^-$) has not been clarified so far. Clarification by an analytical approach to the interfacial reaction as well as a conventional study using the bulk phase concentration is highly desirable. Thus, in this section, we focused on a phase transfer catalytic reaction system with quaternary ammonium bromide, $R_4N^+Br^-$ (R: $n\text{-}C_4H_9$, $n\text{-}C_3H_7$, $n\text{-}C_2H_5$), whose reaction scheme is well known and whose operating conditions are mild. We previously reported that the liquid/liquid interface was the reaction site for formation of ion pairs ($TBA^+C_6H_5O^-$) based on observation of one process in the phase transfer catalysis [7]. The main topic of the present section is to review the dynamic molecular behaviour in the phase transfer catalytic reaction and the mechanism determining the product yield.

3.3.2. Apparatus and Sample Preparation

The reaction scheme between sodium phenoxide (C_6H_5ONa) and diphenylphosphoryl chloride (DPPC) in the water/nitrobenzene (W/NB) system with quaternary ammonium bromide, $R_4N^+Br^-$ (R: $n\text{-}C_4H_9$, $n\text{-}C_3H_7$, C_2H_5), is shown in Figure 3.3. At the beginning of the reaction, $R_4N^+Br^-$ reacts with C_6H_5ONa to form $R_4N^+C_6H_5O^-$ at the interface [7]. Then, this ion pair moves into the organic phase, where it reacts with DPPC to produce triphenyl phosphate ($(C_6H_5O)_3PO$). During this reaction, $R_4N^+Cl^-$ is also formed and transported into the water phase to react again with C_6H_5ONa. $R_4N^+Br^-$ activates the production of triphenyl phosphate by circulating between the two phases.

We investigated the dynamic behaviour of the ion pair ($R_4N^+Cl^-$,$R_4N^+C_6H_5O^-$) formed in the presence of $R_4N^+Br^-$ (1 mM), DPPC (1 mM) and C_6H_5ONa (70 mM).

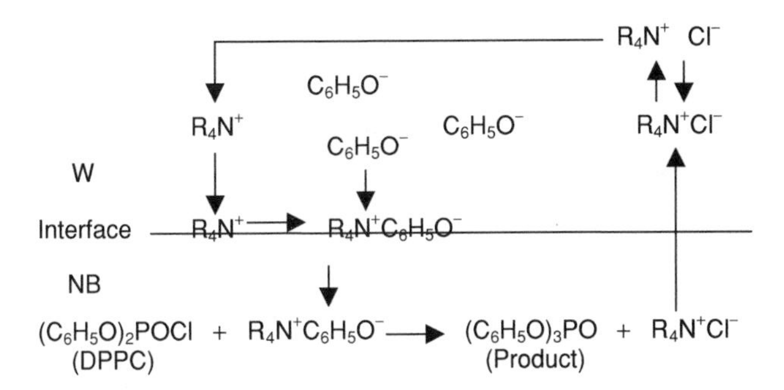

FIGURE 3.3. Proposed phase transfer catalytic reaction scheme between C_6H_5ONa and DPPC in the W/NB system.

We used three kinds of quaternary ammonium bromides, $R_4N^+Br^-$: tetrabutylammonium bromide (TBAB), $(C_4H_9)_4N^+Br^-$; tetrapropylammonium bromide (TPRAB), $(C_3H_7)_4N^+Br^-$; and tetraethylammonium bromide (TEAB), $(C_2H_5)_4N^+Br^-$.

A liquid/liquid interface was prepared by adding 9.8 ml of an aqueous solution of C_6H_5ONa (Aldrich) and NaOH (Kanto Chemical Co.) to 10 ml of nitrobenzene containing DPPC (Kanto Chemical Co.) in the quartz cell. The beam from a YAG laser (CrystaLaser, Model GCL-025S, 532 nm, 25 mW) was passed through a transmitting diffraction grating in front of the cell. The cell was made of quartz glass and had an optically flat bottom. After passing through the sample, one of the diffracted beams, which was mixed with the scattered light, was selected by the aperture in front of the photodiode (Hamamatsu Photonics S1190). Signals from the photodiode were Fourier-transformed and saved by a digital spectrum analyzer (Sony Tektronix Co., Model 3056). The wavelength of the observed capillary wave was 6.6×10^{-2} mm. The quartz cell was placed inside a copper cell that had a thermostat (Tokyo Rikakikai Co., UA-110) to keep the measurement temperature at 283 K since the W/NB interface is most stable at that value. After 0.2 ml of $R_4N^+Br^-$ (Kanto Chemical Co.) was injected into the water phase with a microsyringe, the power spectrum was obtained every 1 second. Ultrapure water (from Millipore Milli-Q system) was used for all sample preparations. NaOH was added to adjust the ionic strength to 0.2 and to prevent the production of phenol or benzoic acid. All chemicals were reagent-grade and were used without further purification.

3.3.3. TR-QELS Measurements of Phase Transfer Catalytic Reactions

First, a typical power spectrum of capillary waves excited at the W/NB interface is shown in Figure 3.4a. The errors on the values of the capillary wave frequency were 0.1 kHz, obtained as the standard deviation of 10 repeated measurements. Capillary wave frequency dependence on C_6H_5ONa is shown in Figure 3.4b. The frequency decreased significantly with increasing C_6H_5ONa concentration. This indicated that interfacial tension was decreased by the interfacial adsorption of C_6H_5ONa.

To elucidate the dynamic molecular behaviour in the phase transfer catalytic reaction, we investigated the time courses of the capillary wave frequencies after the injection of the TEAB, TPRAB and TBAB solutions into the water phase. The time just prior to

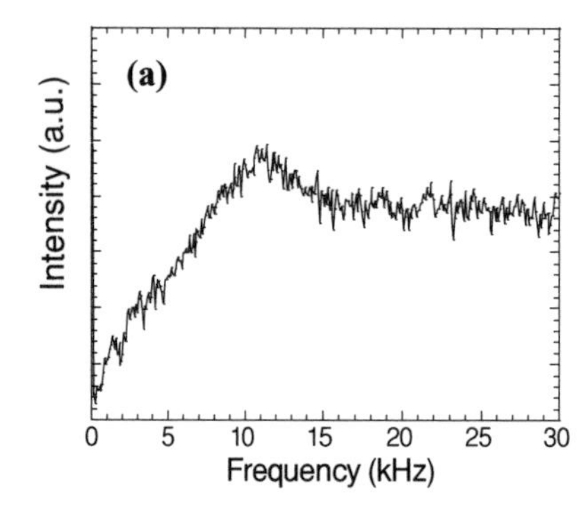

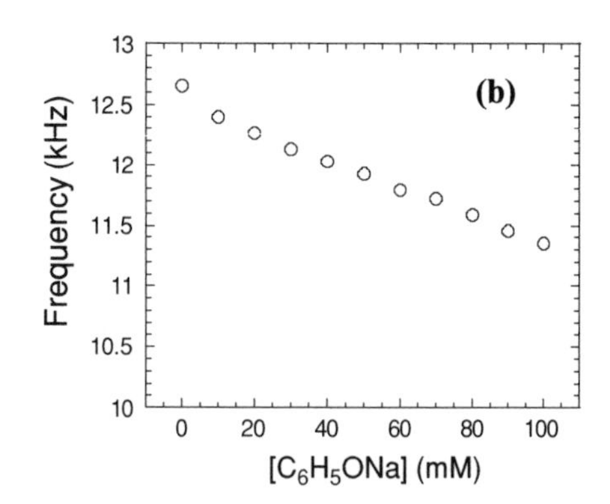

FIGURE 3.4. (a) Power spectrum for capillary waves excited at the W/NB interface (238 K). (b) Capillary wave frequency dependence on the concentrations of C_6H_5ONa (283 K).

the injection was regarded as 0 seconds, and the power spectrum was obtained every 1 second. Capillary wave frequencies after the injection of the TEAB, TPRAB and TBAB solutions are plotted as a function of time in Figures 3.5a, b and c, respectively. These curves were obtained with good reproducibility. No change in capillary wave frequency was observed in each case when only phase transfer catalysis ($R_4N^+Br^-$) was injected (open circles).

Then we measured capillary wave frequencies after the injection of the TEAB, TPRAB and TBAB solutions when the water phase contained C_6H_5ONa. The wave frequency itself was decreased in each case because of the adsorption of C_6H_5ONa onto the interface, but no change in the time course of the frequency after the injection was observed (open triangles). This suggested the following two possibilities. One was that interfacial C_6H_5ONa did not react with $R_4N^+Br^-$ under such conditions. The other was that interfacial C_6H_5ONa reacted with $R_4N^+Br^-$ and formed ion pairs $R_4N^+C_6H_5O^-$ at the interface, but this formation of ion pairs did not affect the capillary wave frequency. On the other hand, when the phase transfer catalyst was injected into the interface between the water containing C_6H_5ONa and the nitrobenzene containing DPPC, the capillary wave frequencies decreased gradually (open squares). After attaining their minimum, they increased gradually and then became constant after 8–15 seconds. These times are summarized in Table 3.1.

TABLE 3.1. Times of the minimum and stationary state frequencies and the change of frequency (Δf) of $R_4N^+Br^-$ estimated in this study.

	TEAB	TPRAB	TBAB
Time of minimum (s)	6	7	6
Time of stationary state (s)	13	14	9
Δf (adsorption) (kHz)	0.43	0.78	0.43
Δf (desorption) (kHz)	0.43	0.47	0.16
Ratio of adsorption onto the interface of stationary state (%)[a]	0	40	64

[a] $(\Delta f(\text{adsorption}) - \Delta f(\text{desorption})) / \Delta f(\text{adsorption})$.

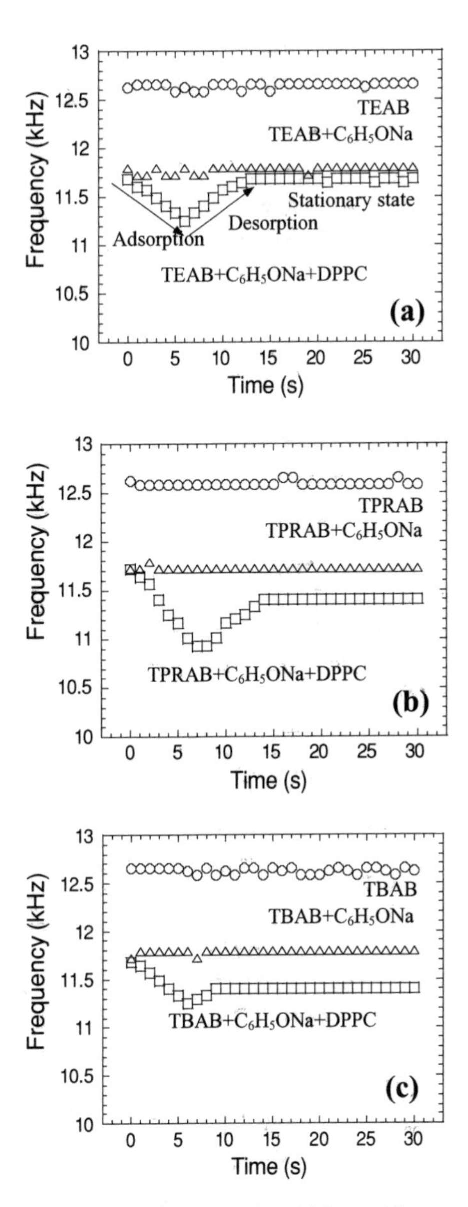

FIGURE 3.5. Time course of capillary wave frequency after an injection of a quaternary ammonium bromide solution (1 mM) at 283K. (a)TEAB, (b) TPRAB and (c) TBAB. The concentration of C_6H_5ONa in the water phase and DPPC in nitrobenzene phase was 70 mM and 1 mM, respectively.

We assigned the frequency changes as follows. As shown within the dotted line rectangle of Figure 3.6a, the decrease in frequency corresponded to the adsorption of $R_4N^+Cl^-$ onto the interface and the increase corresponded to the desorption from the interface, because the change in frequency was observed only when DPPC was added. The time of the minimum frequency was almost the same regardless of the phase transfer

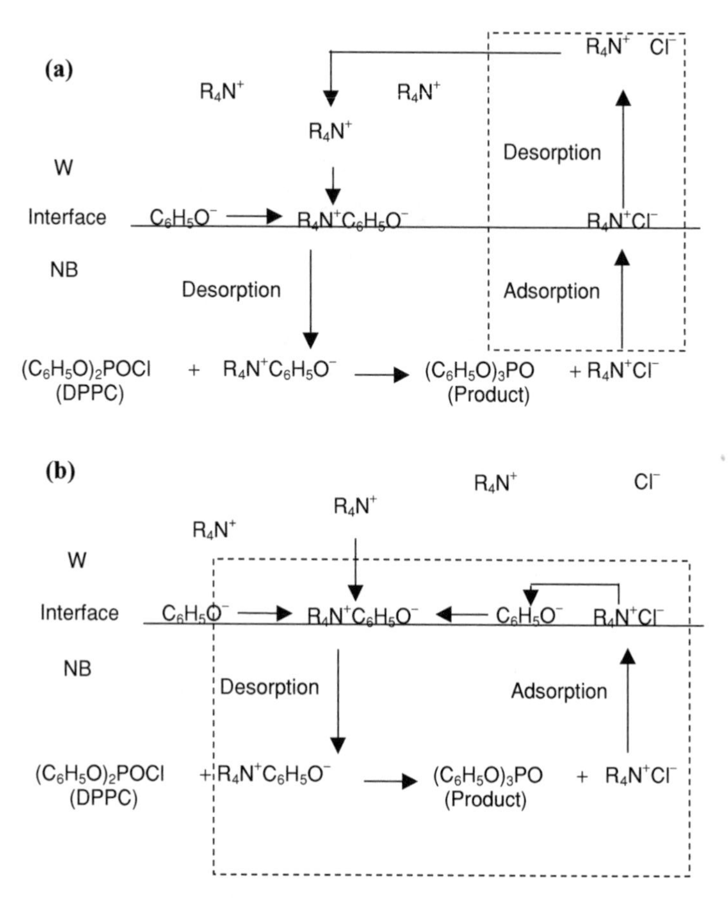

FIGURE 3.6. (a) Reaction scheme that corresponds to the decrease and increase in capillary frequency. (b) Model of the phase transfer catalytic reaction in the stationary state. Reaction proceeds between the adsorbed chemical species at the interface after adsorption from the water phase.

catalyst, whereas the frequency of TBAB became constant much faster than the other two. This meant that the adsorption rates of $R_4N^+Cl^-$ were almost the same for the three phase transfer catalysts, whereas the desorption behaviour of the TBAB ion pair differed from that of the other two. This suggested that the desorption rate of TBAB ion pair was very low and the stationary state could be established quickly at high interfacial molecular density. Furthermore, the ratio of interfacial adsorption of the stationary state is defined and summarized in Table 3.1. This parameter reflected the amount of interfacial adsorbate of the stationary state. This result indicated that the interface accommodated the most adsorbate in the cyclic reaction of TBAB. In short, the TBAB ion pair was difficult to desorb from the interface to the water phase, and thus many interfacial adsorbate molecules were present for the stationary state. This characteristic behaviour of TBAB was considered to be related to the high product yield as a phase transfer catalyst.

Based on the experimental fact that the high interfacial molecular density of the TBAB ion pair was related to the high product yield, we considered the interface was the reaction site controlling the phase transfer catalytic reaction. We proposed a model for the phase transfer catalytic reaction in the stationary state. It is shown within the dotted line rectangle of Figure 3.6b. This model is a further explanation for our previous study, which demonstrated the interfacial reaction of TBAB below 50 mM [7].

We successfully observed dynamic molecular behaviour reflecting the interfacial specificity by focusing on the phase transfer catalyst system using the TR-QELS method. Our results provided new information on the characteristic behaviour of phase transfer catalysis which gives the high yield reaction.

3.4. A CHEMICAL OSCILLATION INDUCED BY ANIONIC SURFACTANT AT A W/NB INTERFACE

3.4.1. Chemical Oscillation at a Liquid/Liquid Interface

Non-linear phenomena accompanied by periodic changes of electrochemical potential have been the subject of many research activities since Dupeyrat and Nakache [39] reported on periodic macroscopic movements of an oil/water interface and generation of electrochemical potential in 1978. These authors found such non-linear behaviour at a W/NB interface with positively charged cationic surfactants. They explained the non-linear behaviour on the basis of formation of ion pairs between the positively charged cationic surfactants in the aqueous phase and negatively charged picrate anions dissolved in the oil phase. The ion pairs formed at a W/NB interface were assumed to be removed from the interface by a phase transfer process and oscillatory behaviour was explained in terms of the Marangoni effect.

Later, Yoshikawa and Matsubara [40] further studied a non-linear system and proposed a mechanism for the periodic behaviour that involved the formation of inverted micelles that suddenly moved to the oil phase after the concentration of adsorbed surfactants reached a critical value. They extended the experiment to a water/oil/water three-phase system in a U-shaped glass tube that gave spontaneous and stable oscillatory behaviour over a long period [41]. Since then, various characteristics of non-linear behaviour have been investigated and several mechanisms for the non-linear behaviour have been proposed by many research groups including ours[2,5,10,42–48]; however, the mechanism at a molecular level has not been clarified yet and no consensus has been achieved. The difficulty in the explanation seems to come from not only the complexity and diversity of the systems, but also limitations of the observation methods that enable us to monitor dynamic molecular behaviour at liquid/liquid interfaces with sufficient interfacial selectivity and time resolution. In this section, the TR-QELS method has been applied to the investigation of W/NB—sodium dodecyl sulfate (SDS) two-phase system [10].

3.4.2. Simultaneous Measurement of an Electric Potential and Interfacial Tension in a Chemical Oscillation

This section reports the simultaneous measurement of an electric potential and interfacial tension to obtain further information on the relation between the generation of an electric potential and the dynamic molecular behaviour of the surfactants at the liquid/liquid interface. For the simultaneous measurement of an electric potential, Ag/AgCl electrodes equipped with salt bridges containing a concentrated aqueous solution of KCl and aqueous solution of 0.01 M TBACl were immersed in water and nitrobenzene phases, respectively. The electric potential of the water phase against the nitrobenzene phase ΔV was measured with a potentiometer. The W/NB interface was prepared by gently pouring 0.1 M LiCl (Wako; special grade) aqueous solution onto the nitrobenzene phase (Kanto Chemical; special grade; 99.5% purity) containing 0.01 M tetra-n-butylammonium tetraphenylborate (TBATPB) in a quartz glass cell. Here, LiCl and TBATPB were supporting electrolytes. TBATPB was obtained by mixing tetra-n-butylammonium bromide (Kanto Chemical; Cica Regent) and sodium tetraphenylborate (Kanto Chemical; Cica Regent) in a water/ethanol/acetone solution and recrystallizing the precipitate of TBATPB in an ethanol/acetone solution. Sodium dodecyl sulfate (SDS) (Kanto Chemical Co., Inc; first grade) solution (10 mM) was injected with a syringe pump (LMS model 210) into the water phase (at a point far from the W/NB interface) at 3 μl/min. Ultrapure water (from Millipore Milli-Q system) was used for all aqueous sample preparations. All experiments were performed at room temperature (298 K).

Figure 3.7 shows the simultaneous measurement of the time courses of the electric potential across the interface and the interfacial tension at a chemical oscillation induced at the W/NB interface by successive introduction of SDS into the water phase. Clearly, the interfacial tension and the electric potential changed simultaneously. No change in the interfacial tension was observed before the first electric potential generation. This result indicated that the electric potential was induced not by desorption of the surfactants from the interface, but by their sudden and corrective adsorption onto the interface. The absolute value of the electric potential at the peaks was almost constant at about 200 mV under our experimental conditions. In contrast, the baseline of the electric potential gradually increased. Corresponding to the increase of the electric potential, a

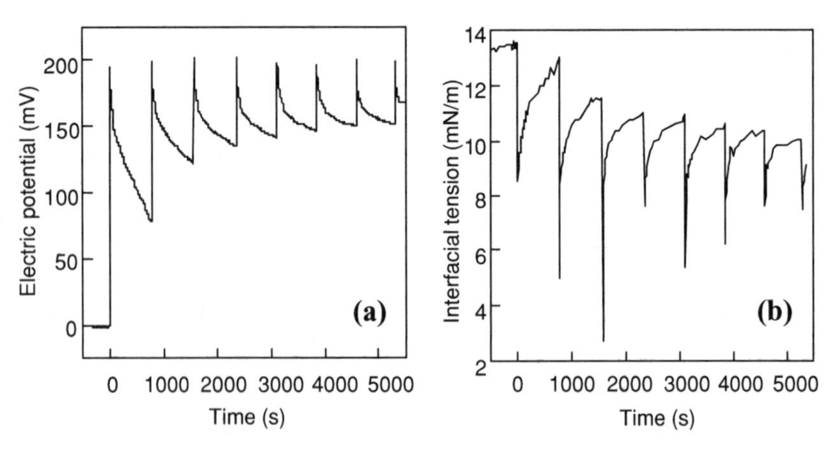

FIGURE 3.7. The time course of electric potential. Time 0 indicates the onset of the first oscillation. (b) The simultaneous measurement of the time course of interfacial tension. The value at the onset of each oscillation fluctuated due to the macroscopic wave-linked interfacial movements and the subsequent fluctuation of the scattered light. A few tens of seconds after the onset of oscillation, we could measure the interfacial tension correctly again.

gradual decrease of the interfacial tension was observed. Unfortunately, the value of the interfacial tension fluctuated at the time the electric potential was generated and it was not correctly measured because of the flip (sudden straining) motion of the W/NB interface. The gradual increase of the baseline of electric potential and corresponding decrease of the interfacial tension indicated that some part of the surfactant DS^- ions at the interface remained in the proceeding oscillation phenomenon and the DS^- ions at the interface were gradually accumulated. Interestingly, though the baseline of the electric potential gradually increased and the interfacial tension decreased correspondingly, the value of the generated electrical potential and periodic time of oscillation were almost constant throughout the oscillation. In short, though the surface coverage of the DS^- ions gradually changed, the periods of the oscillation did not change. In our experimental set-up, the inflow speed of the surfactant was always constant. Thus, these features indicated that the conditions for the onset of the oscillation were determined primarily, not by the surface coverage of the surfactant molecules, but by the flux of the surfactants in the aqueous subphase. This meant that the surfactant concentration in the adjacent water phase to the interface was critical to the onset of the periodic electric potential generation. Thus, it could be considered that when the concentration of the surfactants in water phase adjacent to the interface reached some critical value, the sudden and corrective adsorption of the surfactants occurred. Under our experimental conditions, the period of the oscillation was about 750 seconds. This timescale was extremely long in terms of normal molecular adsorption and desorption (milliseconds) and micelle formation (several tens of seconds at most). This result might indicate that some hindrance effect towards the gradual adsorption of DS^- ions onto the interface was at work in the oscillation system, where no external electric fields were applied and no additional molecules that could assist the ion pair formation were present.

3.4.3. Relaxation Process

In this subsection, we focus on the relaxation process in the chemical oscillation, getting further information on the desorption behaviour of the surfactant from the interface. Figure 3.8 shows simultaneous measurements of the time courses of electric potential and interfacial tension when the concentration of TBAB was 3×10^{-5} M. The decrease of electric potential was reaccelerated at around 250 seconds, where the interfacial tension recovered to the baseline. The recovery of the interfacial tension indicated that there were few surfactant molecules at the interface. Thus, some other factor must also contribute to generation of electrical potential with a long lifetime besides the electrical double layer that was composed of the ionic surfactants (DS^- ions) and the counter ions (Na^+) in the water phase. We considered that some of the DS^- ions formed ion pairs with Na^+ in the water phase, while others formed ion pairs with TBA^+ in the oil phase. In the latter case, the remaining counter Na^+ ions in the water phase and supporting electrolyte TPB^- in the oil phase could also form an electric double layer and contribute to the generation of electric potential. After the formation and desorption of the ion pairs of DS^- and TBA^+ from the interface, the relaxation of the electric double layer between Na^+ and TPB^- across the W/NB interface was considered to be accelerated and this led to the slow decay of the electric potential. Since Na^+ and TPB^- favoured water phase and oil phase, respectively, we considered that this relaxation occurred on a long timescale.

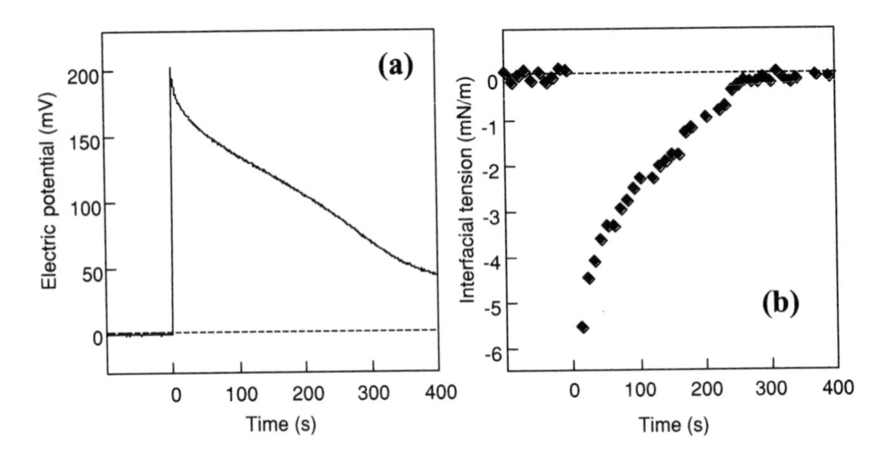

FIGURE 3.8. Simultaneous measurements of (a) electric potential and (b) interfacial tension at the Br^- concentration 3×10^{-5} M under the condition that only the first oscillation was induced.

3.4.4. The Role of Coexisting Ions

Next we report effects of the coexisting ions on the dynamic molecular behaviour across the interface. We investigated the change of relaxation time in the presence of hydrophilic anions in the oil phase that had the same electronic charge as the surfactant molecules. If these anions are present at the interfacial region, we can expect that an ion exchange between them and the charged surfactant molecules would occur and this would accelerate the desorption of the surfactant molecules. Then we measured the chemical oscillation in the presence of a small amount of TBAB that dissociates to the water-insoluble cation TBA^+ and hydrophilic anion Br^- in the oil phase. TBA^+ was already present as a supporting electrolyte at 0.01 M, and thus we could discuss the effect of the presence of the Br^- ions on the relaxation process. The results are shown in Figure 3.9. The relaxation time was drastically accelerated when Br^- was added to the oil phase. This result suggested that the ion exchange between DS^- and Br^- critically affected the relaxation dynamics. In addition, it could be considered that the main origin of the electric potential was derived from the electrical double layer formed by DS^- and the counter ions Na^+ at the interface. In general, the relaxation process has been considered in terms of the ion pair formation at the interface. Our results demonstrated that the ion exchange could drastically accelerate the relaxation process in a chemical oscillation induced by the charged surfactants.

Further experiments to examine the ion exchange mechanism for the acceleration in the relaxation process of the chemical oscillation were also performed. We varied the kind of hydrophilic anions (Cl^-, Br^- and I^-) in the oil phase and investigated the dependence of the desorption rate of DS^- ions on the standard free energy of transfer of those anions. Since the ion exchange should occur at the oil/water interface, we expected that the more the standard free energy of transfer of the anions from the interface to the water phase would decrease, the more efficiently the ion exchange would occur between the hydrophilic anions and DS^- ions, resulting in the acceleration of the desorption

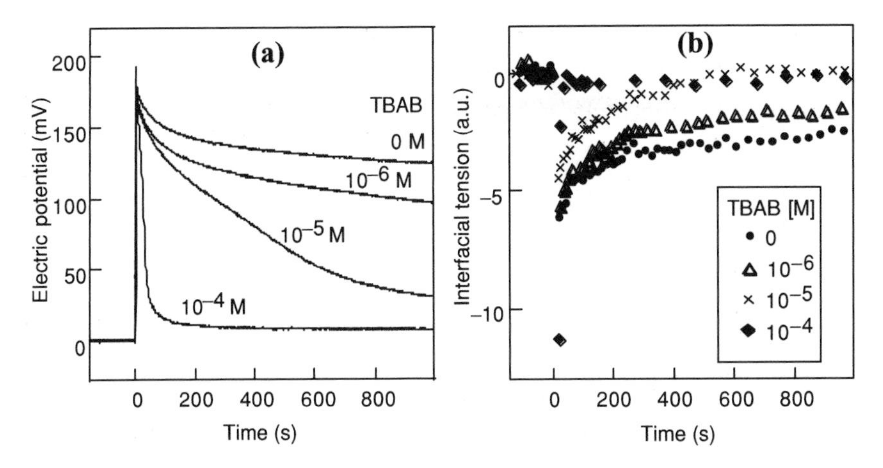

FIGURE 3.9. (a) Time courses of electric potential while varying the concentration of TBAB. (b) Simultaneous measurements of the time courses of interfacial tension.

of the DS^- ions from the interface. When the ion exchange occurs at the interface, the decrease of the standard free energy of the anions is the change of free energy of transfer from the interface to the water phase (ΔG_{i-w}). However, although water and nitrobenzene are immiscible, the interface between them should have some thickness where the water molecules and nitrobenzene molecules are mingled at the molecular level. Thus, we roughly estimated the average of the free energy of the anion at the interface (G_i) as the mean value of the free energy of the anion in the water phase (G_w) and that in the oil phase (G_o), namely $G_i = (G_w + G_o)/2$. Then ΔG_{i-w} value was $G_w - G_i = (G_w - G_o)/2 = \Delta G_{o-w}/2$, where ΔG_{o-w} is the standard free energy of transfer from the oil phase to the water phase. The ΔG_{i-w} ($= \Delta G_{o-w}/2$) values of Cl^-, Br^- and I^- at the W/NB interface were –15.3, −14.3 and − 9.4 $kJ \cdot mol^{-1}$, respectively, as calculated from ΔG_{o-w} values obtained by electrochemical measurements [49].

Figure 3.10a shows the time courses of the interfacial tension when the hydrophilic anions were added into the oil phase. The concentrations of these ions were set at 1.0×10^{-4} M. Here, desorption rate of each case was determined by $1/\Delta t$, where Δt is the period for the interfacial tension to become the initial value again after the oscillation. The Δt values of each ion (Cl^-, Br^- and I^-) were 110, 150 and 1000 seconds, respectively. Figure 3.10b shows the dependence of the desorption rate of the DS^- ions on ΔG_{i-w} of Cl^-, Br^- and I^- from nitrobenzene to water. We can consider that the desorption rate is proportional to the exponential of $-\Delta G_{i-w}/RT$ (R: gas constant, T: temperature). We plotted the dependence of the log value of the desorption rate ($\ln(1/\Delta t)$) on the standard free energy of transfer from the interface to the water phase (ΔG_{i-w}). The solid straight line with the slope $-1/RT(R : 8.314\ J \cdot K^{-1} \cdot mol^{-1}$, T: 298 K) was obtained by the least-square fitting method. The line fitted the experimental results quite well. This result strongly supported the ion exchange model and we could consider that the ΔG_{i-w} value of the coexisting ions in the oil phase strongly affected the desorption rate, namely the relaxation process in the chemical oscillation induced by the adsorption of charged surfactants.

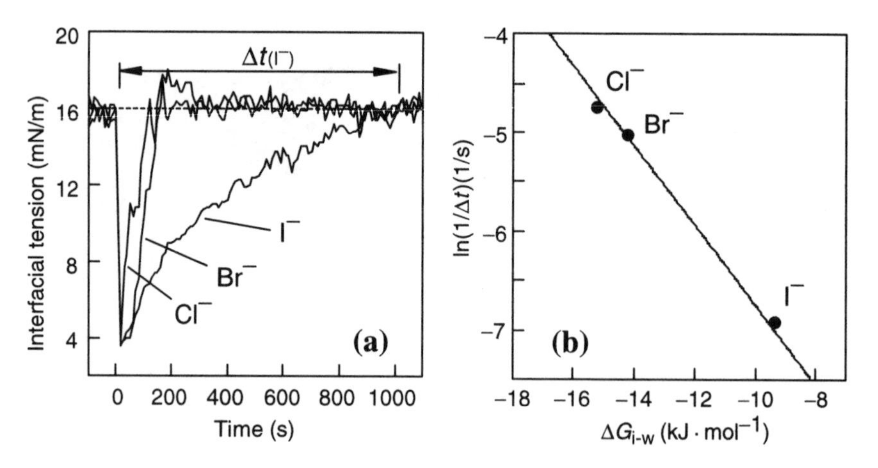

FIGURE 3.10. (a) Time courses of the interfacial tension when hydrophilic anions (Cl^-, Br^- and I^-) were in the oil phase. The concentration of these anions was 1.0×10^{-4} M. (b) Dependence of the desorption rates ($\ln(1/\Delta t)$) on the standard free energy of ion transfer from the W/NB interface to the water phase ($\Delta G_{i-w} = \Delta G_{o-w}/2$). The solid straight line with slope $-1/RT$ (R: gas constant 8.314 $J \cdot K^{-1} \cdot mol^{-1}$, T: temperature 298 K) was obtained by the least-square fitting method.

3.5. CONCLUDING REMARKS

In this chapter, the TR-QELSmethod was introduced and recent applications of this method to the investigations on dynamic molecular behaviours at liquid/liquid interfaces were reviewed [7–11]. Dynamic molecular behaviour of the phase transfer catalysts and the reaction scheme for high yield synthesis was revealed by the TR-QELS method in Section 3.3. The role of the coexisting ions in chemical oscillation system was discussed in Section 3.4. Two-step relaxation processes and acceleration of the relaxation by co-existing ions were described. The related ion-exchange mechanism in the relaxation process and experiments to examine the mechanism in terms of the change of standard free energy of ion transfer from the liquid / liquid interface to the bulk phase were also documented. These results showed the advantages of the TR-QELS method in studying interfacial chemical reactions and molecular transfer phenomena across liquid/liquid interfaces. Further investigation of the mechanism for the onset of the chemical oscillation phenomena is now in progress. We are also applying the TR-QELS method to real-time monitoring of other biologically and/or industrially important systems such as enzymatic hydrolysis reactions in biological membranes and adsoption/desorption phenomena of the self-assembled molecular system such as reversed micelles at liquid/liquid interfaces. These new applications will be reported elsewhere in the near future [50–53].

ACKNOWLEDGEMENTS

We are grateful to Dr. Kensuke Arai (University of Florida) and Dr. Hajime Katano (Fukui Prefectural University) for helpful discussions. Our research results in this chapter were obtained with support by the Grant-in-Aids for Scientific Research (B) (No. 13129203) by the Ministry of Education, Science, Sports, and Culture in Japan.

REFERENCES

1. S. Takahashi, A. Harata, T. Kitamori and T. Sawada, Anal. Sci. **7**, 645 (1991).
2. S. Takahashi, A. Harata, T. Kitamori and T. Sawada, Anal. Sci. **10**, 305 (1994).
3. Z. Zhang, I. Tsuyumoto, S. Takahashi, T. Kitamori and T. Sawada, J. Phys. Chem. A **101**, 4163 (1997).
4. I. Tsuyumoto, N. Noguchi, T. Kitamori and T. Sawada, J. Phys. Chem. B **102**, 2684 (1998).
5. S. Takahashi, I. Tsuyumoto, T. Kitamori and T. Sawada, Electrochim. Acta **44**, 165 (1998).
6. Z. Zhang, I. Tsuyumoto, S. Takahashi, T. Kitamori and T. Sawada, J. Phys. Chem. B **102**, 10284 (1998).
7. Y. Uchiyama, I. Tsuyumoto, T. Kitamori and T. Sawada, J. Phys. Chem. B **103**, 4663 (1999).
8. Y. Uchiyama, T. Kitamori and T. Sawada, Langmuir **16**, 6597 (2000).
9. Y. Uchiyama, M. Fujinami, T. Sawada and I. Tsuyumoto, J. Phys. Chem. B **104**, 4699 (2000).
10. T. Takahashi, H. Yui and T. Sawada, J. Phys. Chem. B **106**, 2314 (2002).
11. H. Yui, Y. Ikezoe, T. Takahashi and T. Sawada, J. Phys. Chem. B **107**, 8433 (2003).
12. H. Lamb, *Hydrodynamics*, 6th ed., Dover, New York, 1945, Chapter IX.
13. M. Sano, M. Kawaguchi, Y.L. Chen, R.J. Skarlupka, T. Chang, G. Zografi and H. Yu, Rev. Sci. Instrum. **57**, 1158 (1986).
14. V.G. Levich, *Physicochemical Hydrodynamics*, Prentice Hall, New York, 1962, Chapter 11.
15. F.C. Goodrich, J. Phys. Chem. **66**, 1858 (1962).
16. R.H. Katyl and U. Ingard, Phys. Rev. Lett. **19**, 64 (1967).
17. R.H. Katyl and U. Ingard, Phys. Rev. Lett. **20**, 248 (1968).
18. J. Lucassen, J. Chem. Soc. Faraday Trans. **64**, 2221 (1968).
19. J. Lucassen, J. Chem. Soc. Faraday Trans. **64**, 2230 (1968).
20. M.A. Bouchiat and J. Meunier, Phys. Rev. Lett. **23**, 752 (1969).
21. J.A. Mann, J.F. Baret, F.J. Dechow and R.S. Hansen, J. Colloid Interface Sci. **37**, 14 (1971).
22. S. Hård, Y. Hamnerius and O. Nilsson, J. Appl. Phys. **47**, 2433 (1976).
23. J.C. Earnshaw, R.C. McGivern, A.C. McLaughlin and P.J. Winch, Langmuir **6**, 649 (1990).
24. J.C. Earnshaw and P.J. Winch, J. Phys.: Condens. Matter **2**, 8499 (1990).
25. J.C. Earnshaw and E. McCoo, Langmuir **11**, 1087 (1995).
26. I. Tsuyumoro and H. Uchikawa, Anal. Chem. **73**, 2366 (2001).
27. A. Trojanek, P. Krtil and Z. Samec, Electrochem. Commun. **3**, 613 (2001).
28. N. Sakamoto, K Sakai and K. Takagi, Jpn. J. Appl. Phys. **41**, 6240 (2002).
29. M. Hosoda, K. Sakai and K. Takagi, Jpn. J. Appl. Phys. **40**, L231 (2001).
30. K. Sakai, D. Mizuno and K Takagi, Phys. Rev. E **63**, 046302 (2001).
31. C.M. Starks, J. Am. Chem. Soc. **93**, 195 (1971).
32. W.P. Weber and G.W. Gokel, *Phase Transfer Catalysis in Organic Synthesis*, Springer-Verlag, Berlin, 1977, Chapters 1 and 5.
33. C.M. Starks and C. Liotta, *Phase Transfer Catalysis*, Academic Press, New York, 1978, Chapters 2, 3, and 4.
34. E.V. Dehmlow and S.S. Dehmlow, *Phase Transfer Catalysis*, Verlag Chemie, Weinheim, 1980, Chapter 2.
35. Y. Kimura and S.L. Regen, J. Org. Chem. **47**, 2493 (1982).
36. D. Mason, S. Magdassi and Y. Sasson, J. Org. Chem. **55**, 2714 (1990).
37. A.W. Herriott and D. Picker, J. Am. Chem. Soc. **97**, 2345 (1975).
38. Y. Kimura and S. L. Regen, J. Org. Chem. **47**, 2493 (1982).
39. M. Dupeyrat and E. Nakache, Bioelectrochem. Bioenerg. **5**, 134 (1978).
40. K. Yoshikawa and Y. Matsubara, J. Am. Chem. Soc. **105**, 5967 (1983).
41. K. Yoshikawa and Y. Matsubara, J. Am. Chem. Soc. **106**, 4423 (1984).
42. K. Toko, K. Yoshikawa, M. Tsukiji, M. Nosaka and K. Yamafuji, Biophys. Chem. **22**, 151 (1985).
43. S. Kihara, M. Suzuki, K. Maeda, K. Ogura, S. Umetani and M. Matui, Anal. Chem. **58**, 2954 (1986).
44. K. Arai, F. Kusu and K. Takamura, Chem. Lett. 1517 (1990).

45. K. Arai, S. Fukuyama, F. Kusu and K. Takamura, Electrochim. Acta **40**, 2913 (1995).
46. K. Maeda, W. Hyogo, S. Nagami and S. Kihara, Bull. Chem. Soc. Jpn. **70**, 1505 (1997).
47. T. Yoshidome, T. Higashi, M. Mitsushio and S. Kamata, Chem. Lett. 855 (1998).
48. V. Pimienta, R. Etchenique and T. Bushe, J. Phys. Chem. A **105**, 10037 (2001).
49. J. Koryta, Electrochim. Acta **29**, 445 (1984).
50. T. Morisaku, H. Yui, M. Iwazumi, Y. Ikezoe, M. Fujinami, and T. Sawada, Anal. Chem. **76**, 3794 (2004).
51. M. Takahashi, H. Yui, Y. Ikezoe, and T. Sawada, Chem. Phys. Lett. **390**, 104 (2004).
52. Y. Ikezoe, S. Ishizaki, T. Takahashi, H. Yui, and T. Sawada, J. Colloid Interf. Sci., **275**, 298 (2004).
53. Y. Ikezoe, S., Ishizaki, H. Yui, and T. Sawada, Anal. Sci., **20**, 435 (2004).

4

Direct Force Measurement at Liquid/Liquid Interfaces

Raymond R. Dagastine and Geoffery W. Stevens
University of Melbourne, Melbourne, Victoria 3010, Australia

4.1. INTRODUCTION

Bulk colloidal dispersion properties such as stability, rheology and phase behaviour are commonly mediated if not controlled by the interaction forces between the colloidal particles. The particle size can range from a few nanometres to approximately 100 μm and the force scale is on the order of 100 nm or less. There is a great body of research on studying colloidal forces, both indirectly and through direct force measurement. The basic principle behind direct force measurement is that by measuring the colloidal force between two particles or surfaces of known geometry, one can study the fundamental colloidal forces present in the dispersion. The results of the study can be compared to theoretical predictions or scaled to describe other geometrical situations because the geometry of the force measurement is well defined.

Dispersion behaviour in systems with liquid/liquid or liquid/gas interfaces (i.e. droplet or bubbles) has traditionally been described in terms of rheological properties, wetting properties, including contact angle and interfacial tensions, or phase behaviour and stability measurements. Direct force measurements provide a means to fundamentally probe the interactions between deformable interfaces that significantly impact the dispersion (or emulsion) behaviour.

The study of forces between deformable interfaces can be broken into two categories, the interactions between two sets of deformable interfaces (e.g., two oil drops in water), or a rigid particle and a single deformable interface. Study of the forces in these systems is motivated by the prevalence of both types of systems (drop–drop or drop–rigid particle) in industrial problems. For example, wetting and adhesion of oil emulsions in porous media are concerns in the petroleum industry for both liquid/liquid separations and oil recovery [1]. An understanding of the interaction forces between

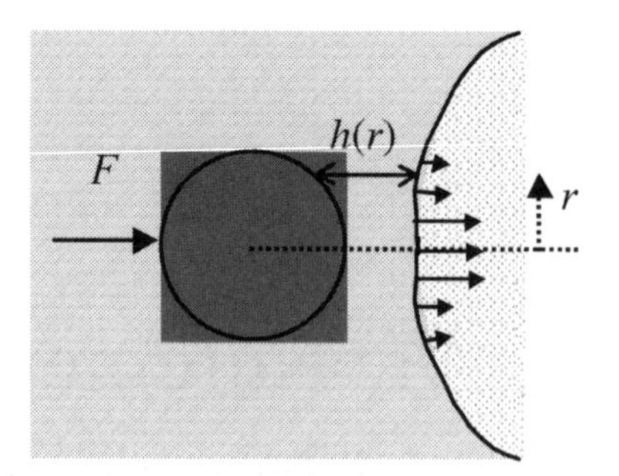

FIGURE 4.1. The interaction between a rigid particle and a liquid/liquid interface. A repulsive colloidal force causes the interface to bend as a function of radial position, r, thus separation distance between the sphere and interface is also a function of radial position, $h(r)$.

liquid/liquid interfaces, where attraction between droplets can lead to coalescence and phase separation in emulsions, is vital in order to predict and model emulsion stability in processing and storage situations (e.g., in the hydrometallurgy, food processing and cosmetic industries [2]). Additionally, wetting and oil deposition in aqueous solutions onto natural and synthetic fibres are of great interest in the textile industry [3]. Also, the motion of a particle near a liquid/liquid interface has applications in particle coatings, adsorption of species onto the particles as well as incorporating macromolecules into biological systems [4]. Furthermore, colloidal forces at the air/water interface are of great interest in floatation processes and sewage treatment [5].

Applying direct force measurement to liquid interfaces is often complicated by interface deformation, making both measurement and the interpretation more difficult than for rigid systems. This is exemplified in Figure 4.1, where a liquid/liquid interface is in close proximity to a rigid particle. A repulsive colloidal force between the particle and the interface causes the interface to bend away from the particle as the two bodies approach. Using modern direct force measurement techniques, the total force acting on the rigid particle can be measured, but this is not a complete characterization of the measurement because the interface geometry is changing with force. The film thickness of water between the oil and sphere is a function of radial distance, but the force measurement alone does not provide a method to determine how this film thickness changes with the measured force. At the same time, curvature of the interface is also crucial to the measurement where liquid/liquid interfaces commonly have absorbed species at the interface, which may rearrange as the two curved surfaces come together. Evidence of the enhanced lateral mobility or rearrangement of absorbed species is demonstrated by the body of literature studying the surface pressure of air/water or oil/water interfaces with absorbed species as a function of surface area [6].

The goal of this chapter is to provide an overview of the measurement of colloidal forces at liquid/liquid interfaces, using predominately atomic force microscopy (AFM). First, some of the types and origins of the relevant colloidal forces are introduced. This is followed by a general description of the operation of AFM at rigid interfaces. The next sections focus on forces at liquid/liquid interfaces, beginning with a discussion of other measuring techniques employed at liquid/liquid interfaces, followed by a summary of the types of systems that have been studied with AFM. The data analysis methods are described and finally a discussion of the types of behaviours seen in these systems is presented with a combination of a summary of both experimental results and modelling calculations. Possible directions of future investigations are discussed in the conclusions.

4.2. TYPES OF COLLOIDAL FORCES

A variety of interaction behaviours can be observed between liquid/liquid interfaces based on the types of colloidal forces present. In general, they can be separated into static and dynamic forces. Static forces include electrostatic, steric, van der Waals and hydrophobic forces, relevant to stable shelf life and coalescence of emulsions or dispersions. Dynamic forces arise from flow in the system, for instance during shear of an emulsion or dispersion. Direct force measurements tend to center on static force measurements, and while there is a large body of work on the study of film drainage between both liquid or solid interfaces, there are very few direct force measurements in the dynamic range between liquid interfaces. Below are general descriptions of some of the types of force observed and brief discussions of their origins.

4.2.1. Electrostatic Forces

A thorough discussion of the basic theory describing electrostatic interactions can be found in [7]; the pertinent points are discussed below. Electrostatic forces arise from the osmotic pressure difference between two charged surfaces as a result of the local increase in the ionic distribution around each charged surface. For a single electrified interface, the local ion distribution is coupled to the potential distribution near that surface and can be described using the Poisson–Boltzmann equation. The solution of this equation shows that for low surface potentials the potential follows an exponential function with distance from the interface, D, given by

$$\psi = \text{const.}\exp(-\kappa D) \tag{1}$$

where ψ is the surface potential, and κ^{-1} is the characteristic decay length of the potential distribution, referred to as the Debye length, defined by

$$\kappa = \left(\frac{1}{\varepsilon \varepsilon_0 k_B T} \sum_i n_i^0 e^2 z_i^2 \right)^{1/2} \tag{2}$$

where $n_i{}^0$ is the bulk concentration of ion i, z_i is the charge number for ion i, e is the charge on an electron, k_B is the Boltzmann constant, T is temperature, ε_0 is the permittivity in free space and ε is the dielectric constant. The solution to the full Poisson–Boltzmann equation (for large surface potentials) is less limited but leads to more complicated expressions given in [7].

The electrostatic interaction between two flat surfaces is an extension of the single electrified interface where the ion distributions overlap at close separations. The two-surface case can also be described according to the Poisson–Boltzmann equation where the ion distributions for each charged interface are coupled. A complete solution to describe the interaction force between these two surfaces requires numerical approach. An efficient method was developed by Chan *et al.* [8], where the bulk electrolyte concentration and the charge or potential at the surface are required.

A simplification to this for low surface potential where the electrostatic double layers only have weak overlap leads to an exponential expression for the force between two flat interfaces (in the case of a symmetric electrolyte and symmetric surfaces)

$$f = 2\varepsilon_0 \varepsilon \kappa^2 \psi_0^2 \exp(-\kappa D) \tag{3}$$

where ψ_0 is the infinite separation surface potential and f is the force per unit area. At infinite separation, the surface charge density σ is related to the surface potential via $\sigma = -\varepsilon_0 \varepsilon \kappa \psi_0$ for low surface potentials. Most electrostatic surfaces are encompassed by two limiting situations. The first is constant charge, where the number of charge sites is preserved with separation. The second is constant surface potential, where ions adsorb or desorb as separation distance changes to keep the surface potential constant. Situations bridging these behaviours are referred to as charge regulation mechanisms that behave as a hybrid of the two limiting cases. The above discussion was outlined for symmetric surfaces (equal surface charge or potential) but the case of dissimilar surface charge or potential is also applicable. For a further discussion of this refer to [7] and [9].

4.2.2. Steric Forces

It is well known that absorbed or grafted polymer layers on colloidal particles are a common method to stabilize dispersions [10]. The repulsive forces originate when two polymer layers overlap and an osmotic pressure or steric force arises repelling the two surfaces. There have been a number of experimental and theoretical investigation into the behaviour of steric forces, summarized in [10–14] to list a few. The results of a scaling theory approach, developed by De Gennes [15,16], describe the force f for two flat surfaces with grafted polymer layers at high surface coverage (polymer brushes) as

$$f \approx k_B T s^{3/2} \left[\left(\frac{2L}{D}\right)^{9/4} - \left(\frac{D}{L}\right)^{3/4} \right] \tag{4}$$

where s is the average distance between detachment points and L is the polymer brush thickness. The first term in Equation (4) accounts for the osmotic pressure increase inside each brush and the second term corresponds to the elastic restoring force of the brush. The above expression is well approximated by an exponential functional form [17,18]

$$f \approx 50 k_B T s^{3/2} \exp\left(-\frac{\pi D}{L}\right) \tag{5}$$

in the region of $0.2 < D/L < 0.9$, below $D/L = 0.2$ the osmotic pressure term in Equation (4) dominates and the exponential description breaks down. The exponential description of the steric force has also been expanded to describe absorbed polymer layers by Luckham [11].

4.2.3. *Van der Waals Forces*

The van der Waals force is ubiquitous in colloidal dispersions and between like materials, always attractive and therefore the most common cause of dispersion destabilization. In its most common form, intermolecular van der Waals attraction originates from the correlation, which arises between the instantaneous dipole moment of any atom and the dipole moment induced in neighbouring atoms. On this macroscopic scale, the interaction becomes a many-body problem where allowed modes of the electromagnetic field are limited to specific frequencies by geometry and the dielectric properties of the system.

The earliest quantitative theory to describe van der Waals forces between two colloidal particles, each containing a statistically large number of atoms, was developed by Hamaker, who used pairwise summation of the atom–atom interactions. This approach neglects the multi-body interactions inherent in the interaction of condensed phases. The modern theory for predicting van der Waals forces for continua was developed by Lifshitz who used quantum electrodynamics [19,20] to account for the many-body molecular interactions and retardation within and between materials. Retardation is a reduction of the interaction because of a phase lag in the induced dipole response that increases with distance.

Hamaker's linear superposition theory provides a single parameter (the Hamaker constant) to describe the interaction between macroscopic bodies. The force per unit area, f_{132}, between infinite flat plates of media 1 and 2, separated by medium 3, is given by

$$f_{132} = -\frac{A_{132}}{6\pi D^3} \tag{6}$$

where A_{132} is the Hamaker constant and D is the separation distance. Lifshitz continuum theory, which includes the effects of many-bodied interactions and retardation, effectively causes the Hamaker constant to vary with separation distance. The Hamaker constant for a number of systems has been tabulated (in [7,18,21] for example), while the Lifshitz analysis requires more information. For a further discussion of van der Waals forces refer to [7,18,21,22].

4.2.4. *DLVO Forces*

Frequently more than one type of force is significant in a colloidal dispersion, and while the above discussions focus on the general behaviour of individual forces, it is also necessary to be concerned about how colloidal forces behave in concert. This is the crux of the most well-known theory in colloidal forces, DLVO theory (named after the work of Derjaguin and Landua [23] and Verwey and Overbeek [24]). DLVO theory defines the total potential energy as a sum of the repulsive and attractive components. The original DLVO theory was developed for colloidal stability incorporating electrostatic and van der Waals forces, but further study has proposed incorporating solvent structuring effects and steric forces [7].

DLVO theory describes the behaviour of energy, yet the discussion of the types of colloidal forces above is in terms of force; it is important to note that the force is the negative derivative of energy. For two flat interfaces with and interaction energy E, the force between the two interfaces is defined as

$$\Pi(D) = -\frac{\mathrm{d}E(D)}{\mathrm{d}D} \tag{7}$$

where $\Pi(D)$ is the force per unit area, also referred as the disjoining pressure.

4.3. ATOMIC FORCE MICROSCOPY AT RIGID INTERFACES

AFM, invented in 1986 [25], has been optimized as a surface imaging tool for a variety of materials from metals to insulators by scanning the surface with a thin metallic cantilever, with a sharp tip on the end, coming to a point with a radius on the order of 10 nm. As the tip is scanned across a surface, the tip-stage position is controlled in three dimensions by a set of oriented piezo crystals (see Figure 4.2). An optical lever comprised of a laser in conjunction with a photodiode is used to measure cantilever deflection in a closed loop feedback system (based on amount of deflection of the cantilever in the simplest mode of operation). As the tip rasters across the surface, a three-dimensional topography map of the surface is created.

4.3.1. Force Measurements at Rigid Surfaces

The use of AFM has been expanded to measure the forces acting on an AFM tip as a function of separation between the tip and surface, but the tip geometry is often difficult to characterize and inconvenient to compare to theoretical predictions of force measurements. By attaching a colloidal size sphere (on the order of 5–10 μm in radius a) to the end of a cantilever (first accomplished by Ducker [26]), it is possible to measure the colloidal interaction between surfaces in system with a well-characterized geometry (shown schematically in Figure 4.2). The measurement of a force–distance cycle is the record of the deflection of the cantilever as the two surfaces approach through motion of the piezo stage. The separation distance between the surfaces is not measured directly

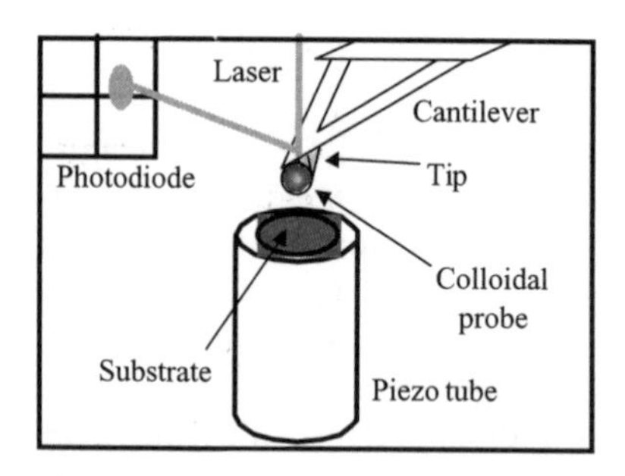

FIGURE 4.2. A schematic of an atomic force microscope, comprising a piezo tube, a cantilever, a sample or substrate to image and an optical lever using a laser and four-quadrant photodiode. A colloidal probe (radius ~5 μm) is added to the end of the cantilever for direct force measurement. For imaging application, the probe is absent and the tip is rastered across the surface.

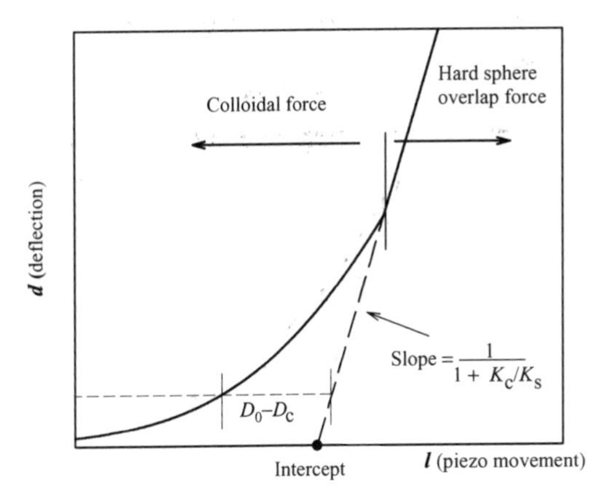

FIGURE 4.3. The extraction of separation distance D_0 from the linear compliance region. Reprinted from Ref. [67], with permission from Elsevier.

and further analysis is required to obtain the force acting on the sphere as a function of axial separation distance D_0. The data analysis process is shown graphically (see Figure 4.3) and is discussed below.

The cantilever deflection d is related to the total force F on the particle via $F = K_c d$, where K_c is the cantilever spring constant. By employing a basic distance balance

$$\Delta l = \Delta d + \Delta D_0 + \Delta z \tag{8}$$

where the change in piezo movement, Δl, is equal to the sum of the change in deflection, Δd, the change in separation distance, ΔD_0, and the substrate deformation, Δz. Assuming the substrate is an elastic body and $F = K_s \Delta z$ (where K_s is the effective spring constant of the surface), it is possible to write the above equation as

$$\Delta d = (\Delta l + \Delta D_0) \Big/ \left(1 + \frac{K_c}{K_s}\right) \tag{9}$$

In the conventional analysis (i.e. for rigid surfaces) of a linearly elastic substrate, the "force curve" is divided into two regions (see Figure 4.3). In the first region, colloidal forces are significant and the cause of deflection of the cantilever. In the second region, a linear behaviour in the force curve is observed because the probe is moving in contact with the surface. The probe surface is effectively at some small cut-off distance D_c from the substrate surface that is of the order of the center-to-center molecular spacing. In this "linear compliance" region, all of the deflection of the cantilever is caused by piezo movement and $\Delta D_0 = D_c - D_0$ may be assumed to be constant. Hence from Equation (9), Δd is linear in Δl with a slope $1 + K_c/K_s$. The intercept of the linear region then gives ΔD_0. For "stiff" substrates $K_s \gg K_c$ and the slope of the linear region on a Δd vs. Δl plot should be unity. For a general K_c/K_s the slope and intercept of the linear regression of the "linear compliance" region is then used to determine the separation distance outside the linear region by rearranging and substituting the experimental slope and intercept into Equation (9) to obtain

$$D_0 = D_c + \left(\frac{d}{\text{slope}}\right) - l + \text{intercept} \tag{10}$$

The cut-off distance D_c for rigid surfaces is sufficiently so small that it is commonly neglected in the display of force versus separation curves. This cut-off distance may not be negligible in the presence of absorbed layers of surfactant or polymers, compressed between the surfaces; thus it is common to report apparent separation distance on AFM measurements [27].

4.4. FORCE MEASUREMENTS AT LIQUID INTERFACES

4.4.1. Other Force Measurement Techniques at Liquid/Liquid Interfaces

Before reviewing the force measurements using AFM at deformable interfaces, a description of other techniques used for force measurement at liquid/liquid interfaces is presented. The classical experiment of measuring disjoining pressure of a thin film between a flat surface and a large flattened drop is well known [28]. The film thickness can then be determined accurately by the use of interferometry. This technique has had great success, but is less relevant when describing colloidal forces. The difference between the two (in the context of this chapter) is a colloidal force measurement between bodies where the curvature of each surface is important. For example, if an absorbed species at oil water drop interface rearranges as a result of the approach of a particle, the effects of the curvature of the particle and drop are critical for this to occur, while this effect is not easily observed with two flat interfaces.

A more recently developed force measurement technique, coined the liquid surface force apparatus (LSFA), brings a drop made from a micropipette close to a flat liquid/liquid interface [29–32]. A piezo electric drive is used to change the position of the micropipette while the deflection of the pipette and the radius of the drop are recorded with piezo motion. The drop radius and thus the film thickness between the two liquid/liquid interfaces are recorded using interferometry. The method requires a calibration of the interferometer, where the drop must come into contact with the other liquid interface. The distance resolution of the film is about ± 1 nm at a 50-nm separation and ± 5 nm at a separation of 10 nm. This is a very robust technique where the authors have proposed attaching a particle to the end of the pipette instead of a drop [29]. In comparing this method to AFM, the only drawback of the LSFA is the weaker distance resolution. It is important point out that both methods required a contact point for distance calibration.

Two similar techniques to probe hydrodynamic behaviour have been recently developed by Horn *et al.* [33] and Gee and co-workers [34–37]. The measurements by Horn *et al.* employ interferometry to measure the shape of a mercury drop at the end of a microcapillary as it approaches a flat interface. The capillaries are much larger than those used in an LSFA, where the drop profile is recorded under hydrodynamic forces. The work by Gee [34–37] is similar but uses hydrocarbon drops and determines the separation using imaging ellipsometry/reflectometry. This method is useful for systems with more practical interest when discussing the drainage between droplets in emulsions with surfactants.

The last two methods discussed measure the interaction with droplets small enough to behave as rigid particles. The first technique is total internal reflection microscopy (TIRM) where Odicahi [38] measured the interaction between silicon oil, bitumen and bromodecane droplets above a flat glass plate. This is a very sensitive technique which measures the forces acting on a Brownian particle suspended above a flat surface (forces on the order of femto-Newtons). The work in [39] uses the methods developed by Odiachi in systems where the droplet is less dense than water. The last method is a magnetic force method where the force between water in oil emulsions is measured [40–45]. The water droplets (on the order of 100–200 nm in diameter) have magnetic particles dispersed in them, making the droplets align in a magnetic field and Bragg scattering is used to determine droplet separation. This is a novel technique, which measures force on the order of pico- to nano-Newtons.

4.4.2. AFM Force Measurement Techniques at Liquid/Liquid Interfaces

4.4.2.1. One Liquid/Liquid Interface. The measurement of forces at deformable interfaces with a rigid probe using AFM is becoming commonplace. Ducker *et al.* [5] measured the forces between a silica particle attached to the AFM cantilever and an air bubble anchored to the piezo-driven stage in aqueous solution. Similar experiments were reported by Fielden *et al.* [46] and Butt and co-workers [47–49]. The above investigations probed forces in the presence of electrolyte and surfactant. In some cases, either one or both surfaces were hydrophobic and a large rapid deflection of the cantilever was observed, which led to either engulfment or formation of a three-phase contact line. This behaviour is typically faster than the sampling rate of the AFM measurement and is referred to as "jump-in", caused from a strong attractive force. With the addition of surfactant, jump-in was observed less frequently and only repulsion was seen in some cases. The presence of jump-in and formation of a three-phase contact line has been proposed as a method to measure the contact angle of water on the particle through initial experimental studies [46,48,50] and followed by a combination of modelling and experimental studies [51–53].

Measurements of forces between probe particle and sessile oil drops in water were first reported by Mulvaney *et al.* [54] and Snyder *et al.* [55], with a number of studies following by Hartley *et al.* [56], Aston *et al.* [57,58], Attard and co-workers [59–61], Nespolo *et al.* [62] and Dagastine *et al.* [63]. As an example of the types of typical forces observed between an alkane droplet and a silica probe in the presence of an anionic surfactant, refer to Figure 4.4 [63]. The general shape of the force curve is similar, at first glance, to the behaviour at rigid surfaces, but as discussed below, this is a product of changes in separation and interface deformation. Also note that at low surfactant concentrations, jump-in does occur, but with an increase in surfactant concentration (which leads to a decrease in interfacial tensions) only repulsive forces are observed.

4.4.2.2. Two Liquid/Liquid Interfaces. Measuring the interactions between two liquid droplets with AFM is a natural extension of the one droplet problem and more relevant to emulsion stability. The first and only such measurement is between alkane droplets in

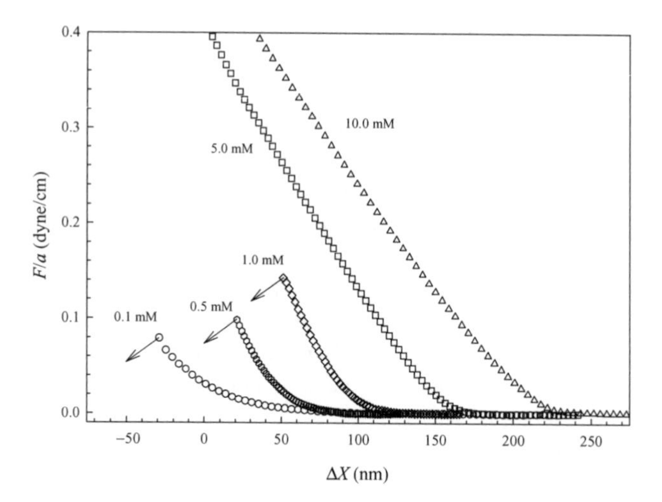

FIGURE 4.4. AFM force curves for a silica sphere and a teradecane oil drop in terms of F/a and ΔX as a function of sodium dodecyl benzene sulfonate concentration. The arrow denotes jump-in for the 0.1, 0.5 and 1.0 mM concentrations. Jump-in was never observed for the 5 and 10 mM concentrations. The force curves have an expanded horizontal axis. Reprinted from Ref. [63], with permission from Elsevier.

an anion surfactant solution reported by Dagastine *et al.* [64]. The effects of modulating electrostatic interactions through the presence of an anionic surfactant and changes in interfacial tension were consistent with the behaviours observed for a single liquid/liquid interface. Also, the range of approach velocities probe both the static and dynamic forces between the droplets, showing a clear definition where hydrodynamic drainage effects have significant impact on the measurement. Similar issues in regards to analysis and interpretation are likely to be encountered.

4.4.3. Analysis and Interpretation

AFM investigations at deformable interfaces between a single liquid/liquid interface and a rigid particle have shown that the force curves exhibit a linear region similar to the behaviour seen for rigid systems (as shown in Figure 4.4), but the slope of the deflection force curve in the linear region is much less than 1. The first analysis proposed by Ducker [5] assumed that the drop behaved as a Hookean spring and that separation distance is constant in the linear region. (This is a similar approach to the case for rigid surfaces.) A number of theoretical studies [57,65–67] have shown that the separation distance is changing in this "linear compliance" region and that the psuedo-linear appearance to the force curve is a combination of the changes in surface separation coupled with surface deformation [57,65–67]. These theoretical analyses also show that the assumption that the interface behaves as a Hookean spring is reasonable for low to moderate deformations. The breakdown of the constant separation distance assumption invalidates the analysis used by Ducker and others [5,46,50,54,56] over the full range of forces.

The experimental method coupled with the appropriate use of theories developed for data interpretation makes it possible to probe the forces between liquid interfaces

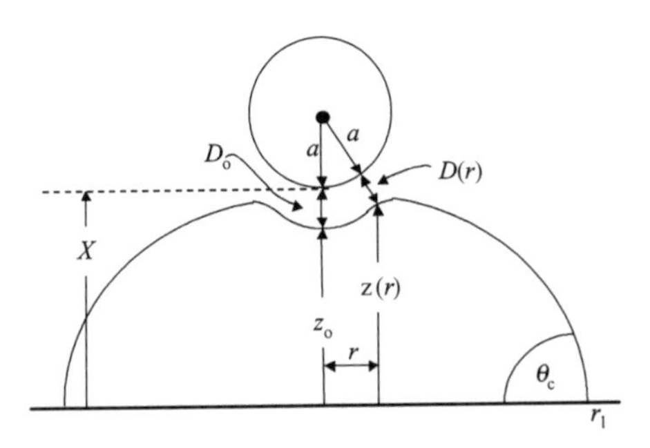

FIGURE 4.5. The geometry of the probe/drop experiment. The drop has an undistorted radius of curvature R_0 and the probe radius a, such that $a \ll R_0 \ll \lambda$ so that gravitational distortion can be neglected. (The deformation is not shown to scale.). Reprinted from Ref. [67], with permission from Elsevier.

in the presence of a large range of absorbed species acting as stabilizers, emulsifiers or flocculants. The general approach of the models developed to interpret the AFM measurements is discussed below. This is followed in the next section by discussions of the general behaviour of the force curves in terms of both modelling and experimental observations.

4.4.3.1. Drop Shape Modelling. This approach is based on modelling the interface shape during the AFM measurement (see Figure 4.5). Once the interface (or drop) shape is determined, the total force on the particle as a function of piezo movement can be calculated, thus effectively predicting the AFM "force curve". The radius R_0 of the drop or bubble forming the liquid/liquid or liquid/gas system is well below the capillary length, $\lambda = (\gamma/\Delta\rho g)^{1/2}$, where $\Delta\rho$ is the density difference and γ is the interfacial tension. Thus, the effects of gravity (g) are negligible and the interfacial tension and the colloidal forces present govern the interface shape.

Most of the model approaches require a number of input parameters including the interfacial tension, γ, the contact angle of the drop on the substrate, θ_c, density, and the interaction forces between the two flat surfaces. With the exception of the last parameter, these are readily available from literature sources or straightforward measurements.

The interaction forces, or disjoining pressures, are calculated from theoretical models for flat interfaces and mapped to curved surfaces using the Deraguin's approximation [7]. The Deraguin's approximation is a geometric scaling limited to the case where the surface separation is much smaller than the radius of either interfaces (i.e. a and $R_0 \gg D_0$ in Figure 4.5). For example, the force $F(h)$ between two spheres of radius a_1 and a_2 is related to the potential energy between two flat plates, $V(h)$, by

$$F(h) = 2\pi\left(\frac{a_1 a_2}{a_1 + a_2}\right) V(h) \tag{11}$$

where h is the separation distance. The force models also require additional information, for example electrostatic models require surface potential or surface charge and electrolyte concentration. Once all the parameters are set, calculation of the drop shape at different surface separations is possible.

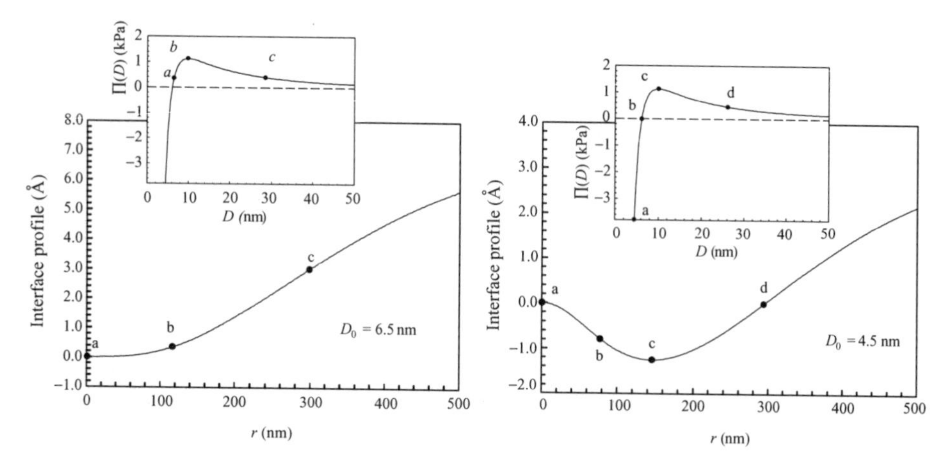

FIGURE 4.6. The interface profile (relative to the interface position at $r = 0$) for an alkane drop in water with a silica probe at an axial separation distance D_0 of 6.5 and 4.5 nm. The dots denote the corresponding disjoining pressure value at the marked radial distance. The inset is the disjoining pressure (electrostatic and van der Waals) for a water/alkane/silica system with a univalent (non-surface-active) electrolyte ($\kappa^{-1} = 30$ nm $\psi_0 = 50$ mV). Reprinted from Ref. [67], with permission from Elsevier.

Model interface profiles for an alkane drop in water with a silica probe at two different axial separation distances are shown in Figure 4.6 from the modelling results in [67]. The only forces used in the model disjoining pressure were an electrostatic repulsion and a van der Waals attraction. As the particle surface separation changes, the colloidal force varies radially, sampling different points on the disjoining pressure curve (the insets in Figure 4.6). For $D_0 = 6.5$ nm, the interface is deformed away from the probe because only positive disjoining pressures are sampled. The dots mark the corresponding disjoining pressure at that radial distance. Note that as the disjoining pressure samples approach being negative, where $\Pi(D)$ becomes negative at 6.1 nm (see inset), the interface profile is reasonably flat at small values of r. A reversal in interface curvature is exhibited for the $D_0 = 4.5$ nm case where the central section of the drop is sampling negative $\Pi(D)$ values. As the axial separation is decreased, more of the negative portion of the disjoining pressure is sampled, leading to a larger peak growth and eventually interface instability. The interface profiles are then used to predict AFM force measurements, discussed in the next section.

4.4.3.2. Predicting the AFM Measurement. A repulsive electrostatic disjoining pressure is used to illustrate how the measurable AFM "force curve" can be calculated, in this case for a silica probe above an alkane drop. The calculation of $F(D_0)$ and $X(D_0)$ as a function of axial separation D_0 was performed by the algorithm developed by Dagastine *et al.* [66,67]. Figure 4.7 shows the plot of $F(D_0)$ and $X(D_0)$ plotted as a function of D_0 and the plot of $F(D_0)$ and $X(D_0)$ parametric in D_0. The general shape of the F versus ΔX curve is agreement with the behaviour seen for the AFM experiments shown in Figure 4.4. The calculation results are comparable to the AFM force curve to within a constant shift in ΔX. This curve shows that the separation distance is changing in the pseudo-linear portion of the $F(\Delta X)$ plot, which verifies the need for a rigorous analysis to data [57,65,66]. For a brief summary of the calculation of these force curves, see appendix.

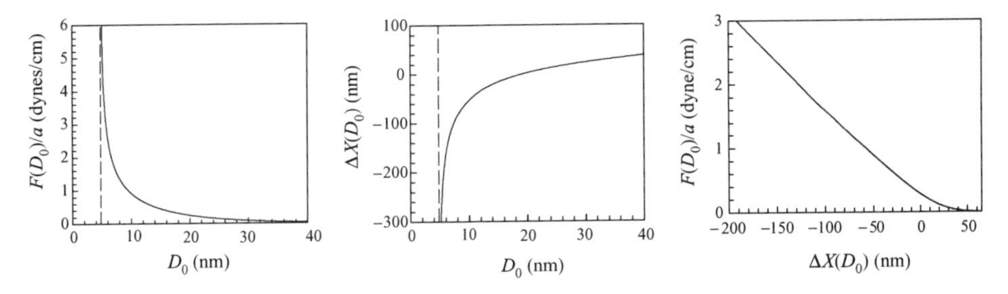

FIGURE 4.7. (a) $F(D_0)/a$ as a function of central separation distance D_0 for $\psi_0 = 50$ mV, $\theta_c = 30°$, $\gamma = 30$ dynes/cm, $\kappa^{-1} = 10$ nm, $a = 2$μm and $R_0 = 0.5$ mm. The dashed vertical line denotes the wrapping distance D_w. (b) $X(D_0)$ as a function of central separation distance D_0. The dashed vertical line denotes the wrapping distance D_w. (c) $F(D_0)/a$ and $\Delta X(D_0)$ plotted parametrically as a function of central separation distance D_0. Reprinted from Ref. [66], with permission from Elsevier.

4.4.3.3. Model Verification and Limitation. There have been several studies that have applied the models to experimental data, the most notable being the independent verification of the model (developed by Chan *et al.* [66] and Dagastine and White [67]) to AFM force measurements between a decane droplet and a silica probe in SDS solutions in a study by Nespolo *et al.* [62]. The wetting properties (interfacial tension and contact angle) were measure-independent of the AFM experiment, and the disjoining pressure required for the model was constructed using attractive van der Waals forces and electrostatic repulsion. The parameters for the electrostatic force model were measured via an in-depth study of the electrokinetic behaviour of both decane droplets and silica particles in SDS solution [68]. The agreement between the data and model predictions was within the error in the AFM measurements.

The series of modelling studies generally followed the same approach discussed in the previous sections, either through analytic expressions [39,59–61,66,67] or numerical calculations [57,65]. All of the theoretical studies, except one, require a general form or type (i.e. electrostatic, hydrophobic, etc.) for the colloidal forces incorporated in the model to predict the drop shape. This is reasonable if the behaviour studies are expected to be analogous to rigid systems, but this is quite limited to probe systems where the forces are not as well understood, or exhibit behaviour unique to liquid interfaces. One approach has shown a method to extract the interaction energy $E(D)$ between half spaces from the AFM force data between a drop and a rigid probe particle [63] without the use of a colloidal force model. This approach was then applied to the set of data in Figure 4.4, where the interaction energies extracted from the AFM force measurements compared favourably to expected electrostatic repulsion behaviour for that system [63].

4.5. GENERAL BEHAVIOUR OF FORCES AT LIQUID INTERFACES

The modelling of the AFM measurement provides a means to examine the general behaviour of the forces in these systems and the sensitivity to different model parameters [63,65–67], in the context of experimental observations [57,60,62,65]. First, the effects of wetting properties on the force behaviour are discussed below. Then the effects of different disjoining pressures are split into two sections, based in the best way to discuss

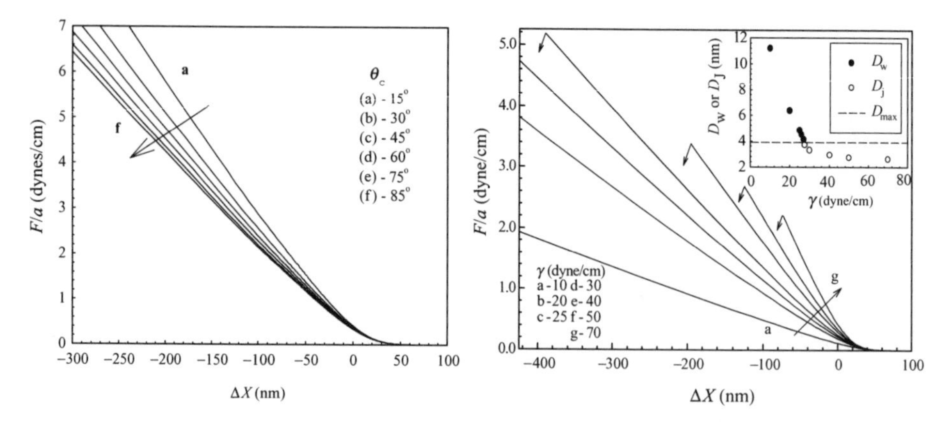

FIGURE 4.8. The effect of changing contact angle θ_c of the drop is plotted as $F(\Delta X)/a$ versus ΔX with $\gamma = 50$ dynes/cm, $\psi_0 = 50$ mV, $\kappa^{-1} = 10$ nm and $a = 2\,\mu$m (left). The effect of changing surface tension γ is shown for a disjoining pressure with both attraction and repulsion ($\theta_c = 80°$, $\psi_0 = 50$ mV, $\kappa^1 = 10$ nm and $a = 2\ \mu$m). The arrows on the curves (right) indicate the point of jump-in, while the no jump-in occurs because of the wrapping of the interface for surface tensions 10, 20 and 25 dynes/cm. The jump-in and wrapping distances for this system are plotted in the inset as a function of surface tension, where the dotted line denotes D_{max} from the position of the disjoining pressure maximum. Reprinted from Ref. [66] (left) and from Ref. [67] (right), with permission from Elsevier.

the termination of an AFM curve. Short-range attractions are typical for many real systems leading to jump-in; yet, in some cases, only a repulsive interaction is observed and zero separation is never reached. In this situation the interface "wraps" the probe particle in the AFM experiment. Both wrapping and the effects of attraction are discussed below.

4.5.1. *Interfacial Properties*

The models [66,67] applied to an alkane drop and a silica probe were used to examine the effect of changing contact angle θ_c; the results are given in Figure 4.8 (left) for a repulsive disjoining pressure $\psi_0 = 50$ mV, $\kappa^{-1} = 10$ nm and $\gamma = 50$ dynes/cm). Note the linear compliance region and a moderate dependence of the compliance (slope) on small contact angles becoming insensitive near 90°. The effects of changing surface tensions are also illustrated in Figure 4.8 (right) for a disjoining pressure with electrostatic repulsion ($\psi_0 = 50$ mV, $\kappa^{-1} = 10$ nm, $\theta_c = 80°$) and van der Waals attraction. The effects on the slope are evident, but the surface tensions also can change the termination behaviour of the force where jump-in is observed only at higher surface tensions ($\gamma = 30, 40, 50, 70$). The absence of jump-in is a manifestation of interface "wrapping".

4.5.2. *Force Curve Termination*

4.5.2.1. Jump-In. Jump-in arises from attractive forces for both rigid and deformable systems alike, caused by a mechanical instability when the gradient of the force–distance curve is equal to the spring constant of the cantilever. The process occurs quickly, typically faster than the data-sampling rate of the AFM, so the initial and final positions of the jump are readily observable, but not all of the dynamic behaviour. The occurrence of jump-in in deformable systems is differentiated from rigid system behaviour because the interface bends towards the probe particle as shown in Figure 4.6 (right). This results in termination of the AFM curve and the modelling of the measurement as well. Examples of jump-in can be seen in modelling results in Figure 4.8 (right) for the higher surface tension cases and the experimental data in Figure 4.4, but a number of investigators have seen jump-in behaviour in both liquid/liquid and liquid/gas systems [5,48,49,54–57].

The origins of the attractive forces are sometimes expected to be van der Waals attraction, acting over the range of a few nanometers. While force measurements with hydrophobic surfaces have proposed the presence of a longer-range attraction from a hydrophobic interaction [5]. Both are possible, while other origins of attractive forces could be charge reversals for the interaction of dissimilar charged surfaces or a combination of forces giving rise to a total interaction with multiple minima in the disjoining pressure. A common observation for the deformable measurements is that addition of surfactants tends to limit or remove jump-in entirely because of the interface wrapping.

4.5.2.2. Wrapping. In [66] it was shown that the interface may "wrap" around the sphere when the surface tension of the interface is decreased sufficiently and/or the repulsive force is sufficiently strong that a separation distance D_w can be found satisfying

$$\Pi(D_w) = 2\gamma\left(\frac{1}{a} + \frac{1}{R_0}\right) \tag{12}$$

where D_w is the solution to Equation (14) describing the film thickness between the interfaces for all radial distances. When $D_0 > D_w$, the interaction area is on the order of the product of a andD_0 ($O(aD_0)$), but as $D_0 \to D_w$, the interaction area is $O(a^2)$, resulting in a very much larger repulsive force in the system than would be observed in a rigid system where the interaction area is always $O(aD_0)$. If the maximum in $\Pi(D)$ occurs at the distance D_{max} and $D_{max} < D_w$, then the attractive part of $\Pi(D)$ is never sampled and no jump-in can occur.

The results in Figure 4.7 (left) show an example of wrapping where the force asymptotically approaches the wrapping distance D_w never reaching zero separation on the $F(D_0)$ versus D_0 curve. The $F(D_0)$ versus $\Delta X(D_0)$ curve in Figure 4.7 (right) exhibits only repulsion where wrapping occurs well after the onset of the linear region in the $F(D_0)$ versus $\Delta X(D_0)$. Another manifestation of wrapping is illustrated in Figure 4.8 (right) where the same disjoining pressure, $\Pi(D)$ (electrostatic repulsion and van der Waals attraction, $\psi_0 = 50$ mV, $\kappa^{-1} = 10$ nm), is used for all the force curves and the surface tension is varied. There is a critical surface tension γ_c for a given $\Pi(D)$ such that

$$\Pi(D_{max}) = 2\gamma_c\left(\frac{1}{a} + \frac{1}{R_0}\right) \tag{13}$$

For $\gamma > \gamma_c$, no wrapping distance D_w can be found and D_0 can be decreased below D_{max} by advancing the stage. Thus the attractive component of $\Pi(D)$ can be sampled in this case and a jump-in is observed. For higher surface tensions the jump-in is marked by the arrows on the curves in Figure 4.8 (right), but as the surface tension decreases and the interface begins to wrap, the jump-in distance decreases, and for $\gamma < \gamma_c$ (27.25 dyne/cm in this case) jump-in no longer occurs. For $\gamma < \gamma_c$, a wrapping distance exists and marks the limiting distance of approach of the surfaces. Moving the stage further towards the probe results in causing more deformation to the drop and the interaction area growing even if $\Pi(D)$ has become essentially constant ($\Pi(D_w)$). In the inset in Figure 4.8 (right), D_w for the wrapping cases and D_J for the jump-in cases are plotted as a function of γ. From this plot, γ_c is evident and the changeover from jump-in to wrapping at D_{max} clear.

In practice, the changeover can be achieved by adding a surfactant to the water/oil or water/air interface. It enhances both interfacial surface potentials and tends to make them equal (i.e. $\Pi(D)$ for similar surfaces > $\Pi(D)$ for dissimilar surfaces) and it lowers the surface tension of the drop. This type of behaviour has been observed in a variety of systems (see Figure 4.4), including oil/water [54,56,62,63] systems where the addition of surfactant prevented jump-in and in air/water systems where the addition of surfactant decreased (and sometimes eliminated) jump-in distances [5,46,48,49].

4.6. CONCLUDING REMARKS

The AFM is a useful tool for probing behaviours at liquid/liquid or liquid/gas interfaces. The studies and modelling of the interaction between a single deformable interface and rigid probe particle have demonstrated the presence of a variety of colloidal forces ranging from electrostatic repulsion to van der Waals and hydrophobic attractions. The deformable nature of these systems frequently causes the interaction area to change with separation, leading to a more complicated behaviour than forces in analogous rigid systems where the interface may wrap the probe particle producing greater forces due to the larger interaction area. Future measurements are likely to probe the behaviours of steric forces from absorbed polymers and other amphiphilic species that have not yet been explored. Extending these techniques to directly probe the interactions between two oil droplets with AFM has been undertaken and may offer a useful method to probe colloidal forces in emulsions in the presence of a variety of stabilizers and flocculants.

ACKNOWLEDGEMENTS

This study was supported by the Particulate Fluids Processing Centre, a special research centre of the Australian Research Council and the United States of America National Science Foundation international research fellowship program.

APPENDIX

Briefly, the method of calculation of $F(\Delta X)$ is as follows [66]. For a given inter-surface central separation distance D_0, we solve the differential equation

$$D'' + \frac{1}{t}D' - \left(2\left(1 + \frac{a}{R_0}\right) - \frac{a\Pi(D)}{\gamma}\right)D_0 = 0 \tag{A1}$$

where $D(0) = D_0$ and $D'(0) = 0$ for the intersurface separation profile $D(t)$. Here t is a scaled transverse radial coordinate (see Figure 4.5).

$$t = \frac{r}{(aD_0)^{1/2}} \tag{A2}$$

Then compute the integrals

$$\left.\begin{aligned} G(D_0) &= \frac{aD_0}{\gamma}\int_0^\infty dt\, t\,\Pi\,(D\,(t)) \\ H(D_0) &= \frac{aD_0}{\gamma}\int_0^\infty dt\, t\,\ln t\,\Pi\,(D\,(t)) \end{aligned}\right\} \tag{A3}$$

and hence the force $F(D_0)$ and separation distance $X(D_0)$ from

$$F\,(D_0) = 2\pi\gamma G\,(D_0) \tag{A4}$$

$$X\,(D_0) = z_0^\infty + D_0 + H\,(D_0) + G\,(D_0)\left(\frac{1}{2}\ln D_0 + B\right) \tag{A5}$$

where B is given in [67]

$$B = 1 + \frac{1}{2}\ln\left[\frac{1+\cos\theta_c}{1-\cos\theta_c}\right] + \ln\left(\frac{a^{1/2}}{2R_0}\right) \tag{A6}$$

In Equation (18), z_0^∞ is the central height above the stage of the undistorted drop ($D_0 = \infty$). Although Equation (18) is absolute, the lack of accurate information about z_0^∞ requires X to be expressed as

$$\Delta X\,(D_0) = \Delta X^{\circ} + D_0 + H\,(D_0) + G\,(D_0)\left(\frac{1}{2}\ln D_0 + B\right) \tag{A7}$$

where ΔX° is an unknown constant set by the choice of origins. Thus a problem of reference frame arises where the existence of ΔX° prevents one from obtaining an absolute separation distance determination. For rigid interfaces, ΔX° may be determined by AFM measurements in hard contact, the constant compliance region of the AFM force curve. For the liquid drop system there is no hard contact region [66] and ΔX° is not immediately extractable for the $F(\Delta X)$ data alone. This problem of absolute separation distance measurement does not have a general solution at this time. For further discussion of methods of absolute separation see [61] or [63].

REFERENCES

1. N.V. Churaev, A.P. Ershov, N.E. Esipova, G.A. Iskandarjan, E.A. Madjarova, I.P. Sergeeva, V.D. Sobolev, T.F. Svitova, M.A. Zakharova, *et al.*, Colloids Surf., A **91**, 97 (1994).
2. R. Douillard, Colloids Surf., A **91**, 113 (1994).
3. A. Perwuelz, T. Novais De Olivera and C. Caze, Colloids Surf., A **147**, 317 (1999).
4. S. R. Dungan and T. A. Hatton, J. Colloid Interface Sci. **164**, 200 (1994).

5. W.A. Ducker, Z.G. Xu and J.N. Israelachvili, Langmuir **10**, 3279 (1994).
6. A.W. Adamson and A.P. Gast, *Physical Chemistry of Surfaces*, John Wiley & Sons, New York, 1997, Chapter 4.
7. R.J. Hunter, *Foundations of Colloid Science*, Clarendon Press, Oxford, 1995, Chapter 6.
8. D.Y.C. Chan, R.M. Pashley and L.R. White, J. Colloid Interface Sci. **77**, 283 (1980).
9. R.J. Hunter, *Modern Aspects of Electrochemistry*, Plenum Press, New York, 1975, Vol. 11.
10. D.H. Napper, *Polymeric Stabilization of Colloidal Dispersions*, Academic Press, New York, 1983.
11. P.F. Luckham, Adv. Colloid Interface Sci. **34**, 191 (1991).
12. P.F. Luckham, Curr. Opin. Colloid Interface Sci. **1**, 39 (1996).
13. G.J. Fleer, M.A. Cohen Struart, J.M.H.M. Scheutjens, T. Cosgrove and B. Vincent, *Polymers at Interfaces*, Chapman and Hall, New York, 1993.
14. H.-J. Butt, M. Kappl, H. Mueller, R. Raiteri, W. Meyer and J. Ruehe, Langmuir **15**, 2559 (1999).
15. P.G. De Gennes, Macromolecules **15**, 492 (1982).
16. P.G. De Gennes, Adv. Colloid Interface Sci. **27**, 189 (1987).
17. S.J. O'Shea, M.E. Welland and T. Rayment, Langmuir **9**, 1826 (1993).
18. J.N. Israelachvili, *Intermolecular and Surface Forces*, Academic Press, New York, 1992 Chapter 15.
19. I.E. Dzyaloshinskii, E.M. Lifshitz and L.P. Pitaevskii, Adv. Phys. **10**, 165 (1961).
20. E.M. Lifshitz, Sov. Phys. JETP **2**, 73 (1956).
21. D.B. Hough and L.R. White, Adv. Colloid Interface Sci. **14**, 3 (1980).
22. J. Mahanty and B.W. Ninham, *Dispersion Forces*, Academic Press, London, 1976.
23. B.V. Derjaguin and L. Landua, Acta Physicochim. URSS **14**, 633 (1941).
24. E.J.W. Verwey and J.T.G. Overbeek, in *Theory of Stability of Lyophobic Colloids*, Elsevier, Amsterdam, 1948.
25. G.K. Binnig and C.F. Quate, Phys. Rev. Lett. **56**, 930 (1986).
26. W.A. Ducker, T.J. Senden and R.M. Pashley, Nature **353**, 239 (1991).
27. T.J. Senden, Curr. Opin. Colloid Interface Sci. **6**, 95 (2001).
28. B.V. Derjaguin, N.V. Churaev and V.M. Muller, *Surface Forces*, Consultant Bureau, New York, 1987, Chapter 1.
29. R. Aveyard, B.P. Binks, W.G. Cho, L.R. Fisher, P.D.I. Fletcher and F. Klinkhammer, Langmuir **12**, 6561 (1996).
30. R. Aveyard, B.P. Binks, S. Clark, P.D.I. Fletcher and I.G. Lyle, Colloids Surf., A **113**, 295 (1996).
31. B.P. Binks, W.G. Cho and P.D.I. Fletcher, Langmuir **13**, 7180 (1997).
32. W.-G. Cho and P.D.I. Fletcher, J. Chem. Soc., Faraday Trans. **93**, 1389 (1997).
33. R.G. Horn, D.J. Bachmann, J.N. Connor and S.J. Miklavcic, J. Phys.: Condens. Mater. **8**, 9483 (1996).
34. D.G. Goodall, M.L. Gee and G.W. Stevens, Langmuir **17**, 3784 (2001).
35. D.G. Goodall, M.L. Gee, G. Stevens, J. Perera and D. Beaglehole, Colloids Surf., A **143**, 41 (1998).
36. D.G. Goodall, G.W. Stevens, D. Beaglehole and M.L. Gee, Langmuir **15**, 4579 (1999).
37. M.L. Gee, D.G. Goodall and G.W. Stevens, Oil Gas Sci. Technol. **56**, 33 (2001).
38. P.C. Odiachi. Carnegie Mellon University, Pittsburgh, PA, 2001, Chapter 6.
39. R.R. Dagastine, Ph.D. thesis, Carnegie Mellon University, Pittsburgh, PA, 2002, Chapter 5.
40. O. Mondain-Monval, A. Espert, P. Omarjee, J. Bibette, F. Leal-Calderon, J. Philip and J.F. Joanny, Phys. Rev. Lett. **80**, 1778 (1998).
41. P. Omarjee, P. Hoerner, G. Riess, V. Cabuil and O. Mondain-Monval, Eur. Phys. J. E **4**, 45 (2001).
42. P. Omarjee, A. Espert, O. Mondain-Monval and J. Klein, Langmuir **17**, 5693 (2001).
43. A. Espert, P. Omarjee, J. Bibette, F.L. Calderon and O. Mondain-Monval, Macromolecules **31**, 7023 (1998).
44. J. Philip, G.G. Prakash, T. Jaykumar, P. Kalyanasundaram, O. Mondain-Monval and B. Raj, Langmuir **18**, 4625 (2002).
45. J. Philip, T. Jaykumar, P. Kalyanasundaram, B. Raj and O. Mondain-Monval, Phys. Rev. E. **66**, 011406/1 (2002).
46. M.L. Fielden, R.A. Hayes and J. Ralston, Langmuir **12**, 3721 (1996).
47. H.-J. Butt, J. Colloid Interface Sci. **166**, 109 (1994).
48. M. Preuss and H.-J. Butt, Langmuir **14**, 3164 (1998).
49. M. Preuss and H.-J. Butt, Int. J. Miner. Process. **56**, 99 (1999).
50. M. Preuss and H.-J. Butt, J. Colloid Interface Sci. **208**, 468 (1998).

51. G.E. Yakubov, O.I. Vinogradova and H.-J. Butt, J. Adv. Sci. Technol. **14**, 1783 (2000).
52. G.E. Yakubov, O.I. Vinogradova and H.J. Butt, Colloid J. (Trans. Kolloid. Z.) **63**, 518 (2001).
53. A.V. Nguyen, J. Nalaskowski and J.D. Miller, J. Colloid Interface Sci. **262**, 303 (2003).
54. P. Mulvaney, J.M. Perera, S. Biggs, F. Grieser and G.W. Stevens, J. Colloid Interface Sci. **183**, 614 (1996).
55. B.A. Snyder, D.E. Aston and J.C. Berg, Langmuir **13,** 590 (1997).
56. P.G. Hartley, F. Grieser, P. Mulvaney and G.W. Stevens, Langmuir **15**, 7282 (1999).
57. D.E. Aston and J.C. Berg, J. Colloid Interface Sci. **235**, 162 (2001).
58. D.E. Aston and J.C. Berg, Ind. Eng. Chem. Res. **41**, 389 (2002).
59. P. Attard and S.J. Miklavcic, Langmuir **17**, 8217 (2001).
60. P. Attard and S.J. Miklavcic, J. Colloid Interface Sci. **247**, 255 (2002).
61. G. Gillies, C.A. Prestidge and P. Attard, Langmuir **17**, 7955 (2001).
62. S.A. Nespolo, D.Y.C. Chan, F. Grieser, P.G. Hartley and G.W. Stevens, Langmuir **19**, 2124 (2003).
63. R.R. Dagastine, D.C. Prieve and L.R. White, J. Colloid Interface Sci., **273**, 339–342 (2004).
64. R.R. Dagastine, G.W. Stevens, D.Y.C. Chan and F. Grieser, J. Colloid Interface Sci., **269**, 84–96 (2004).
65. D. Bhatt, J. Newman and C.J. Radke, Langmuir **17**, 116 (2001).
66. D.Y.C. Chan, R.R. Dagastine and L.R. White, J. Colloid Interface Sci. **236**, 141 (2001).
67. R.R. Dagastine and L.R. White, J. Colloid Interface Sci. **247**, 310 (2002).
68. S.A. Nespolo, M.A. Bevan, D.Y.C. Chan, F. Grieser and G.W. Stevens, Langmuir **17**, 7210 (2001).

5

A Molecular Theory of Solutions at Liquid Interfaces

Andriy Kovalenko[1*] and Fumio Hirata[2]
[1]*National Institute for Nanotechnology, National Research Council of Canada, and Department of Mechanical Engineering, University of Alberta, W6-025, ECERF Bldg., 9107-116 Str., Edmonton, Alberta, T6G 2V4, Canada*
[2]*Institute for Molecular Science, Myodaiji, Okazaki 444-8585, Japan*

5.1. INTRODUCTION

Liquid interfaces are of paramount importance in a variety of fields and technological applications of modern chemistry, biology and pharmacology, such as electrochemistry, photosynthesis, biocatalysis, protein encapsulation and membranes, drugs and gene delivery [1]. Interfaces between immiscible or partially miscible liquids are frequently encountered in chemical and biological systems, and determine their electrokinetics, stability and functioning. An understanding of the structure of a liquid/liquid interface and the mechanisms of charge separation and transfer across the interface is a fundamental problem of modern chemistry and chemical technology. Starting from the pioneering work of Benjamin [2], computer simulations were used to study the mechanism and dynamics of transfer of single ions across a realistic liquid/liquid interface modelled with molecular force fields to take into account chemical functionalities of solution species (see literature in [3]). However, molecular dynamics of realistic liquid interfaces with a finite concentration of ions is not yet feasible [4], whereas interfacial ion transfer experiments and technological processes involve a current of ions in the presence of an external electric field applied across the liquid/liquid interface. Micelles are an example of a ternary liquid interface where amphiphilic molecules of surfactant coating the boundaries of nanoheterogeneous oil phases determine their stability in aqueous solution [5]. The structural properties of realistic micelles and other

* On leave from: Institute for Molecular Science, Myodaiji, Okazaki 444-8585, Japan.

self-assembling systems were studied by molecular dynamics simulations of a pre-assembled surfactant nanophase [6–10] and its spontaneous formation [11,12] up to nanometre and nanosecond scales. Unfortunately, molecular simulation at the microscopic level to describe their composition, stability and mass balance in solution above the critical micelle concentration is virtually impossible because of the necessity to consider tens of millions of water molecules per micellar aggregate as small as 100 surfactant molecules and their characteristic time as long as milliseconds [13]. Microscopic description of such systems can be achieved by employing statistical mechanical theory of interfaces of molecular liquids and electrolyte solutions [14,15].

One of the methods employed to describe liquid/vapour and liquid/liquid interfaces of simple fluids was density functional theory (DFT) [16–19]. A site–site DFT formalism has been developed for liquid interfaces of molecular species with the hard sphere and 12-6 Lennard–Jones-type potentials between their interaction sites [20–22], and applied to spherical and cylindrical micelles and vesicles formed in a model system of two-, three- and polyatomic amphiphiles in single-sphere solvents [20–24]. In the site–site DFT approach, the free energy functional is usually split up into the term because of the core repulsion that is replaced with that of a reference system of purely repulsive (hard body) particles, and the contribution of attraction taken into account by using the random phase approximation [14] perturbation scheme. Such a subdivision works well for fluids of molecular species with weakly attractive site–site interactions like the 12-6 Lennard–Jones potential. However, it is problematic for realistic systems of charged polyatomic species with strong association and orientational ordering of species, like hydrogen bonding. A more consistent approach should provide proper coupling between the short-range solvation structure and the long-range correlations of molecules with partial site charges. A substantial step in this direction had been the RISM (reference interaction site model) integral equation theory of inhomogeneous polyatomic fluids, derived within the density functional formalism by Chandler *et al.* [25,26]. Unfortunately, this approach cannot be applied to liquid interfaces because it approximates the direct correlation function in the inhomogeneous fluid with that in the bulk, and thus does not allow for a change of the direct correlation function between the fluid phases.

A successful approach to fluids and solutions of molecular species is provided by the RISM theory, which was pioneered by Chandler and Andersen for non-polar molecular fluids [27], and extended by Hirata *et al.* to charged species [28]. Kovalenko and Hirata proposed a new closure [29] (below referred to as the KH approximation) that enabled description of associating molecular fluids and solutions in the whole density range from gas to liquid [30,31]. It is a non-trivial optimized coupling of the so-called hypernetted chain (HNC) approximation and the mean spherical approximation (MSA). The former is switched on in the repulsive core region, whereas the latter is automatically applied to long-range enhancement tails of the distribution functions in the critical regime as well as to their high association peaks wherever they arise. The KH closure approximation is adequate for description of charged polyatomic fluids and mixtures over a wide range of density with proper account for their chemical specificities. The RISM–KH theory qualitatively reproduces the liquid/vapour phase diagrams of associating molecular fluids such as water, methanol and hydrogen fluoride, and their structure in the vapour as well as liquid phases, including hydrogen bonding [30,31]. It appropriately describes the association structures that sequentially change in a mixture of water and amphiphilic *tert*-butyl alcohol (TBA) with their molar fraction, including micromicelles [32, 33].

Consideration of a liquid interface that fluctuates requires proper account of mechanical and chemical balance between the phases. The description of spatially inhomogeneous fluids can be performed on the basis of Born–Green–Yvon (BGY) integro-differential equation [14,15]

$$\nabla_1 \log \rho_a(\mathbf{r}_1) = -\nabla_1 \beta u_a^{\text{ext}}(\mathbf{r}_1) + \sum_b \int d\mathbf{r}_2\, \rho_b(\mathbf{r}_2)\, g_{ab}(\mathbf{r}_1, \mathbf{r}_2; [\rho]) \nabla_2\, \beta u_{ab}(\mathbf{r}_1, \mathbf{r}_2) \tag{1}$$

or the Lovett–Mou–Buff–Wertheim (LMBW) equation [15, 34, 35]

$$\nabla_1 \log \rho_a(\mathbf{r}_1) = -\nabla_1\, \beta u_a^{\text{ext}}(\mathbf{r}_1) + \sum_b \int d\mathbf{r}_2\, c_{ab}(\mathbf{r}_1, \mathbf{r}_2) \nabla_2\, \rho_b(\mathbf{r}_2) \tag{2}$$

for the one-particle density distributions $\rho_a(\mathbf{r}_1)$, where $u_a^{\text{ext}}(\mathbf{r}_1)$ is the external field potential acting on a particle of sort a and $u_{ab}(\mathbf{r}_1, \mathbf{r}_2)$ is the interaction potential between two particles of sort a and b in the presence of the surface, and $\beta = 1/(k_B T)$, i.e., the inverse temperature with the Boltzmann constant k_B. As an input, the BGY equation requires the inhomogeneous two-particle distribution function $g_{ab}(\mathbf{r}_1, \mathbf{r}_2)$, that is, essentially a three-particle distribution function. The LMBW equation requires an input of the inhomogeneous two-particle direct correlation function $c_{ab}(\mathbf{r}_1, \mathbf{r}_2)$. Both the BGY and LMBW equations should yield identical results for the density distributions, provided the inhomogeneous two-particle correlation functions are known exactly. Quite complex in nature, these functions are usually constructed of bulk correlations by employing approximations appropriate to the system physics. Fischer and Methfessel [36] used the BGY equation to describe the gas–liquid interface of simple fluid of Lennard–Jones atomic species. They replaced the inhomogeneous distribution function $g(\mathbf{r}_1, \mathbf{r}_2; [\rho])$ by the distribution of homogeneous hard-sphere fluid $g_{\text{hs}}(|\mathbf{r}_1 - \mathbf{r}_2|; \bar{\rho}_{\text{hs}})$ at number density $\bar{\rho}_{\text{hs}}((\mathbf{r}_1 + \mathbf{r}_2)/2)$ determined by coarse-graining of the density distribution $\rho(\mathbf{r})$. The attractive force term of $g(\mathbf{r}_1, \mathbf{r}_2; [\rho])$ was neglected. Taylor and Lipson [37] developed a site–site BGY equation with a set of Kirkwood-type superposition approximations to the site–site distribution functions for fluids of flexible polyatomic chains with the hard-sphere and square-well potentials. Attard [38] derived a site–site BGY equation with proper triplet superposition approximation for fluids of hard-sphere polymers with soft bonds. Even for homogeneous fluids with short-range interactions, the superposition approximations expressing the triplet distribution functions in terms of products of the binary distributions are tedious and rather specific to the particular systems. As distinct, it is much easier to approximate the direct correlation function because it has an additive asymptotics of the interaction potential. Iatsevitch and Forstmann [39] developed a description of a planar interface of immiscible Lennard–Jones atomic liquids by using the LMBW equation (2) with the inhomogeneous direct correlation function constructed as

$$c_{ab}(\mathbf{r}_1, \mathbf{r}_2) = \frac{1}{2}\Big[c_{ab}(|\mathbf{r}_1 - \mathbf{r}_2|; \bar{\rho}(\mathbf{r}_1)) + c_{ab}(|\mathbf{r}_1 - \mathbf{r}_2|; \bar{\rho}(\mathbf{r}_2))\Big] \tag{3}$$

from the homogeneous direct correlation functions $c_{ab}(r;\bar{\rho})$ at the set of densities of species equal to the coarse-grained local densities $\bar{\rho} \equiv \{\bar{\rho}_a\}$. The direct correlation functions $c_{ab}(r;\bar{\rho})$ at the fluid density $\bar{\rho}$ varying through the instability region were in turn calculated by linear interpolation between the homogeneous direct correlation functions $c_{ab}^{\text{blk}}(r;\rho)$ in the coexisting bulk phases. The latter were obtained from the reference HNC theory [14].

Of the BGY and LMBW approaches, the latter is most suitable for the description of interfaces of electrolyte solutions with polar/non-polar polyatomic species in molecular solvents exhibiting an associating structure, such as hydrogen bonding. An approximate theory constructing the inhomogeneous two-particle correlations as discussed below consists of two stages: determination of the chemical and mechanical equilibrium between the coexisting fluid phases, and then calculation of the density distribution describing the interfacial structure. Below we address these questions.

5.2. MOLECULAR THEORY OF FLUID PHASE EQUILIBRIA

A description of molecular fluids can employ the molecular Ornstein–Zernike integral equation for the six-dimensional correlation functions between molecular species. The orientational dependence in the correlations can be treated by expanding the correlation functions in a reasonably simple basis of the generalized spherical harmonics [14,40], recapitulating essential features of the solutions. This method is computationally very intensive even for relatively simple molecules, and becomes increasingly cumbersome with molecular asphericity. It is extremely time-consuming for calculation of phase diagrams requiring evaluation of a number of thermodynamic states. Moreover, it is not easy to extend the molecular Ornstein–Zernike approach to flexible molecular models. A cost-efficient description of complex polyatomic species can be achieved by partial contraction of their configurational space, averaging out the orientational degrees of freedom.

Chandler and Andersen introduced the site–site Ornstein–Zernike (SSOZ) integral equation for the radial correlation functions between pairs of interaction sites of polyatomic species [14,27],

$$h^{ab}_{\alpha\gamma}(r) = \omega^{a}_{\alpha\mu} * c^{ab}_{\mu\nu} * \omega^{b}_{\nu\gamma} + \omega^{a}_{\alpha\mu} * c^{ac}_{\mu\nu} * \rho^{c} h^{cb}_{\nu\gamma}, \tag{4}$$

where $h^{ab}_{\alpha\gamma}(r)$ and $c^{ab}_{\alpha\gamma}(r)$ are respectively the total and direct correlation functions between interaction site α of solution species a and site γ of species b, the intramolecular matrix $\omega^{a}_{\alpha\gamma}(r)$ determines the geometry of molecules of species a, ρ^a is their number density and "$*$" means convolution in the direct space and summation over repeating indices. For rigid particles with site separations $l^{a}_{\alpha\gamma}$, the intramolecular matrix comprises delta functions: $\omega^{a}_{\alpha\gamma}(r) = \delta_{\alpha\gamma}\delta(\mathbf{r}) + (1-\delta_{\alpha\gamma})\delta(r - l^{a}_{\alpha\gamma})/(4\pi(l^{a}_{\alpha\gamma})^2)$. In reciprocal space, it is written as

$$\omega^{a}_{\alpha\gamma}(k) = \frac{\sin(kl^{a}_{\alpha\gamma})}{kl^{a}_{\alpha\gamma}}, \tag{5}$$

implying that $l^{a}_{\alpha\alpha} = 0$. Approximate closures to the SSOZ equation (4) are constructed analogously to atomic fluids [14]. Complemented with the Percus–Yevick type closure relation, it has been known as the RISM theory. Schweizer and Curro [41] generalized the RISM theory to flexible polymer chains with the intramolecular distributions $\omega^{a}_{\alpha\gamma}(k)$ in a smeared form peculiar to chains of freely tangent segments. Another generalization to deformable molecules determines the intramolecular distribution functions self-consistently within the DFT by including them into the appropriately constructed functional of the free energy [42].

The RISM–HNC, or extended RISM theory [28], has an advantage of being able to readily handle the description of ion-molecular solutions for realistic models of polyatomic species at liquid density, with proper account for such chemical specificities as hydrogen bonding [43]. However, as association in fluid of charged species increasingly strengthens in a vapour phase and near critical lines, the RISM–HNC equations strongly overestimate clustering and become divergent [44]. This drawback is eliminated in the KH closure approximation [29–31], which couples in a new way the HNC functional form applied to the regions of density depletion $g^{ab}_{\alpha\gamma}(r) < 1$, and the linearization reducing to the MSA form applied to the regions of enrichment $g^{ab}_{\alpha\gamma}(r) > 1$,

$$g^{ab}_{\alpha\gamma}(r) = \begin{cases} \exp\left(\psi^{ab}_{\alpha\gamma}(r)\right) & \text{for } \psi^{ab}_{\alpha\gamma}(r) \leq 0, \\ 1 + \psi^{ab}_{\alpha\gamma}(r) & \text{for } \psi^{ab}_{\alpha\gamma}(r) > 0, \end{cases} \tag{6}$$

with

$$\psi^{ab}_{\alpha\gamma}(r) = -\beta u^{ab}_{\alpha\gamma}(r) + h^{ab}_{\alpha\gamma}(r) - c^{ab}_{\alpha\gamma}(r),$$

where $g^{ab}_{\alpha\gamma}(r) = h^{ab}_{\alpha\gamma}(r) + 1$, the site–site distribution function. A modification to the KH approximation (6) includes a non-linear, quadratic term in addition to the linearized MSA form,

$$g^{ab}_{\alpha\gamma}(r) = \begin{cases} \exp\left(\psi^{ab}_{\alpha\gamma}(r)\right) & \text{for } \psi^{ab}_{\alpha\gamma}(r) \leq 0, \\ 1 + \psi^{ab}_{\alpha\gamma}(r) + \dfrac{\tau}{2}\left(\psi^{ab}_{\alpha\gamma}(r)\right)^2 & \text{for } \psi^{ab}_{\alpha\gamma}(r) > 0. \end{cases} \tag{7}$$

The parameter τ can be optimized for thermodynamic consistency between the virial and compressibility routes to the free energy of the system. As discussed in Section 5.3.2, the KHM closure (7) brings about an additional long-range term in the direct correlation functions in the critical regime.

The RISM integral equations in the KH approximation lead to closed analytical expressions for the free energy and its derivatives [29–31]. Likewise, the KHM approximation (7) possesses an exact differential of the free energy. Note that the solvation chemical potential for the MSA or PY closures is not available in a closed form and depends on a path of the thermodynamic integration. With the analytical expressions for the chemical potential and the pressure, the phase coexistence envelope of molecular fluid can be localized directly by solving the mechanical and chemical equilibrium conditions,

$$P(\rho_{\text{vap}}, T) = P(\rho_{\text{liq}}, T), \tag{8a}$$

$$\mu(\rho_{\text{vap}}, T) = \mu(\rho_{\text{liq}}, T), \tag{8b}$$

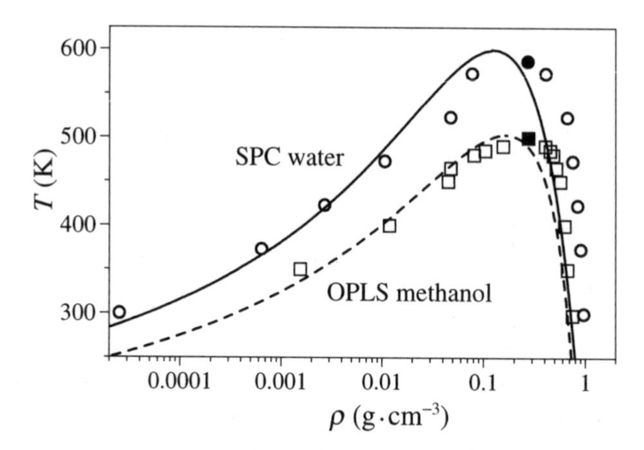

FIGURE 5.1. Liquid/vapour coexistence curves of SPC water (solid line) and OPLS methanol (dashed line) following from the RISM–KH theory. Molecular simulation results for water [45] and methanol [46] (open circles and squares, respectively) and critical point extrapolations (filled symbols).

for the vapour and liquid phase densities ρ_{vap} and ρ_{liq}, whether or not the solution exists for all intermediate densities. Provided the isotherm is continuous across all the region of liquid/vapour phase coexistence, Equations (8) are exactly equivalent to the Maxwell construction on either pressure or chemical potential isotherm.

The KH closure approximation has been shown to be adequate for treatment of phase transitions of associating molecular fluids [30,31]. Figure 5.1 shows the liquid/vapour coexistence curves of water and methanol predicted by the RISM–KH theory against the results of simulations [45,46]. A logarithmic scale is used to resolve gas phase densities. The theory qualitatively reproduces the phase diagrams of SPC water and OPLS methanol. Remarkable is a good fit for the vapour branch of the binodals. The critical temperature of SPC water following from the RISM–KH description, $T_c = 600$ K, is close to 587 K obtained in the simulations. The methanol critical temperature $T_c = 503$ K versus 500 K in the simulation and its vapour branch are again reproduced quite well. The liquid branch of both water and methanol are shifted to smaller densities. As discussed in Refs.[30,31], this disadvantage can be alleviated by further modifying the KH closure like in the KHM approximation (7) and proceeding to the three-dimensional (3D-RISM) description. The theory gives the critical density to be larger for methanol, which is similar to the trend observed in the simulation results. Figure 5.2 draws a comparison of the coexistence envelopes of water, methanol, hydrogen fluoride and dimethylsulfoxide. The theory reproduces the relative positions of their critical points as well as the difference in their density at ambient conditions in qualitative agreement with the experimental data.

Much as for pure substances, the RISM–KH theory yields qualitatively good results for mixtures, in particular, for such a complex system as aqueous solution of TBA, which is an amphiphile substance. It reproduces in good agreement with experiment the structural changes from dilute TBA molecules encaged by the water tetrahedral hydrogen bonding to micromicelles of TBA molecules formed with rising the TBA concentration and finally to dilute water molecules incorporated into the zigzag alcohol hydrogen bonding in the alcohol-rich regime [32]. The predictions of the RISM–KH theory are in very

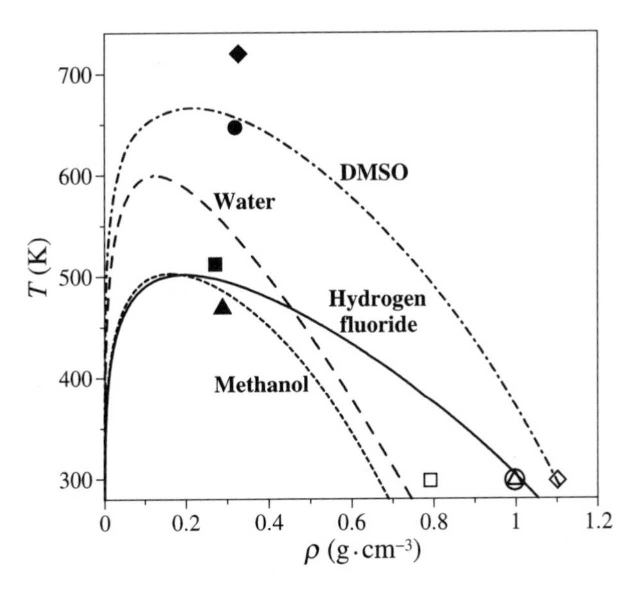

FIGURE 5.2. Liquid/vapour coexistence curves of hydrogen fluoride (solid line), methanol (short-dashed line), water (long-dashed line) and dimethylsulfoxide (dash-dotted line) following from the RISM–KH theory. Experimental data for their critical points (filled triangle, square, circle and rhomb, respectively), and at ambient conditions (corresponding open symbols).

good agreement also with the experimental data for the compressibility of TBA–water mixtures [33]. As is seen in Figure 5.3, the theory well reproduces the compressibility anomalies with the temperature and molar fraction of the mixture, including the isosbestic point and the minimum at low TBA concentrations corresponding to formation of micromicelles.

5.3. MICROSCOPIC DESCRIPTION OF LIQUID/LIQUID AND LIQUID/VAPOUR INTERFACES

Using the BGY and LMBW integro-differential equations (1) and (2) for single-particle density distributions $\rho_a(\mathbf{r}_1)$ require an input of the inhomogeneous two-particle distribution function $g_{\alpha\gamma}(\mathbf{r}_1, \mathbf{r}_2)$ or the direct correlation functions $c_{\alpha\gamma}(\mathbf{r}_1, \mathbf{r}_2)$. They can be obtained by solving the inhomogeneous Ornstein–Zernike (IOZ) equation [15] complemented with an appropriate closure. This inhomogeneous approach leads to the description of a liquid interface without making any phenomenological assumptions on the structure of the inhomogeneous two-particle pair correlations. However, its generalization to fluids of charged polyatomic particles runs into significant difficulties of dealing with multidimensional integral equations. Long-range tails of the distribution functions corresponding to phase transitions and phase separations in such liquids require large domains of discretization of the integral equations, whereas high peaks describing specific association structures need fine grid resolution. As an affordable alternative, we propose a site–site LMBW equation for the single-site density distributions, which is a generalization of the LMBW equation (2) to polyatomic fluids [47]. It uses the inhomogeneous site–site direct correlation functions of molecular fluid non-linearly interpolated

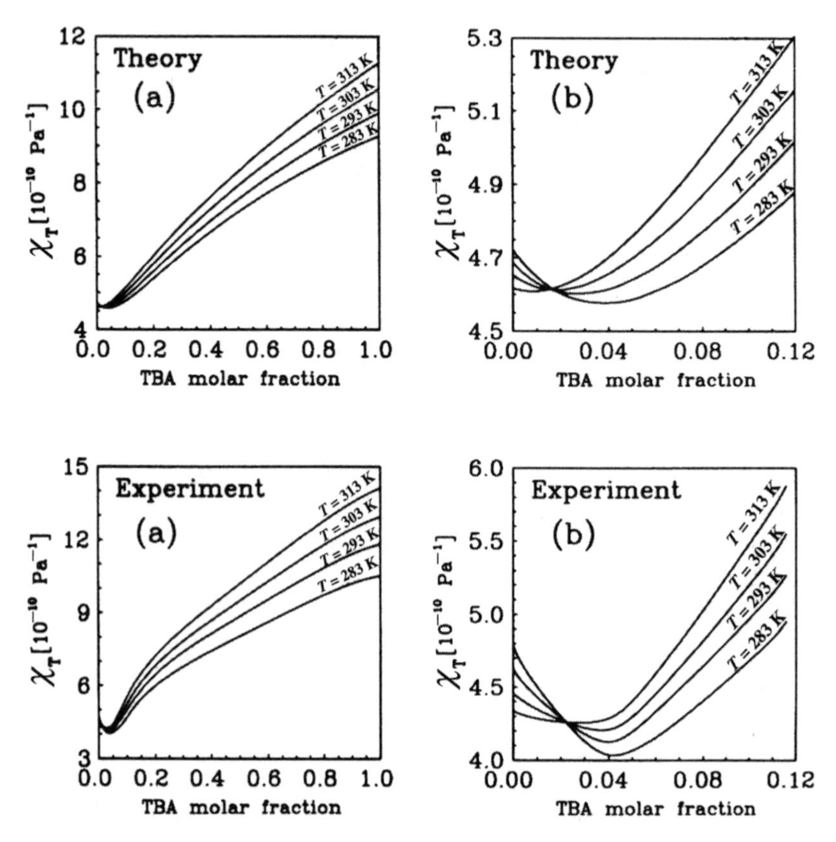

FIGURE 5.3. Isothermal compressibility of TBA–water mixture as a function of TBA molar fraction at four temperatures $T = 283; 293; 303; 313$ K, (a) in the whole concentration range, and (b) within the water-rich region with the anomalous behaviour of χ_T shown in more detail. Predictions of the RISM–KH theory in comparison with experimental data.

between the homogeneous ones. On the other hand, development of the inhomogeneous scheme for the liquid/vapour interface is of crucial importance even for simple fluid of Lennard–Jones particles, as it allows one to obtain physical features of the interface [48]. Below we develop a methodology of these two approaches to liquid interfaces.

5.3.1. Molecular Theory of Liquid Interfaces of Polyatomic Fluids

Consider an inhomogeneous fluid mixture of polyatomic species with the density distributions $\rho_\alpha(\mathbf{r})$ of interaction sites α in an external field with the site potential $u_\alpha^{\text{ext}}(\mathbf{r})$. Minimization of the Helmholtz free energy with respect to variations $\delta\rho_\alpha$ performed analogously to Refs. [34,35] yields the relation

$$\nabla_1 \rho_\alpha(\mathbf{r}_1) = -\,\rho_\alpha(\mathbf{r}_1)\nabla_1\,\beta u_\alpha^{\text{ext}}(\mathbf{r}_1) + \rho_\alpha(\mathbf{r}_1)\sum_\gamma \int d\mathbf{r}_2\,\tilde{c}_{\alpha\gamma}(\mathbf{r}_1,\mathbf{r}_2;[\rho])\nabla_2\,\rho_\gamma(\mathbf{r}_2), \tag{9}$$

where the inhomogeneous site–site direct correlation function $\tilde{c}_{\alpha\gamma}(\mathbf{r}_1, \mathbf{r}_2)$ is defined in terms of the functional inverse of the susceptibility or density–density correlation function [25,26]

$$\chi_{\alpha\gamma}^{-1}(\mathbf{r}_1, \mathbf{r}_2) = \rho_\alpha^{-1}(\mathbf{r}_1)\delta_{\alpha\gamma}\delta(\mathbf{r}_1 - \mathbf{r}_2) - \tilde{c}_{\alpha\gamma}(\mathbf{r}_1, \mathbf{r}_2). \tag{10}$$

The susceptibility $\chi_{\alpha\gamma}$ can be subdivided into the *intramolecular* ideal gas part $\chi_{\alpha\gamma}^{(0)}$ and the *intermolecular* term,

$$\chi_{\alpha\gamma}(\mathbf{r}_1, \mathbf{r}_2) = \chi_{\alpha\gamma}^{(0)}(\mathbf{r}_1, \mathbf{r}_2) + \rho_\alpha(\mathbf{r}_1)\rho_\gamma(\mathbf{r}_2)h_{\alpha\gamma}(\mathbf{r}_1, \mathbf{r}_2). \tag{11}$$

In the limit of a uniform fluid, the intramolecular part of the susceptibility is translationally invariant and is expressed through the intramolecular matrix simply as

$$\chi_{\alpha\gamma}^{(0)}(\mathbf{r}_1, \mathbf{r}_2)\Big|_{\text{uniform}} = \rho_\alpha\omega_{\alpha\gamma}(|\mathbf{r}_1 - \mathbf{r}_2|). \tag{12}$$

Defining the *intramolecular* site–site direct correlation function as

$$c_{\alpha\gamma}^{(0)}(\mathbf{r}_1, \mathbf{r}_2) = \rho_\alpha^{-1}(\mathbf{r}_1)\delta_{\alpha\gamma}\delta(\mathbf{r}_1 - \mathbf{r}_2) - \left[\chi_{\alpha\gamma}^{(0)}(\mathbf{r}_1, \mathbf{r}_2)\right]^{-1}, \tag{13}$$

the whole direct correlation function can be subdivided as

$$\tilde{c}_{\alpha\gamma}(\mathbf{r}_1, \mathbf{r}_2) = c_{\alpha\gamma}^{(0)}(\mathbf{r}_1, \mathbf{r}_2) + c_{\alpha\gamma}(\mathbf{r}_1, \mathbf{r}_2), \tag{14}$$

where the latter term is identified as the *intermolecular* site–site direct correlation function. In the limit of uniform fluid, $c_{\alpha\gamma}(\mathbf{r}_1, \mathbf{r}_2)$ satisfies the SSOZ integral equation (4) introduced by Chandler and Andersen [27]. Inserting the approximation (12) for the intramolecular susceptibility into the relations (13) and (14) yields the direct correlation function in the form

$$\tilde{c}_{\alpha\gamma}(\mathbf{r}_1, \mathbf{r}_2) = \delta_{\alpha\gamma}(\mathbf{r}_1, \mathbf{r}_2)\,\rho_\gamma^{-1}(\mathbf{r}_1) - \omega_{\alpha\gamma}^{-1}(\mathbf{r}_1, \mathbf{r}_2)\,\rho_\gamma^{-1}(\mathbf{r}_1) + c_{\alpha\gamma}(\mathbf{r}_1, \mathbf{r}_2). \tag{15}$$

Substituting the relation (15) into Equation (9) leads on simple evaluation to the equation

$$\begin{aligned}\nabla_1\,\rho_\alpha(\mathbf{r}_1) = &-\rho_\alpha(\mathbf{r}_1)\nabla_1\,\beta u_\alpha^{\text{ext}}(\mathbf{r}_1) + \nabla_1\,\rho_\alpha(\mathbf{r}_1) - \omega_{\alpha\gamma}^{-1}(r_{12}) * \nabla_2\,\rho_\gamma(\mathbf{r}_2)\\ &+ \rho_\alpha(\mathbf{r}_1)\,c_{\alpha\gamma}(\mathbf{r}_1, \mathbf{r}_2) * \nabla_2\,\rho_\gamma(\mathbf{r}_2).\end{aligned} \tag{16}$$

On cancellation of the term $\nabla_1\,\rho_\alpha(\mathbf{r}_1)$ in the two sides and functional multiplication of Equation (16) by the intramolecular matrix $\omega_{\alpha\gamma}(r_{12})$, we obtain

$$\begin{aligned}\nabla_1\,\rho_\alpha(\mathbf{r}_1) = \omega_{\alpha\gamma}(r_{12}) * \big(&-\rho_\gamma(\mathbf{r}_2)\nabla_2\,\beta u_\gamma^{\text{ext}}(\mathbf{r}_2)\\ &+ \rho_\gamma(\mathbf{r}_2)\,c_{\gamma\mu}(\mathbf{r}_2, \mathbf{r}_3) * \nabla_3\,\rho_\mu(\mathbf{r}_3)\big),\end{aligned} \tag{17}$$

where "$*$" means integration over repeating spatial variables and summation over repeating site indices. Notice that since the intramolecular matrix $\omega_{\alpha\gamma}(r_{12})$ precedes the density $\rho_\gamma(\mathbf{r}_2)$ in the right-hand side, the left-hand side of Equation (17) contains the gradient of the density rather than of the density logarithm as in the original LMBW equation (2). Therefore, Equation (17) admits solutions with non-physical regions of a negative site density $\rho_\alpha(\mathbf{r}) < 0$ at the gas side of the interface. In fact, this is caused by the choice of the intramolecular susceptibility in the form (12) with the intramolecular matrix $\omega_{\alpha\gamma}(r_{12})$ performing uniform averaging over the orientation of molecular species, which is more appropriate for the case of a weakly inhomogeneous system. Indeed, when the site density $\rho_\alpha(\mathbf{r}_1)$ is vanishing to the gas phase and its derivative in the left-hand side of Equation (17) should vanish too, the convolution through $\omega_{\alpha\gamma}(r_{12})$ in a system with strong alignment of molecules can keep the derivative non-zero, coupling it with other sites γ through improper orientations of molecules. A simple approximation alleviating this shortcoming of Equation (17) can be constructed by changing the order of terms to multiply with factor $\rho_\alpha(\mathbf{r}_1)$ on the orientational averaging [convolving with $\omega_{\alpha\gamma}(r_{12})$] rather than prior to it. This immediately leads to

$$\nabla_1 \ln \rho_\alpha(\mathbf{r}_1) = \omega_{\alpha\gamma}(r_{12}) * \left(-\nabla_2\, \beta u_\gamma^{\text{ext}}(\mathbf{r}_2) + c_{\gamma\mu}(\mathbf{r}_2, \mathbf{r}_3) * \nabla_3\, \rho_\mu(\mathbf{r}_3)\right). \tag{18}$$

This is our new equation for the single-site density distributions which generalizes the LMBW equation (2) to polyatomic fluids, below called a site–site LMBW (SS-LMBW) equation [47]. As distinct from the site–site DFT approaches [20–24], the SS-LMBW equation properly treats the short- and long-range correlations coupled in the site–site direct correlation function $c_{\gamma\mu}(\mathbf{r}_2, \mathbf{r}_3)$. The SS-LMBW theory also differs from the RISM approach to inhomogeneous polyatomic fluids derived within the density functional formalism by Chandler *et al.* [25,26]. In Equations (3.7) and (3.13) of Ref. [26] for the site density profiles $\rho_\alpha(\mathbf{r})$, the orientational averaging with the intramolecular matrix $\omega_{\alpha\gamma}(r_{12})$ is applied to the Boltzmann exponent, with further superposition of the contributions to the density profile from the different sites of the solvent molecule in a Kirkwood-type approximation,

$$\rho_\alpha(\mathbf{r}_1) = \prod_\gamma \omega_{\alpha\gamma}(r_{12}) * \exp\Big(-\beta u_\gamma^{\text{ext}}(\mathbf{r}_2) + \sum_\gamma c_{\gamma\mu}^{\text{blk}}(r_{23}) * (\rho_\mu(\mathbf{r}_3) - \rho_\mu^{\text{blk}})\Big),$$

where $c_{\gamma\mu}^{\text{blk}}(r)$ is the homogeneous site–site direct correlation function of bulk solvent, and ρ_μ^{blk} is its density. As distinct, the SS-LMBW equation (18) is for the gradient $\nabla\rho_\alpha(\mathbf{r})$ rather than for the site density profile $\rho_\alpha(\mathbf{r})$, and involves the inhomogeneous two-particle direct correlation function $c_{\gamma\mu}(\mathbf{r}_2, \mathbf{r}_3)$ rather than the bulk homogeneous one $c_{\gamma\mu}^{\text{blk}}(r_{23})$. (Notice that in the approximation $c(\mathbf{r}_1, \mathbf{r}_2) \approx c^{\text{blk}}(r_{12})$, the LMBW equation (2) reduces to the Ornstein–Zernike equation with the HNC closure.) Furthermore, the SS-LMBW equation (18) applies the orientational averaging with the intramolecular matrix $\omega_{\alpha\gamma}(r_{12})$ to an expression in the argument of the Boltzmann exponent, that is to a part of the site potentials of mean force rather than to the site distributions.

To solve Equation (18), we construct the inhomogeneous site–site direct correlation function $c_{\alpha\gamma}(\mathbf{r}_1, \mathbf{r}_2)$ by the following procedure [47]. In the spirit of the approximation (3) for simple atomic fluid, we span it between the homogeneous site–site direct correlation functions $c_{\alpha\gamma}^{\text{blk}}(r_{12}; \boldsymbol{\rho}^{(i)})$ of coexisting bulk phases i with site densities $\boldsymbol{\rho}^{(i)}$. (Here vector $\boldsymbol{\rho}$ denotes a set of densities $\{\rho_\alpha\}$ for all sites α.) For an interface of two immiscible/partially miscible liquids, phase 1 with density $\boldsymbol{\rho}^{(1)} = \{\rho_{\text{I}}^{\text{rich}}, \rho_{\text{II}}^{\text{poor}}\}$ is rich in component I and poor in component II, whereas phase 2 with density $\boldsymbol{\rho}^{(2)} = \{\rho_{\text{I}}^{\text{poor}}, \rho_{\text{II}}^{\text{rich}}\}$ is poor in component I and rich in component II. The inhomogeneous site–site direct correlation function is represented by an average of two homogeneous terms parametrically dependent on the coarse-grained density profiles $\bar{\boldsymbol{\rho}}(\mathbf{r}_1)$ and $\bar{\boldsymbol{\rho}}(\mathbf{r}_2)$ at the two given positions,

$$c_{\alpha\gamma}(\mathbf{r}_1, \mathbf{r}_2) = \frac{1}{2}\Big[\tilde{c}_{\alpha\gamma}\big(r_{12}; \bar{\boldsymbol{\rho}}(\mathbf{r}_1)\big) + \tilde{c}_{\alpha\gamma}\big(r_{12}; \bar{\boldsymbol{\rho}}(\mathbf{r}_2)\big)\Big]. \tag{19}$$

The effective homogeneous direct correlations $\tilde{c}_{\alpha\gamma}(r; \bar{\boldsymbol{\rho}})$ are in turn interpolated over the site densities $\bar{\boldsymbol{\rho}}$ between the homogeneous direct correlation functions $c^{\mathrm{blk}}_{\alpha\gamma}(r;\boldsymbol{\rho}^{(i)})$ of the bulk liquid phases $i = 1$ and 2 in equilibrium at the densities $\boldsymbol{\rho}^{(1)}$ and $\boldsymbol{\rho}^{(2)}$, as well as of a gas phase $i = 0$ with all species at a vanishing density $\boldsymbol{\rho}^{(0)}$,

$$\tilde{c}_{\alpha\gamma}(r;\boldsymbol{\rho}) = \sum_{i=0}^{2} \tilde{w}_i(\boldsymbol{\rho})\, c^{\mathrm{blk}}_{\alpha\gamma}(r;\boldsymbol{\rho}^{(i)}). \tag{20}$$

Unlike the form (3) with linearly interpolated $c_{ab}(r;\bar{\rho})$ the interpolation procedure (20) is inherently non-linear, which is suggested by the following considerations. The SS-LMBW equation (18) can be rewritten in the form of a linear equation for the density gradient $\rho'_\alpha(\mathbf{r}) \approx \boldsymbol{\nabla}\rho_\alpha(\mathbf{r})$,

$$A_{\alpha\gamma}(\mathbf{r}_1,\mathbf{r}_2) * \rho'_\gamma(\mathbf{r}_2) = B_\alpha(\mathbf{r}_2), \tag{21}$$

with the matrix coefficients

$$\begin{aligned} A_{\alpha\gamma}(\mathbf{r}_1,\mathbf{r}_2) &= \delta_{\alpha\gamma}(\mathbf{r}_{12}) - \omega_{\alpha\mu}(r_{13}) * \rho_\mu(\mathbf{r}_3)\, c_{\mu\gamma}(\mathbf{r}_3,\mathbf{r}_2) \\ B_\alpha(\mathbf{r}_2) &= -\omega_{\alpha\gamma}(r_{12}) * \rho_\gamma(\mathbf{r}_2) \boldsymbol{\nabla}_2\, \beta u^{\mathrm{ext}}_\gamma(\mathbf{r}_2) \end{aligned}$$

containing the density distribution $\rho_\alpha(\mathbf{r})$ and thus introducing non-linearity in the SS-LMBW equation. (The same applies to the conventional LMBW equation (2) as well.) As discussed below, in the case of the liquid interface in the external gravity field the free term B_α in Equation (21) is very small compared to the components of the matrix $A_{\alpha\gamma}$. This leads to the condition that the determinant of this matrix should be as small,

$$\left|\, \delta_{\alpha\gamma}(\mathbf{r}_{12}) - \omega_{\alpha\mu}(r_{13}) * \rho_\mu(\mathbf{r}_3)\, c_{\mu\gamma}(\mathbf{r}_3,\mathbf{r}_2) \,\right| \ll 1. \tag{22}$$

Notice that it is intimately related to the condition determining the spinodal instability of bulk liquid, $\left|\, \delta_{\alpha\gamma}(\mathbf{r}_{12}) - \omega_{\alpha\mu}(r_{13}) * \rho^{\mathrm{blk}}_\mu c^{\mathrm{blk}}_{\mu\gamma}(r_{32}) \,\right| = 0$. Indeed, the latter specifies the region of the density where the isothermal compressibility of the bulk liquid (or liquid mixture) becomes negative, making the homogeneous phase with the intermediate densities unstable and causing its separation into phases with gas and liquid densities (or high and low ratio of mixture components). The relation (22) thus describes the existence of an inhomogeneous density distribution of the fluid. If the approximation to the inhomogeneous direct correlation function $c_{\alpha\gamma}(\mathbf{r}_1,\mathbf{r}_2)$ linearly interpolates over the local coarse-grained density $\rho_\alpha(\mathbf{r})$ between the homogeneous functions $c^{\mathrm{blk}}_{\alpha\gamma}(r;\boldsymbol{\rho}^{(i)})$, the condition (22) is not accurately met and there may be no solution to the SS-LMBW equation (18), which is non-linear in $\rho_\alpha(\mathbf{r})$. Such a behaviour was, in fact, observed in the case of the conventional LMBW equation for simple atomic fluid by Iatsevitch and Forstmann [39]. Using linear interpolation between the bulk direct correlation functions, they had to replace the coarse-grained density distribution as $\rho_\alpha(z) \to \rho_\alpha(z+\Delta)$, empirically shifting it by Δ across the interface in order to obtain converging solutions to the LMBW equation.

Since the approximation (19)–(20) to the inhomogeneous direct correlation function $c_{\alpha\gamma}(\mathbf{r}_1,\mathbf{r}_2)$ spans it between the homogeneous ones of the bulk phases $c^{\mathrm{blk}}_{\alpha\gamma}(r;\boldsymbol{\rho}^{(i)})$, a suitable way to ensure the existence of solutions to the SS-LMBW equation (18) is to optimize the expansion coefficients $\tilde{w}_i(\boldsymbol{\rho})$. First we span the vector of site densities

$\boldsymbol{\rho}$ between the densities $\boldsymbol{\rho}^{(i)}$ of the bulk liquid phases $i = 1, 2$ and a gas phase $i = 0$. The weight coefficients should satisfy the constraint $\sum_i w_i = 1$. Projecting $\boldsymbol{\rho}$ onto the density differences $\boldsymbol{\rho}^{(i)} - \boldsymbol{\rho}^{(0)}$ gives the weight factors

$$w_i(\boldsymbol{\rho}) = \frac{(\boldsymbol{\rho} - \boldsymbol{\rho}^{(0)}) \cdot (\boldsymbol{\rho}^{(i)} - \boldsymbol{\rho}^{(0)})}{(\boldsymbol{\rho}^{(i)} - \boldsymbol{\rho}^{(0)}) \cdot (\boldsymbol{\rho}^{(i)} - \boldsymbol{\rho}^{(0)})} \quad \text{for } i = 1, 2, \tag{23}$$

or explicitly evaluating the scalar products,

$$w_i(\boldsymbol{\rho}) = \sum_{\alpha} (\rho_\alpha - \rho_\alpha^{(0)})(\rho_\alpha^{(i)} - \rho_\alpha^{(0)}) \Big/ \sum_{\alpha} \left(\rho_\alpha^{(i)} - \rho_\alpha^{(0)}\right)^2 \quad \text{for } i = 1, 2.$$

Then we obtain the weights $\tilde{w}_i(\boldsymbol{\rho})$ in the superposition (20) by "distorting" the linear dependencies $w_i(\boldsymbol{\rho})$. We assume the non-linear functional form as

$$\tilde{w}_i = w_i\left(f_i w_i (w_i - 1)^2 + 1\right) \quad \text{for } i = 1, 2 \tag{24}$$

and get the last "distorted" weight $\tilde{w}_0(\boldsymbol{\rho})$ for the projection onto the gas phase density vector $\boldsymbol{\rho}^{(0)}$ from the constraint

$$\sum_{i=0}^{2} \tilde{w}_i = 1. \tag{25}$$

The form (24) is so constructed that the distortion $\tilde{w}_i - w_i$ quadratically vanishes as $w_i \to 1$ or $w_i \to 0$ in the bulk phases. The correction is maximal when the coarse-grained density $\boldsymbol{\rho}(\mathbf{r})$ most deviates from the bulk densities of the coexisting phases $\boldsymbol{\rho}^{(1)}$ and $\boldsymbol{\rho}^{(2)}$ and enters the instability region on the phase diagram. For a given inhomogeneous direct correlation function $\boldsymbol{c}(\mathbf{r}_1, \mathbf{r}_2)$ appearing in the kernel of the SS-LMBW equation, the latter can be considered as an eigenvalue problem for the density distribution values at a distance from the interface, $\boldsymbol{\rho}(\mathbf{r} \to \pm\infty)$. The additional relations determining the "distortion" factors f_i for $i = 1, 2$ are obtained from the boundary conditions of each site density profile tending to its bulk value in the liquid phase rich in the corresponding component,

$$\rho_{\mathrm{I}}(\mathbf{r}) \to \rho_{\mathrm{I}}^{\mathrm{rich}} \quad \text{as } \mathbf{r} \to -\infty\,, \tag{26a}$$

$$\rho_{\mathrm{II}}(\mathbf{r}) \to \rho_{\mathrm{II}}^{\mathrm{rich}} \quad \text{as } \mathbf{r} \to +\infty\,, \tag{26b}$$

where phases $i = 1$ and 2 occupy respectively the left and right half-spaces $\mathbf{r} < 0$ and $\mathbf{r} > 0$. In the process of solution, the SS-LMBW equation (18) for the site density profiles $\rho_\alpha(\mathbf{r})$ are being converged simultaneously with the relations (26). The limiting values of the density profiles in the opposite phases poor in the corresponding components,

$$\rho_{\mathrm{I}}(\mathbf{r}) \quad \text{as } \mathbf{r} \to +\infty\,, \tag{27}$$

$$\rho_{\mathrm{II}}(\mathbf{r}) \quad \text{as } \mathbf{r} \to -\infty\,, \tag{28}$$

are eigenvalues of Equation (18), functionally dependent on its kernel factor $c_{\alpha\gamma}(\mathbf{r}_1, \mathbf{r}_2)$. Their consistency with the bulk densities of the depleted components $\rho_{\mathrm{I}}^{\mathrm{poor}}$ and $\rho_{\mathrm{II}}^{\mathrm{poor}}$ depends on the precision of the approximation (19)–(20) and (23)–(26) to the inhomogeneous direct correlation function, as well as on the accuracy of the site–site direct correlation functions of bulk phases $c^{\text{blk}}_{\alpha\gamma}(r;\boldsymbol{\rho}^{(i)})$ yielded by the RISM equations (4) with the KHM closure (7).

Because Equation (21) is written for the density derivative $\rho'_\alpha(\mathbf{r})$, the density distribution $\rho_\alpha(\mathbf{r})$ to be inserted into its matrix coefficients $A_{\alpha\gamma}$ and B_γ is determined up to the integration constant

$$\rho^{\text{av}}_\alpha = V^{-1}\int_V d\mathbf{r}\,\rho_\alpha(\mathbf{r}), \tag{29}$$

which gives the average density of fluid contained in the volume $V \to \infty$ of the integral equation domain. To numerically integrate the density derivative $\rho'_\alpha(\mathbf{r})$, it is convenient to use the fast Fourier transform and the relation in the reciprocal space

$$\imath\mathbf{k}\,\rho_\alpha(\mathbf{k}) = \rho'_\alpha(\mathbf{k}), \tag{30a}$$

$$\rho_\alpha(\mathbf{k}=0) = \rho^{\text{av}}_\alpha. \tag{30b}$$

Rewriting the SS-LMBW equation (18) for the density distribution $\rho_\alpha(\mathbf{r})$ requires expressing it through the derivative $\rho'_\alpha(\mathbf{r})$ and thus involves the integration constant (29) anyway. Specifying its value between the bulk densities of the coexisting phases as

$$\rho^{\text{av}}_\alpha = \eta\rho^{(1)}_\alpha + (1-\eta)\rho^{(2)}_\alpha \quad \text{with } 0<\eta<1 \tag{31}$$

determines the interface position subdividing the whole volume into the portions ηV and $(1-\eta)V$ occupied respectively by phases 1 and 2.

The constraint (25) is important because all the homogeneous direct correlation functions in the expansion (20) have the long-range asymptotics of the intermolecular site–site interaction potential,

$$c^{\text{blk}}_{\alpha\gamma}(r_{12}) \to -\beta u^{ab}_{\alpha\gamma}(r_{12}) \quad \text{for } r_{12}\to\infty,$$

which has to be exactly the same for the inhomogeneous direct correlation function as well,

$$c_{\alpha\gamma}(\mathbf{r}_1,\mathbf{r}_2) \to -\beta u^{ab}_{\alpha\gamma}(r_{12}) \quad \text{for } r_{12}\to\infty.$$

For charged sites, it is the asymptotics of the site–site Coulomb interaction potential, and the constraint (25) ensures the condition of local electroneutrality is obeyed.

Following the familiar recipe for a weighted local density of fluid in DFT [15], the coarse-grained local density $\bar{\rho}_\alpha(\mathbf{r})$ of site α is determined by weighting the single-site density distribution $\rho_\alpha(\mathbf{r})$ within a sphere of radius r^{w}_α,

$$\bar{\rho}_\alpha(\mathbf{r}) = \frac{3}{4\pi(r^{\text{w}}_\alpha)^3}\,\Theta\Big(r^{\text{w}}_\alpha - |\mathbf{r}-\mathbf{r}'|\Big) * \rho_\alpha(\mathbf{r}'). \tag{32}$$

In the common case of the site–site interaction potential comprising the Lennard–Jones and Coulomb terms, it is reasonable to choose the weighting radius for site α as half its Lennard–Jones size, $r^{\text{w}}_\alpha = \sigma^{\text{LJ}}_\alpha/2$. The convolution (32) is conveniently calculated in the reciprocal space as

$$\bar{\rho}_\alpha(\mathbf{k}) = 3\frac{\sin(kr^{\text{w}}_\alpha) - kr^{\text{w}}_\alpha\cos(kr^{\text{w}}_\alpha)}{(kr^{\text{w}}_\alpha)^3}\,\rho_\alpha(\mathbf{k}).$$

The site–site direct correlation functions $c_{\alpha\gamma}(r;\boldsymbol{\rho}^{(1)})$ and $c_{\alpha\gamma}(r;\boldsymbol{\rho}^{(2)})$ of coexisting phases 1 and 2 with densities $\boldsymbol{\rho}^{(1)} = \{\rho_{\rm I}^{\rm rich}, \rho_{\rm II}^{\rm poor}\}$ and $\boldsymbol{\rho}^{(2)} = \{\rho_{\rm I}^{\rm poor}, \rho_{\rm II}^{\rm rich}\}$ are obtained by solving the RISM integral equation (4) with the KHM closure (7). The RISM–KH equations are also solved for $c_{\alpha\gamma}(r;\boldsymbol{\rho}^{(0)})$ of a gaseous mixture with both components I and II at vanishing densities $\boldsymbol{\rho}^{(0)} = \{\rho_{\rm I}^{\rm gas}, \rho_{\rm II}^{\rm gas}\}$.

In the case of a liquid/vapour interface it is straightforward to reduce the above procedure of approximating the inhomogeneous direct correlation functions $c_{\alpha\gamma}(\mathbf{r}_1,\mathbf{r}_2)$ given by Equations (19)–(20) and (23)–(26) to the non-linear interpolation between two phases, vapour and liquid.

For a planar interface the interpolation (19)–(20) of the inhomogeneous site–site direct correlation function in terms of the homogeneous ones simplifies the SS-LMBW equation (18) to the form

$$
\begin{aligned}
\frac{\mathrm{d}\ln\rho_\alpha(z_1)}{\mathrm{d}z_1} = \sum_\gamma \int_{-\infty}^{\infty} \mathrm{d}z_2\, \omega_{\alpha\gamma}(r_{12}) \Bigg(-\beta \frac{\mathrm{d}u_\gamma^{\rm ext}(z_2)}{\mathrm{d}z_2} + \frac{1}{2}\sum_{i=0}^{2} \\
\times \Bigg[\tilde{w}_i(\bar{\rho}(z_2)) \int_{-\infty}^{\infty} \mathrm{d}z_3\, c_{\alpha\gamma}^{\rm blk}\big(p=0, |z_2-z_3|; \boldsymbol{\rho}^{(i)}\big) \frac{\mathrm{d}\rho(z_3)}{\mathrm{d}z_3} \\
+ \int_{-\infty}^{\infty} \mathrm{d}z_3\, c_{\alpha\gamma}^{\rm blk}\big(p=0, |z_2-z_3|; \boldsymbol{\rho}^{(i)}\big)\, \tilde{w}_i(\bar{\rho}(z_3)) \frac{\mathrm{d}\rho(z_3)}{\mathrm{d}z_3} \Bigg] \Bigg)
\end{aligned}
\tag{33}
$$

involving the zeroth planar harmonics of the homogeneous direct correlation functions in bulk fluid phase i,

$$
c_{\alpha\gamma}^{\rm blk}(p=0, z_1, z_2; \boldsymbol{\rho}^{(i)}) = 2\pi \int_0^{\infty} s\, \mathrm{d}s\, c_{\alpha\gamma}^{\rm blk}(s, z_1, z_2; \boldsymbol{\rho}^{(i)}).
$$

All the three integrals in Equation (33) are simply convolutions of one-dimensional functions and can be sequentially evaluated by using the fast Fourier transform technique.

The external gravitational field $u_\gamma^{\rm ext}(z) = m_\gamma g z$ restricts the size of capillary waves at the liquid interface and thus shapes its average density profile [49,50]. As the external field vanishes, the capillary waves of large length slowly wash out the interface. (See the discussion in Section 5.3.2.) The potential energy of typical organic molecules with a mass of interaction sites $m_\gamma \lesssim 100$ atomic units in the gravitational field with the acceleration $g = 9.8\ \mathrm{m\cdot s^{-2}}$ constitutes a very small value of order of 10^{-13}, compared to the other terms in the SS-LMBW equation (33), and the interface width appears to be on macroscopic scale. Solving the equation on a smaller finite domain introduces cut-off of capillary waves at the domain size, much shorter than the interface width that would result from the gravity field. Therefore, the gravity potential term $\mathrm{d}u_\gamma^{\rm ext}/\mathrm{d}z = m_\gamma g$ can be neglected as its effect is shaded by the cut-off of capillary waves. Notice that such a macroscopic width results purely from statistical averaging of the "local" microscopic density profile bent by long capillary waves. Solving the equation on a smaller finite domain is equivalent to considering a local portion at the surface of a long capillary wave, neglecting its curvature and oscillations with respect to the average interface position. The long capillary wave can be taken into account by adding the long-range analytical asymptotics to the solution on the finite domain.

To illustrate the SS-LMBW methodology, we calculated the structure of the planar liquid/vapour as well as liquid/liquid interfaces of non-polar and polar molecular fluids: hexane and methanol at ambient conditions. The site–site interaction potentials to appear in the closure (7) were specified in the form comprising the Coulomb and 12-6 Lennard–Jones terms,

$$u_{\alpha\gamma}(r) = \frac{q_\alpha q_\gamma}{r} + 4\varepsilon_{\alpha\gamma}\left[\left(\frac{\sigma_{\alpha\gamma}}{r}\right)^{12} - \left(\frac{\sigma_{\alpha\gamma}}{r}\right)^{6}\right]$$

with the standard Lorentz–Berthelot mixing rules for cross-terms, $\sigma_{\alpha\gamma} = (\sigma_\alpha + \sigma_\gamma)/2$ and $\varepsilon_{\alpha\gamma} = (\varepsilon_\alpha\varepsilon_\gamma)^{1/2}$. The geometry $l^a_{\alpha\gamma}$ to appear in the intramolecular matrix (5) as well as the site charges q_α and Lennard–Jones sizes σ_α and energies ε_α of the hexane and methanol molecules were taken from the OPLS force field [51]. The densities of liquid n-hexane $\rho^{\rm liq} = 0.655\ \mathrm{g\cdot cm^{-3}}$ and liquid methanol $\rho^{\rm liq} = 0.791\ \mathrm{g\cdot cm^{-3}}$ at temperature $T = 300$ K were taken from literature [52]. For the n-hexane–methanol binary mixture [53], the densities and methanol mole fractions of two phases coexisting at $T = 300$ K are $\rho^{(1)} = 0.6528\ \mathrm{g\cdot cm^{-3}}$, $x^{(1)}_{\rm MeOH} = 0.28$ and $\rho^{(2)} = 0.7217\ \mathrm{g\cdot cm^{-3}}$, $x^{(2)}_{\rm MeOH} = 0.80$. The site–site direct correlation functions of the bulk phases in equilibrium were obtained by solving the RISM–KHM equations (4) and (7) with the parameter $\tau = 0.25$ for vapour and $\tau = 0.5$ for liquid phases.

The SS-LMBW equation (33) was discretized in the interval $-409.6 \le z \le 409.6$ Å on a linear grid with a resolution of 0.05 Å. To position the interface at $z = 0$, the parameter in Equation (31) setting the average density of the coexisting phases was chosen as $\eta = 0.5$. The SS-LMBW equation (33) with the boundary conditions (26) were converged by using the modified direct inversion in the iterative subspace (MDIIS) method [54].

Figure 5.4 shows the results of the SS-LMBW/RISM–KHM theory for the site density profiles of the liquid/vapour interface of ambient n-hexane. The density profiles are very similar for all the interaction sites (from the end CH_3 to inner CH_2 groups). There is a tiny enhancement of the inner sites of the n-hexane molecule compared to the end ones (lower part of Figure 5.4). This shows some tendency of preferential orientation of n-hexane molecules at the liquid/vapour interface in normal to its plane. The density perturbation quickly fades into the vapour phase at a distance of 20 Å but decays considerably slower into the liquid up to about 60 Å. No oscillations in the liquid/vapour site density profiles are observed for this fluid characterized with a relatively weak attraction of the Lennard–Jones interaction potential. Iatsevitch and Forstmann [39] obtained oscillating profiles of the liquid/vapour interface of Lennard–Jones atomic fluid at the comparably lower temperature and hence higher density. This might be an artifact of their procedure of weighting the local density with a spatial bias they used in construction of the inhomogeneous direct correlation function $c(\boldsymbol{r}_1, \boldsymbol{r}_2)$ to be inserted into the LMBW equation (2). For Lennard–Jones, non-associating fluid, no oscillations are obtained even in the inhomogeneous OZ-KHM/LMBW approach, which is a higher order theory accurately yielding the inhomogeneous two-particle direct correlation function (see Section 3.2).

The structure of the liquid/vapour interface of ambient methanol obtained in the SS-LMBW/RISM–KHM approach is depicted in Figure 5.5. Compared to the non-polar n-hexane, the decay of the inhomogeneity away from the interface is noticeably quicker for the polar methanol: 10 Å to the gas and 30 Å to the liquid phase. The coarse-grained

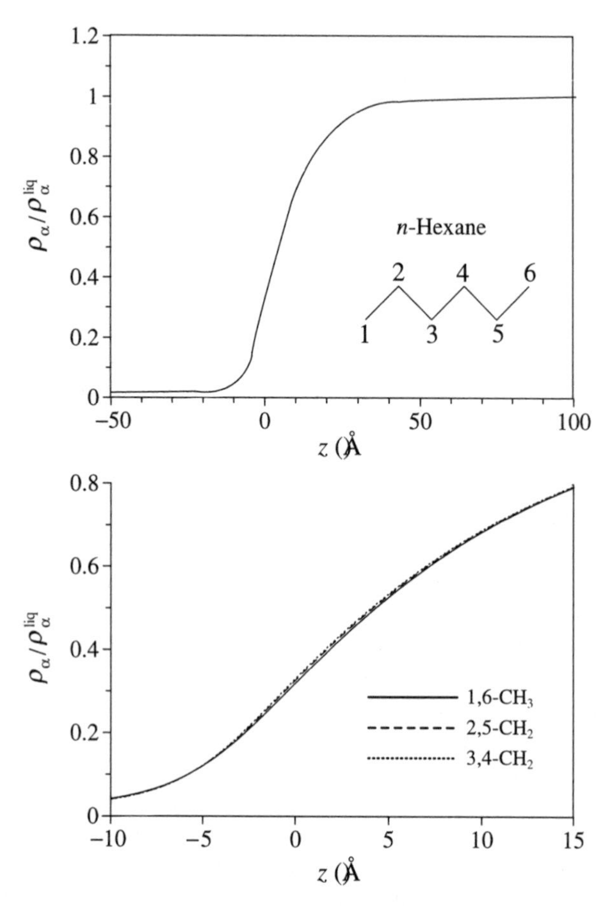

FIGURE 5.4. Site density profiles of the planar liquid/vapour interface of n-hexane at ambient temperature, following from the SS-LMBW/RISM–KHM theory. Outer 1,6-CH_3, middle 2,5-CH_2 and inner 3,4-CH_2 groups of the n-hexane molecule are represented by solid, dashed and dotted lines, respectively. The density profiles are normalized to the side densities in the liquid phase, ρ_α^{liq}. The lower part zooms in the interface region.

density of methanol reaches the bulk liquid value already at a distance of 10 Å from the interface. However, considerable oscillations of the site density profiles indicative of layering of methanol molecules spread quite deeply to the liquid phase, up to 30 Å. As seen in the lower part of Figure 5.5, the methyl density profile oscillates almost in phase with that of hydroxyl hydrogen and opposite in phase to that of oxygen. This alternating layering can be attributed to hydrogen bonding into zigzag chains orientationally ordered in normal to the interface. Methanol molecules at the interface turn with the oxygens by about 45° towards the liquid bulk, somewhat distorting the chains. Possible orientations and layering of methanol molecules corresponding to these oscillations of the site density profiles are drawn in Figure 5.5. The three lines mark the positions of the methyl group, oxygen and hydroxyl hydrogen sites of molecules layered in the first plane with the density approaching the liquid phase value (the first maxima of the density profiles at

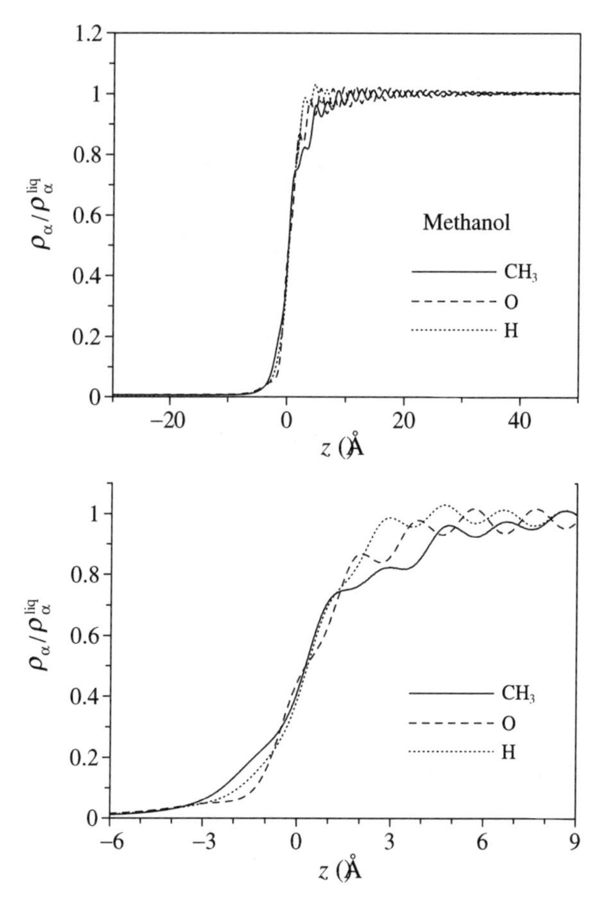

FIGURE 5.5. Site density profiles of the planar liquid/vapour interface of methanol at ambient temperature, following from the SS-LMBW/RISM–KHM theory. Methyl group, oxygen and hydroxyl hydrogen sites of the methanol molecule are represented by solid, dashed and dotted lines, respectively. The density profiles are normalized to the side densities in the liquid phase, ρ_α^{liq}. The lower part zooms in the interface structure.

respectively $z = 2.8$, $z = 3.8$ and $z = 2.9$ Å in Figure 5.5). Termination of zigzag chains by a plane perpendicular to their direction forms a corrugated surface of the liquid with a part of molecules protruding to the vapour phase. Along with capillary waves this shapes the microscopic structure of the methanol liquid/vapour interface. Methanol molecules at the surface of the liquid are oriented with their methyl groups towards the vapour phase. This leads to the first peak of the methyl density profile at $z_{\mathrm{Me}} = 1$ Å, followed by the first peaks of hydroxyl oxygen and hydrogen at $z = 1.2$ and $z = 2$ Å in Figure 5.5. Besides, the first pre-peak of the methyl density profile at $z = -1.5$ Å at the slope of the density profiles descending into the vapour phase coincides with a hollow at the slope of the oxygen density profile. Similar small oscillations of microscopic length are present in the simulation data for the density profile of the methanol liquid/vapour interface [55]. (However, it is difficult to draw definite conclusions from that data because it could be

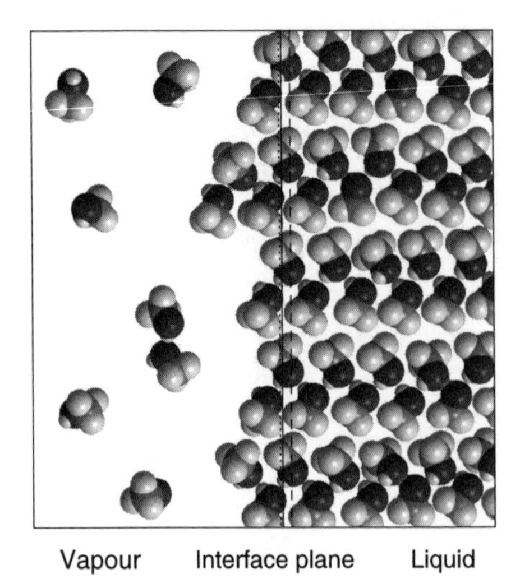

FIGURE 5.6. Molecular orientations and ordering at the methanol liquid/vapour interface, illustrating the site density profiles in Figure 5.5. The solid, dashed and dotted lines show the positions of methyl groups, oxygens and hydroxyl hydrogens, respectively, corresponding to the maxima of the density profiles in the liquid phase.

statistical noise of the simulation.) Besides, the orientations of methanol molecules at the liquid surface, derived above from the site density profiles in Figure 5.5, agree with those observed in the simulation [55].

Figure 5.7 exhibits the prediction of the SS-LMBW/RISM–KHM theory for the site density profiles of the liquid/liquid interface in the binary system with partially miscible n-hexane and methanol at ambient conditions. The bulk limiting values of the n-hexane density $\rho_{n\text{-hex}}^{\text{rich}}$ in the phase rich in n-hexane ($z \to -\infty$) and of the methanol density $\rho_{\text{MeOH}}^{\text{rich}}$ in the phase rich in methanol ($z \to +\infty$) are enforced by the boundary conditions (26). As discussed above, the bulk limits of the density profiles on the depletion side yielded by the SS-LMBW equation (18) using the approximate inhomogeneous direct correlation function (19) can differ from the values appearing in the RISM–KHM equations (4) and (7) for bulk phases. The latter are laid off with bold marks in Figure 5.7. The consistency for the bulk density $\rho_{n\text{-hex}}^{\text{poor}}$ of n-hexane in the phase rich in methanol is very good. The bulk density $\rho_{\text{MeOH}}^{\text{poor}}$ of methanol in the phase rich in n-hexane is appropriately small, although overestimated by a factor of 1.8. This discrepancy can be treated by optimizing the parameter τ in the KHM closure (7) for the thermodynamic consistency between the virial and compressibility routes to the free energy of the bulk phases, as well as by improving the functional form (24) used in the approximation procedure for the inhomogeneous direct correlation function (19).

Oscillations of the site density profiles develop at both sides of the liquid/liquid interface to a depth of 50 Å, showing significant layering at the interface. The n-hexane and methanol density profiles oscillate in opposite phase to each other. Their period of about 20 Å is three to four times longer than the size of the methanol and n-hexane molecules. This renders the oscillations as a collective, "hydrostatic" effect of a varying

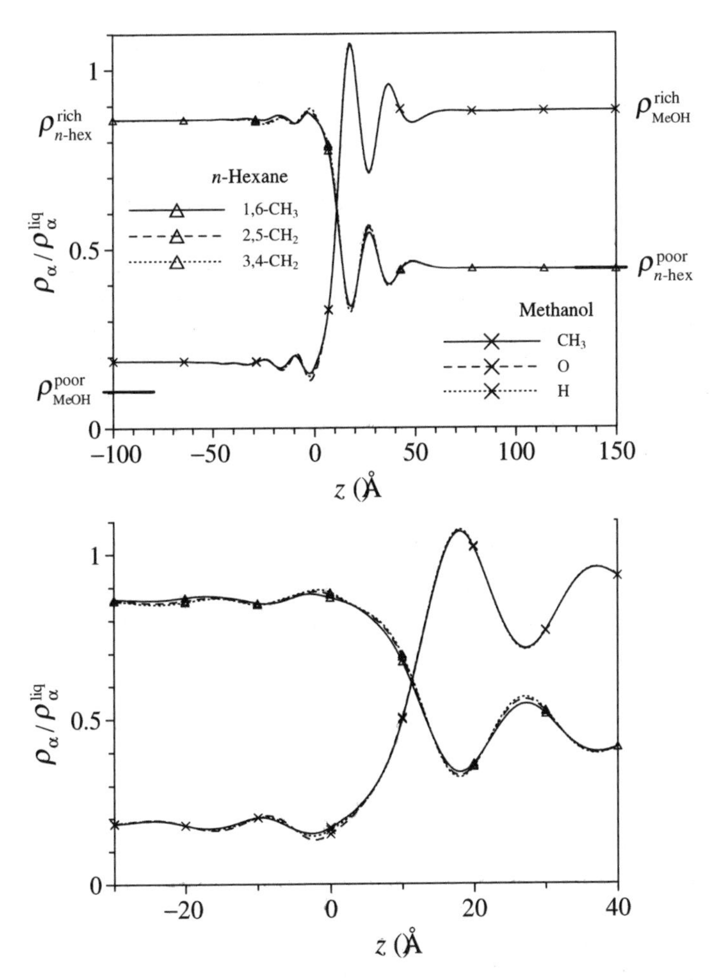

FIGURE 5.7. Site density profiles of the planar liquid/liquid interface of n-hexane and methanol at ambient conditions, following from the SS-LMBW/RISM–KHM theory. Outer 1,6-CH_3, middle 2,5-CH_2 and inner 3,4-CH_2 groups of n-hexane (solid, dashed and dotted lines with triangles, respectively). Methyl group, oxygen and hydroxyl hydrogen sites of methanol are represented by solid, dashed and dotted lines with crosses, respectively. The density profiles of hexane and methanol are normalized to the site densities in the corresponding pure liquids, ρ_α^{liq}. The bold segments at the left and right edges of the plot mark the bulk densities of depleted components $\rho_{\text{MeOH}}^{\text{poor}}$ and $\rho_{n\text{-hex}}^{\text{poor}}$ that appear in the RISM–KHM equations (4) and (7) for the correlations of bulk phases. The lower part zooms in the interface region.

local coarse-grained density related to capillary waves rather than the short-range solvation structure of the molecular species in the system. The n-hexane site density profiles oscillate in phase, but with the amplitude somewhat smaller for the inner 3,4-CH_2 and middle 2,5-CH_2 groups compared to the outer 1,6-CH_3 ones. This can be interpreted as noticeable orientational ordering of n-hexane molecules in perpendicular to the interface. Indeed, for such molecules the density profiles of the two end groups 1-CH_3 and 6-CH_3 are of the same amplitude as those of 3,4-CH_2 but shifted to the right and left by the corresponding intramolecular distances. Superposition of these two shifted oscillations yields the density profile of 1,6-CH_3 with oscillations in phase but of a smaller amplitude than that of 3,4-CH_2. As distinct from the methanol liquid/vapour interface (Figure 5.5), the methyl group, oxygen and hydroxyl hydrogen density profiles of methanol in Figure 5.6 oscillate in phase. They practically coincide on the methanol-rich side of the interface, which is indicative of no preferential orientation of methanol molecules and no zigzag hydrogen bonding chains. At the n-hexane-rich side of the interface, the oscillations of the methyl density profile are somewhat smaller in amplitude than those of oxygen. In two strands of a hydrogen-bonded zigzag chain, methyl groups of methanol molecules are directed outwards the chain. Similarly to the considerations for n-hexane, the amplitude of the methyl density profile oscillations smaller than that of oxygen can be interpreted as orientational ordering of methanol zigzag chains in parallel to the interface, with the methyl groups directed towards and outwards the interface. It persists within a 25 Å layer in the n-hexane-rich phase.

5.3.2. *Inhomogeneous Integral Equation Theory of a Liquid/Vapour Interface*

Consider fluid of particles interacting through the Lennard–Jones potential $\varphi(r) = 4\varepsilon[(\sigma/r)^{12} - (\sigma/r)^{6}]$ with the size and energy parameters σ and ε. It is convenient to use cylindrical coordinates and position the interface in the plane $z = 0$. The IOZ equation has the form [15]

$$h(s, z_1, z_2) = c(s, z_1, z_2) + \int \mathrm{d}\mathbf{s}'\, \mathrm{d}z_3\, c(|\mathbf{s} - \mathbf{s}'|, z_1, z_3)\, \rho(z_3)\, h(\mathbf{s}', z_3, z_2), \tag{34}$$

where the inhomogeneous total and direct correlation functions for a pair of particles, $h(s, z_1, z_2)$ and $c(s, z_1, z_2)$, respectively, depend on the distances z_1 and z_2 of the particles to the interface as well as the separation s between their projections onto the interface. The integration over the projected position $\mathbf{s}'$ of the "third" particle in Equation (34) can be evaluated by applying the Hankel transform of a function $f(s)$,

$$f(p) = 2\pi \int_0^\infty s\, \mathrm{d}s\, J_0(ps)\, f(s),$$

where $J_0(ps)$ is the zeroth-order Bessel function of the first kind. In the reciprocal p-space, Equation (34) transforms to

$$h(p, z_1, z_2) = c(p, z_1, z_2) + \int_{-\infty}^{\infty} \mathrm{d}z_3\, c(p, z_1, z_3)\, \rho(z_3)\, h(p, z_3, z_2)\,. \tag{35}$$

It is important to complement the integral equation (34) with an appropriate closure relation for the inhomogeneous two-particle correlations $h(s, z_1, z_2)$ and $c(s, z_1, z_2)$. We extended the KH approximation (6) to the inhomogeneous case,

$$g(s, z_1, z_2) = \begin{cases} \exp\left(\psi(s, z_1, z_2)\right) & \text{for } \psi(s, z_1, z_2) \le 0, \\ 1 + \psi(s, z_1, z_2) & \text{for } \psi(s, z_1, z_2) > 0, \end{cases} \tag{36}$$

where

$$\psi(s, z_1, z_2) = -\beta u(r_{12}) + h(s, z_1, z_2) - c(s, z_1, z_2),$$

and $g(s, z_1, z_2) = h(s, z_1, z_2) + 1$ is the inhomogeneous pair distribution function. For comparison, we also used the inhomogeneous generalization of the modified Verlet (VM) closure. Application of the inhomogeneous HNC closure approximation to the IOZ integral equation (34) results in divergency of the equations.

As an additional relation between the one-particle density profile $\rho(z)$ and the inhomogeneous two-particle correlations completing the IOZ equation (35) with the closure (36), we use the LMBW equation (2). For the planar interface geometry, it takes the form

$$\frac{d\rho(z_1)}{dz_1} = -\beta \frac{du_{ext}(z_1)}{dz_1} + \rho(z_1) \int_{-\infty}^{\infty} dz_2 \, c(p = 0, z_1, z_2) \frac{d\rho(z_2)}{dz_2}, \tag{37}$$

where

$$c(p = 0, z_1, z_2) = 2\pi \int_0^{\infty} s \, ds \, c(s, z_1, z_2) .$$

It is much less sensitive to boundary effects and thus more convenient from the numerical point of view than the BGY equation (1) for the density profile $\rho(z)$. The asymptotics of the density profile of the interface with width W has the form [49]

$$\rho(z) \sim \mathrm{erf}(z/W).$$

It decays outwards the interface to the vapour and liquid bulk much slower than its derivative

$$\rho'(z) \sim \exp(-z^2/W^2).$$

Therefore, the numerical integration in the LMBW equation can be performed on a much smaller z-domain compared to the BGY equation. This significantly reduces computational load for the equations having two z-dimensions. Moreover, the LMBW equation does not involve the assumption of a pairwise interaction potential inherent in the BGY equation [14], which constitutes a benefit for further extensions to systems with non-additive potentials.

Solution of the three coupled integro-differential equations (35)–(36) for the functions $h(s, z_1, z_2)$, $c(s, z_1, z_2)$ and $\rho(z)$ at a given temperature T yields directly the coexisting vapour and liquid phases. Specifying the averaged density of the system ρ_0 in the interval between those of the vapour and liquid phases $\rho_{vap} < \rho_0 < \rho_{liq}$ determines the interface position along z-axis. Above the critical temperature T_c, one obtains a uniform density profile $\rho(z) = \rho_0$. The equations are discretized on the uniform finite grids of M points in the intervals $0 \leq s \leq s_{max}$ and $0 \leq p \leq p_{max}$, and N points in $-z_{max} \leq z \leq z_{max}$. The IOZ equation (35) turns into the set of linear equations

$$\mathbf{H}(p) = \mathbf{C}(p) + \mathbf{C}(p)\,\mathcal{R}\,\mathbf{H}(p) \tag{38}$$

at each of M values of the reciprocal planar vector p, with the matrices $\mathbf{H}$ and $\mathbf{C}$ of size $N \times N$ representing the two-particle correlation functions and the diagonal matrix $\mathcal{R}$ corresponding to the term $\rho(z_1)\,\delta(z_1 - z_2)$. The iterated vector $\mathbf{C}(s)$ is Hankel-transformed to the reciprocal space, and the solution of the IOZ equation for $\mathbf{H}(p)$ is obtained as

$$\mathbf{H}(p) = [\mathbf{1} - \mathbf{C}(p)\mathcal{R}]^{-1}\,\mathbf{C}(p). \tag{39}$$

It is then Hankel-transformed backwards to the direct space by

$$\mathbf{H}(s) = (2\pi)^{-1}\int_0^\infty p\,\mathrm{d}p\,\mathbf{H}(p)J_0(ps),$$

and inserted into the KH closure (36) to calculate the residual of the IOZ–KH equations as

$$\mathbf{R}(s) = \mathbf{G}(s) - \mathbf{H}(s) - 1,$$

where the matrix $\mathbf{G}(s)$ replaces the inhomogeneous two-particle distribution function $g(s, z_1, z_2)$ in Equation (36). Both forward and backward Hankel transforms are evaluated by using the trapezoidal quadrature. The LMBW equation (2) for the density profile derivative $\rho'(z)$ turns into the matrix form

$$[\mathbf{1} - \mathbf{C}_0\,\mathcal{R}\,]\,\boldsymbol{\rho}' = -\mathbf{U}'_{\mathrm{ext}}\mathcal{R} \tag{40}$$

non-linear due to the density profile matrix $\mathcal{R}$ appearing in the kernel, where the matrix $\mathbf{C}_0$ represents the function $c_0(z_1, z_2)$ and the diagonal matrix $\mathbf{U}'_{\mathrm{ext}}$ corresponds to $\delta(z_1 - z_2)\mathrm{d}u_{\mathrm{ext}}/\mathrm{d}z_1$. The density profile is obtained by numerical integration of its derivative,

$$\rho(z) = \rho(0) + \int_0^z \mathrm{d}z_1\,\rho'(z_1), \tag{41}$$

using the trapezoidal rule.

A special case is a system with the gravitational field $u_{\mathrm{ext}}(z) = mgz$. Weeks [49] showed within the capillary wave theory that the width of the planar liquid/vapour interface described by the density profile $\rho(z)$ logarithmically diverges as the external field vanishes with $mg \to 0$. This constitutes no contradiction with the idea of an intrinsic interface width that applies to a small region of the interface having size of the bulk correlation length. The average density profile is widened by the capillary waves, and is slowly washed out when the external field damping the waves vanishes. In fact, the external field is the only factor that holds the liquid/vapour interface in place [49]. As distinct, the separation of spatial scales of an intrinsic profile and capillary waves can be clearly made in the case of an interface of two immiscible liquids [50]. The capillary wave theory predicts the asymptotics of the inhomogeneous two-particle total correlation function as [49,50]

$$h_{\mathrm{cw}}(p, z_1, z_2) = \frac{\rho(z_1)\rho(z_2)}{\epsilon^2 + Ap^2}, \tag{42}$$

where $\epsilon^2 = \beta mg\rho(0)/\rho'(0)$ and $A = \beta\gamma$ for the interface tension γ. In the direct space, it is expressed as

$$h_{\mathrm{cw}}(s, z_1, z_2) = (2\pi)^{-1}\rho(z_1)\rho(z_2)K_0(\epsilon s)$$

with the Bessel function of the second kind K_0, and has the long-range asymptotics

$$h_{\mathrm{cw}}(s, z_1, z_2) \sim \frac{\exp(-\epsilon s)}{\sqrt{\epsilon s}} \quad \text{for } s \to \infty. \tag{43}$$

Notice that this capillary wave term alone cannot represent the whole correlation because it logarithmically diverges as $K_0(\epsilon s) \sim -\log(\epsilon s)$ for $s \to 0$. The singularity $1/p^2$ can also be derived from the condition of existence of an inhomogeneous solution to the LMBW equation (2). Indeed, the matrix equation (40) has non-trivial solutions $\rho' \neq 0$ at $\boldsymbol{u}_{\mathrm{ext}} \to 0$ if and only if

$$\det\left[\mathbf{1} - \mathbf{C}_0 \mathcal{R}\right] \to 0.$$

Accordingly, the solution (38) to the IOZ equation (35) acquires a singularity at $p \to 0$. That is,

$$h(p = 0, z_1, z_2) = 2\pi \int_0^\infty s\, \mathrm{d}s\, h(s, z_1, z_2) \to \infty$$

for particles located at some z_1 and z_2 near the interface, which means that the inhomogeneous two-particle total correlation function $h(s, z_1, z_2)$ has to fade along the interface slower than $\mathcal{O}(s^{-2})$. A particular form of the singularity is determined by the closure approximation used. In the enhancement tail $h > 0$, the KH approximation (36) enforces the asymptotics of the inhomogeneous two-particle direct correlation function to be just the interaction potential between the particles, $c(s, z_1, z_2) \sim -\beta u(r_{12})$. For the 12-6 Lennard–Jones potential, the Hankel transform of the direct correlation function has the analytic form

$$\mathbf{C}(p) = \mathbf{C}_0 + \mathbf{C}_2 p^2 + \mathbf{C}_4 p^4 + \cdots \tag{44}$$

Truncation of this expansion at the second term,

$$\mathbf{C}_{\mathrm{cw}}(p) = \mathbf{C}_0 + \mathbf{C}_2 p^2, \tag{45}$$

leads to the total correlation function with the capillary wave asymptotics

$$\mathbf{H}(p) \sim \frac{1}{\epsilon^2 + Ap^2} \quad \text{for } p \to 0$$

transforming in the direct space into the expression (43).

For the gravity acceleration $g = 9.8\ \mathrm{m \cdot s^{-2}}$ and typical molecular masses m of ambient organic liquids, the external potential term $\beta u'_{\mathrm{ext}}(z_1) = \beta m g$ is of the order of 10^{-13} compared to the other parts of the LMBW equation (2). It becomes important only at extremely small values of wavevector $p \to 0$, when the other terms vanish. This determines the maximal length of capillary waves and hence the width of the liquid/vapour interface shaped by the gravity field to be of microscopic order up to millimetres [49]. We restrict the domain of the numerical solution $s_{\max}$ to the microscopic scale, which cuts off long capillary waves and thus acts as an artificial external field damping them. Accordingly, the range of wavevector p is limited by a minimal value $p_{\min} = 2\pi/s_{\max}$, and therefore the gravity term $\beta m g$ is negligible compared to the cut-off effect and can be omitted. In this approximation, the interface is seen as a local portion

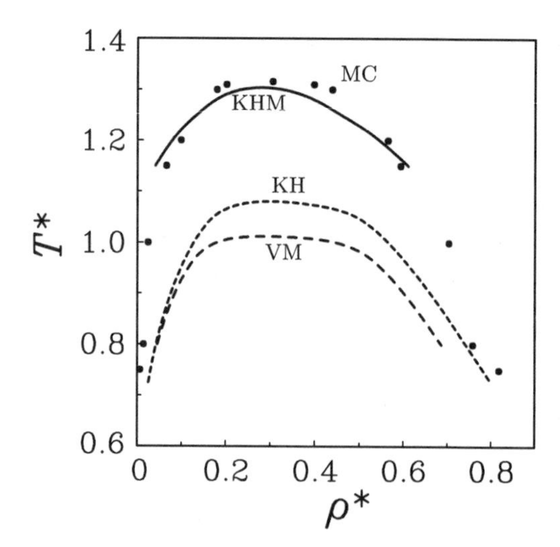

FIGURE 5.8. Liquid/vapour coexistence curve of the Lennard–Jones fluid predicted by the IOZ/LMBW theory in the KHM, KH and VM approximation (solid, short-dash and long-dash lines, respectively). Monte Carlo (MC) simulations are shown by the open circles.

on the surface of a long capillary wave, neglecting its curvature. The long capillary waves beyond the cut-off can be taken into account by adding the long-range analytical asymptotics for $s \to \infty$ (or $p \to 0$) to the solution on the finite s-domain.

The integro-differential equations (35)–(36) for $\mathbf{C}(s)$ and ρ' have been discretized on the linear grid of $N_{s,p} = 100$ and $N_z = 250$ points in the intervals $0 \le s \le 5\sigma$, $0 \le q \le 15\sigma$ and $-12.5\sigma \le z \le 12.5\sigma$. Further increase of the grid resolution has little effect. The resulting set of non-linear equations has been converged by using the MDIIS method [54] to a root mean square accuracy of 10^{-6} in 100–200 MDIIS steps. To further accelerate calculations, the solutions converged at a current temperature are used as an initial guess for the subsequent point with the temperature changed in 5% decrements.

The LMBW equation satisfies the variational principle for the free energy [34,35], and hence *implicitly* provides the chemical and mechanical balance between the liquid and vapour phases. The precision of the phase diagram is determined by the accuracy of the inhomogeneous two-particle direct correlation function related in turn to the quality of a closure to the IOZ equation. Figure 5.8 shows the the liquid/vapour coexistence curve obtained from the IOZ–LMBW equations complemented with the VM closure, KH closure and its modification (KHM approximation) against the Monte Carlo (MC) simulation data [46]. The KH approximation yields the coexistence curve in qualitative agreement with simulations, considerably better than the VM one. Still it underestimates the critical temperature ($T_c \approx 1.3$) by 18%. The agreement for the critical point can be improved by adding a quadratic term to the linearized form, much as in the KHM closure (7),

$$g(s, z_1, z_2) = \begin{cases} \exp\left(\psi(s, z_1, z_2)\right) & \text{for } \psi(s, z_1, z_2) \le 0, \\ 1 + \psi(s, z_1, z_2) + \dfrac{\tau}{2}\psi^2(s, z_1, z_2) & \text{for } \psi(s, z_1, z_2) > 0. \end{cases} \tag{46}$$

The improvement of the inhomogeneous KHM approximation (46) over (36) is at the same level as that of the generalized MSA closure (GMSA). The latter introduces a deviation from the MSA asymptotic behaviour of the direct correlation functions, which is parameterized in terms of Yukawa functions and optimized for thermodynamic self-consistency of the calculated free energy [14]. Good agreement for the upper portion of the coexistence curve as well as for the critical point has been reached at the parameter value $\tau = 0.55$ in the KHM closure (46). The agreement can be achieved for the whole coexistence curve by optimizing the parameter τ at each temperature for the thermodynamic consistency between the compressibility and virial routes to the chemical potential.

With the quadratic non-linearity introduced in the KHM approximation (46), the direct correlation function acquires the long-range asymptotics

$$\begin{aligned} c(s, z_1, z_2) &\to -\beta u(r_{12}) + \frac{\tau}{2} h^2(s, z_1, z_2) \\ &\sim \frac{\exp(-2\epsilon s)}{\epsilon s} \quad \text{for } s \to \infty. \end{aligned} \tag{47}$$

Its Hankel transform has no singularity at $p \to 0$, and so the expansion of the DCF at $p = 0$ keeps the analytic form (44). Accordingly, the total correlation function keeps the asymptotics (43). However, the matrices of the expansion coefficients $\mathbf{C}_0, \mathbf{C}_2, \mathbf{C}_4, \ldots$ in (44) have other, modified values. Through Equation (40) this, in general, changes the profile ρ' and hence results in a modified inverse decay length ϵ appearing in the asymptotics (42), (43) and (46).

Figure 5.9 depicts the scaled density profiles $\rho^*(z) = \rho(z)\sigma^3$ predicted by the IOZ–KHM/LMBW theory [48]. By choosing the average system density ρ_0, all the curves have

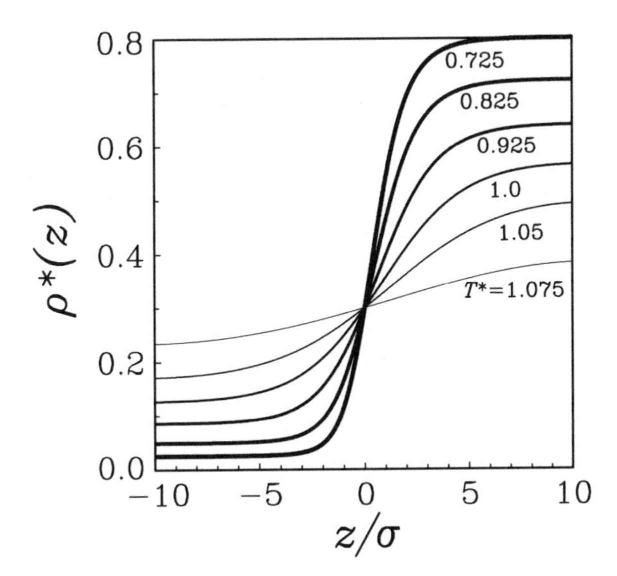

FIGURE 5.9. Reduced density profiles of the planar liquid/vapour interface of the Lennard–Jones fluid, obtained from the IOZ–KH/LMBW theory at reduced temperature $T^* = k_B T/\varepsilon = 0.725, 0.825, 0.925, 1.000, 1.050, 1.075$. Bolder lines correspond to lower temperatures.

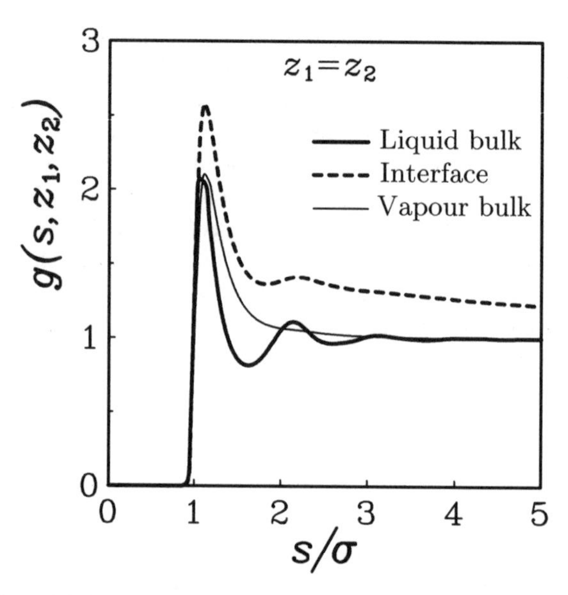

FIGURE 5.10. Sections of the inhomogeneous two-particle distribution function $g(s, z_1, z_2)$ of the Lennard–Jones fluid along the liquid/vapour interface for reduced temperature $T^* = 1.15$ at the distance $z = z_1 = z_2 = -10\sigma$; 0; $+10\sigma$ from the interface, respectively in the vapour phase (thin solid line) at the interface (bold solid line) and in the liquid phase (dashed line). Predictions of the IOZ–KHM/LMBW theory.

been deliberately set to pass through the same point $\rho^*(z) = 0.3$ at $z = 0$. The interface becomes wider as the temperature approaches the critical value. All the density profiles have a monotonous shape. No oscillations were identified in the behaviour of $\rho(z)$ even at low temperatures. This is in contrast with the visible oscillations with a period of about σ obtained from the LMBW equation in the simplified approach constructing the inhomogeneous direct correlation function for LMBW input by interpolation between the homogeneous ones [39]. The latter yields the density profile oscillating in the liquid at the interface with a period of about σ, which can be an artifact of the construction. It is reasonable that for simple Lennard–Jones fluid not forming associative structures, oscillations shorter than the interface width would be smeared out by the capillary waves. Notice that this should not be confused with the oscillations of the inhomogeneous two-particle density distribution either across or along the interface.

Figure 5.10 presents sections of the inhomogeneous two-particle distribution function $g(s, z_1, z_2)$ along the interface plane, at $z_1 = z_2 = z$, predicted by the IOZ–KH/LMBW theory. In the bulk of the vapour ($z \to -\infty$) and liquid ($z \to \infty$) phases, the two-particle distribution goes into the pair distributions of the bulk phase that behave in the usual way. The pair distributions quickly vanish with distance, $g \to 1$ for $s \to \infty$, with several oscillations in the liquid phase. However, at the interface ($z \sim 0$) the pattern is qualitatively different: the distribution function $g(s, z_1, z_2)$ acquires a long-range tail with the asymptotics in the form (43). Figure 5.11 shows sections of the inhomogeneous two-particle direct correlation function $c(s, z_1, z_2)$ along the interface and its Hankel transform $c(p, z_1, z_2)$. In the KHM approximation (46), $c(s, z_1, z_2)$ acquires the long-range tail (47) which is clearly seen on the inset in Figure 5.11a. The direct correlation

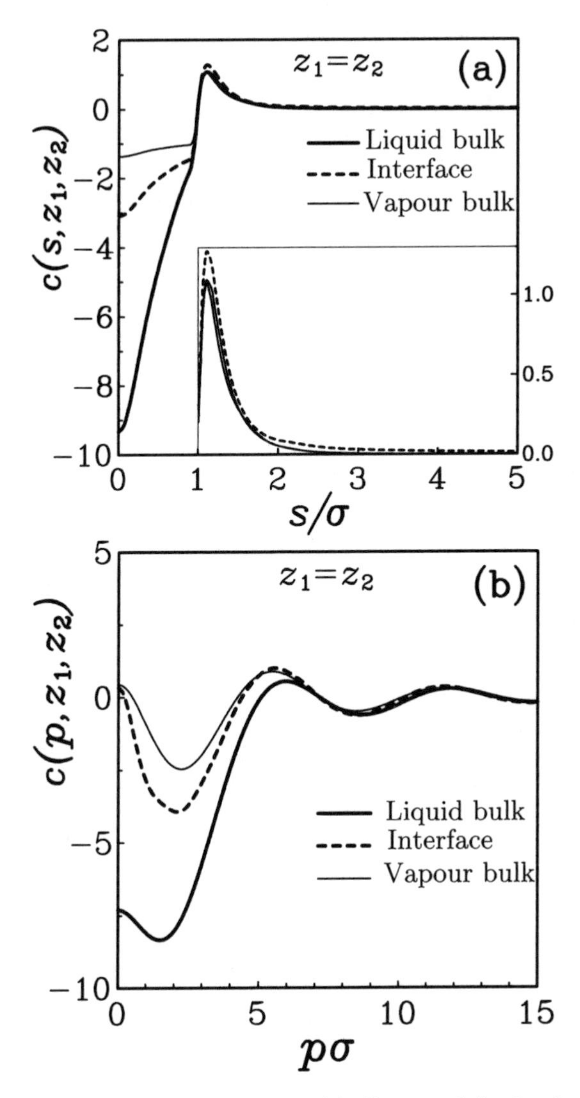

FIGURE 5.11. Sections of the inhomogeneous two-particle direct correlation function $c(s, z_1, z_2)$ of the Lennard–Jones fluid along the liquid/vapour interface (part a), and its Hankel transform $c(p, z_1, z_2)$ (part b). Line nomenclature is the same as in Figure 5.10. Results of the IOZ–KHM/LMBW theory. The inset in part a zooms in the long-range tail of $c(s, z_1, z_2)$ with the asymptotics (47).

function $c(p, z_1, z_2)$ in the reciprocal p-space is analytic at $p = 0$. As is obvious from Figure 5.11b, a significant contribution to its expansion comes from the terms $\mathbf{C}_4 p^4$ and higher, beyond the capillary-wave quadratic form (45).

An important consequence of the above-discussed asymptotics is that from the physical point of view the liquid/vapour interface behaves very similarly to two-dimensional critical fluid. The long-range tail of the inhomogeneous two-particle distributions $g(s, z_1, z_2)$ along the interface is indicative of large density fluctuations. The origin of these two-dimensional fluctuations is the same as in bulk critical fluid: the compressibility diverges as the values of the local density at the interface fall into the instability region on the phase diagram. The external field constrains the fluctuations in a two-dimensional interfacial layer where they appear as capillary waves. The existence of the capillary-wave singularity in the inhomogeneous two-particle distribution function is inherent in the LMBW and IOZ equations, whereas both the asymptotic behaviour and the short-range structure of the inhomogeneous two-particle direct correlation function are determined by a closure to the IOZ equation.

ACKNOWLEDGEMENTS

Financial support from the Japanese Ministry of Education, Culture, Sports, Science and Technology (Monbukagakusho), Grant for Scientific Research of Priority Areas (B) No. 757 on Nanochemistry at Liquid–Liquid Interfaces is gratefully acknowledged.

REFERENCES

1. A.G. Volkov, Ed., *Liquid Interfaces in Chemical, Biological and Pharmaceutical Applications*, Marcel Dekker, New York, 2001, Surfactant Science Series, Vol. 95.
2. I. Benjamin, Science **261**, 1558 (1993).
3. R.A. Marcus, J. Chem. Phys. **113**, 1618 (2000).
4. K. Schweighofer and I. Benjamin, J. Phys. Chem. A **103**, 10274 (1999).
5. J.N. Israelachvili, *Intermolecular and Surface Forces*, 2nd ed., Academic Press, London, 1992.
6. J.C. Shelley and M.Y. Shelley, Curr. Opin. Colloid Interface Sci. 5 **101**, (2000).
7. D.P. Tieleman, D. van der Spoel and H.J.C. Berendsen, J. Phys. Chem. B **104**, 6380 (2000).
8. S. Garde, L. Yang, J.S. Dordick and M.E. Paulaitis, Mol. Phys. **100**, 2299 (2002).
9. H. Domńguez, J. Phys. Chem. B **106**, 5915 (2002).
10. C.D. Bruce, S. Senapati, M.L. Berkowitz, L. Perera and M.D.E. Forbes, J. Phys. Chem. B **106**, 10902 (2002).
11. S.J. Marrink, D.P. Tieleman and A.E. Mark, J. Phys. Chem. B **104**, 12165 (2000).
12. S.J. Marrink, E. Lindahl, O. Edholm and A.E. Mark, J. Am. Chem. Soc. **123**, 8638 (2001).
13. W.M. Gelbart and A. Ben-Shaul, J. Phys. Chem. **100**, 13169 (1996).
14. J.P. Hansen and I.R. McDonald, *Theory of Simple Liquids*, 2nd ed., Academic Press, London, 1986.
15. D. Henderson, Ed., *Fundamentals of Inhomogeneous Fluids*, Marcel Dekker, New York, 1992, Vol. 10.
16. H.T. Davis, *Statistical Mechanics of Phases, Interfaces, and Thin Films*, VCH Publishers, New York, 1996.
17. S. Toxvaerd, J. Chem. Phys. **64**, 2863 (1976).
18. A.M. Somoza, E. Chacon, L. Mederos and P. Tarazona, J. Phys.: Condens. Matter **7**, 5753 (1995).
19. V. Talanquer and D.W. Oxtoby, J. Chem. Phys. **103**, 3686 (1995).
20. T.A. Cherepanova and A.V. Stekolnikov, Chem. Phys. **154**, 41 (1991).
21. T.A. Cherepanova and A.V. Stekolnikov, Mol. Phys. **82**, 125 (1991).
22. V. Talanquer and D.W. Oxtoby, J. Chem. Phys. **113**, 7013 (2000).
23. T.A. Cherepanova and A.V. Stekolnikov, Mol. Phys. **87**, 257 (1996).
24. P.S. Christopher and D.W. Oxtoby, J. Chem. Phys. **117**, 9502 (2002).
25. D. Chandler, J.D. McCoy and S.J. Singer, J. Chem. Phys. **85**, 5971 (1986).
26. D. Chandler, J.D. McCoy and S.J. Singer, J. Chem. Phys. **85**, 5977 (1986).
27. D. Chandler and H.C. Andersen, J. Chem. Phys. 57 (1972) 1930; D. Chandler, J. Chem. Phys. **59**, 2742 (1973).
28. F. Hirata and P.J. Rossky, Chem. Phys. Lett. 83 (1981) 329; F. Hirata, B.M. Pettitt and P.J. Rossky, J. Chem. Phys. **77**, 509 (1982); F. Hirata, P.J. Rossky and B.M. Pettitt, J. Chem. Phys. **78**, 4133 (1983).

29. A. Kovalenko and F. Hirata, J. Chem. Phys. **110**, 10095 (1999).
30. A. Kovalenko and F. Hirata, Chem. Phys. Lett. **349**, 496 (2001).
31. A. Kovalenko and F. Hirata, J. Theor. Comput. Chem. **1**, 381 (2002).
32. K. Yoshida, T. Yamaguchi, A. Kovalenko and F. Hirata, J. Phys. Chem. B **106**, 5042 (2002).
33. I. Omelyan, A. Kovalenko and F. Hirata, J. Theor. Comput. Chem. **2**, 193 (2003).
34. R. Lovett, C.Y. Mou and F.P. Buff, J. Chem. Phys. **65**, 570 (1976).
35. M.S. Wertheim, J. Chem. Phys. **65**, 2377 (1976).
36. J. Fischer and M. Methfessel, Phys. Rev. A **22**, 2836 (1980).
37. M.P. Taylor and J.E.G. Lipson, J. Chem. Phys. **100**, 518 (1994); **102**, 2118 (1995); **102**, 6272 (1995); **104**, 4835 (1996).
38. P. Attard, J. Chem. Phys. **102**, 5411 (1995).
39. S. Iatsevitch and F. Forstmann, J. Chem. Phys. **107**, 6925 (1997).
40. L. Blum and A.J. Torruella, J. Chem. Phys. **56**, 303 (1972).
41. K.S. Schweizer and J.G. Curro, Phys. Rev. Lett. **58**, 246 (1987); J. Chem. Phys. 87 (1987) 1842; Macromolecules **21**, 3070 (1988).
42. T. Munakata, S. Yoshida and F. Hirata, Phys. Rev. E **54**, 3687 (1996).
43. F. Hirata, Bull. Chem. Soc. J. **71**, 1483 (1998).
44. L. Lue and D. Blankschtein, J. Chem. Phys. **102**, 5427 (1995).
45. J.J. de Pablo, J.M. Prausnitz, H.J. Strauch and P.T. Cummings, J. Chem. Phys. **93**, 7355 (1990).
46. M.E. van Leeuwen and B. Smit, J. Phys. Chem. **99**, 1831 (1995).
47. A. Kovalenko and F. Hirata, Chem. Phys. Chem., submitted.
48. I. Omelyan, A. Kovalenko and F. Hirata, Chem. Phys. Lett. **397**, 368 (2004).
49. J.D. Weeks, J. Chem. Phys. **67**, 3106 (1977).
50. J. Stecki, J. Chem. Phys. **107**, 7967 (1997).
51. W.L. Jorgensen, J. Phys. Chem. 90 (1986) 1276; G. Kaminsky, E.M. Duffy, T. Matsui and W.L. Jorgensen, J. Phys. Chem. **98**, 13077 (1994).
52. D.R. Lide, Ed., *CRC Handbook of Chemistry and Physics*, 81st ed., CRC Press, New York, 2000.
53. S. Abbas, J. Satherley and R. Penfold, J. Chem. Soc., Faraday Trans., **93**, 2083 (1997).
54. A. Kovalenko, S. Ten-no and F. Hirata, J. Comput. Chem. **20**, 928 (1999).
55. M. Matsumoto and Y. Kataoka, J. Chem. Phys. **90**, 2398 (1989).

6

Electrochemical Processes at Aqueous/Organic Solution or Aqueous/Membrane Interfaces

Sorin Kihara
Kyoto Institute of Technology, Sakyo, Kyoto 606-8585, Japan

6.1. INTRODUCTION

When an ion transfers at the interface between water, W, and an organic solution, O, a current flows across the interface, and the potential difference at the interface varies depending on the amount of the ion transferred, since the ion carries a charge similarly to an electron. Therefore, the ion transfer reaction can be regarded as an electrochemical process. Another electrochemical process at the W/O interface is the electron transfer, which proceeds when a reductant or an oxidant in W comes into contact with an oxidant or a reductant, respectively, in O at the interface, resulting in an interfacial redox reaction.

The understanding of the electrochemical processes at the W/O interface has been improved drastically during the last 25 years, since the method and concept of voltammetry were expanded to the field of the W/O interface. The new voltammetry is called as *voltammetry for charge transfer at an interface between two immiscible electrolyte solutions* and abbreviated as *VCTIES*. The *VCTIES* has the unmatched advantage that the number of charges (ion or electron) transferred and the transfer energy can be measured simultaneously as the function of current and Galvani potential difference at the W/O interface (voltammogram). The *VCTIES* has been probed also very useful for the elucidation of membrane transport processes of ions or electrons when O was regarded as the membrane.

Typical subjects investigated with the aid of *VCTIES* are the quantitative determination of various ions, the relation between transfer energy of an ion and characteristics of the ion in solutions, the facilitation of the ion transfer by complexing agents or surfactants, the mechanism of the electron transfer, the kinetics of the ion or electron transfer, the process of solvent extraction, the constitution of a membrane potential including ion-selective electrode potential, the oscillation mechanism of the membrane potential or membrane current, the coupling of ion transfer with electron transfer at a W/O interface, processes of respiration mimetic reactions accompanied by the energy transformation, new types of membrane transports, etc. References [1–6] are examples of recent reviews or books relating to *VCTIES* and its application.

In this chapter, typical electrochemical processes at the W/O interface and membrane transport phenomena elucidated mainly by the group of the author based on the concept and methodologies concerning *VCTIES* are introduced.

6.2. VOLTAMMOGRAM FOR ION TRANSFER AT THE W/O INTERFACE

The transfer free energy of an ion i^z, $\Delta G_{tr,i}$, from W to O is the difference between the chemical potential of i^z in W, $\mu_{i,W}$, and that in O, $\mu_{i,O}$.

$$\Delta G_{tr,i} = \mu_{i,O} - \mu_{i,W} \tag{1}$$

When i^z forms an ion-pair $i^z j^{z-}$ with a counter ion j^{z-} or a complex $(iY_p)^z$ with p of neutral ligands Y in O, $\mu_{i,O}$ can be expressed as Equation (4) or (5), respectively, under the condition that the formation constant of $i^z j^{z-}$, $K_{ip,ij}$, or the stability constant of $(iY_p)^z$, $K_{st,(iYp)}$, is much larger than unity.

$$i^z + j^{z-} \overset{K_{ip,ij}}{\rightleftharpoons} i^z j^{z-} \tag{2}$$

$$i^z + pY \overset{K_{st,(iYp)}}{\rightleftharpoons} (iY_p)^z \tag{3}$$

$$\mu_{i,O} = \mu_{i,O}{}^0 + RT \ln a_{i,O} + RT \ln K_{ip,ij} a_{j,O} \tag{4}$$

$$\mu_{i,O} = \mu_{i,O}{}^0 + RT \ln a_{i,O} + RT \ln K_{st,(iYp)} c_{Y,O}{}^p \tag{5}$$

Here, μ_i^0 is the standard chemical potential of i^z, a_i or a_j the activity of i^z or j^{z-}, $c_{Y,O}$ concentration of Y, R the gas constant and T temperature. The subscript O or W means the phase where the species exists.

If the ion-pair and complex formations are negligible in W, $\mu_{i,W}$ is as Equation (6). In this connection, the ion-pair formation in a solvent of large relative dielectric constant such as W is usually not very serious.

$$\mu_{i,W} = \mu_{i,W}{}^0 + RT \ln a_{i,W} \tag{6}$$

When W and O are made to come in contact, the Galvani potential difference, $\Delta_O^W \phi_{tr,i}$, evolves. Here, $\Delta_O^W \phi_{tr,i}$ measured as the potential of W versus O can be related to $\Delta G_{\mathrm{tr,i}}$ through the following equation.

$$\Delta_{\mathrm{O}}^{\mathrm{W}}\phi_{\mathrm{tr,i}} = \frac{\Delta G_{\mathrm{tr,i}}}{zF} \tag{7}$$

The $\Delta_{\mathrm{O}}^{\mathrm{W}}\phi_{\mathrm{tr,i}}$ in the presence of an ion-pair or a complex formation is expressed as Equation (8) or (9), respectively, taking into account Equations (1)–(7).

$$\Delta_{\mathrm{O}}^{\mathrm{W}}\phi_{\mathrm{tr,i}} = \Delta_{\mathrm{O}}^{\mathrm{W}}\phi_{\mathrm{tr,i}}^{0} + \frac{RT}{zF}\ln\frac{a_{\mathrm{i,O}}}{a_{\mathrm{i,W}}} + \frac{RT}{zF}\ln K_{\mathrm{ip,ij}}a_{\mathrm{j,O}} \tag{8}$$

$$\Delta_{\mathrm{O}}^{\mathrm{W}}\phi_{\mathrm{tr,i}} = \Delta_{\mathrm{O}}^{\mathrm{W}}\phi_{\mathrm{tr,i}}^{0} + \frac{RT}{zF}\ln\frac{a_{\mathrm{i,O}}}{a_{\mathrm{i,W}}} + \frac{RT}{zF}\ln K_{\mathrm{st,(iY}p)}c_{\mathrm{Y,O}}{}^{p} \tag{9}$$

where $\Delta_{\mathrm{O}}^{\mathrm{W}}\phi_{\mathrm{tr,i}}^{0}$ is the standard Galvani potential difference.

The ion transfer at the W/O interface can be observed as a voltammogram, when the potential difference E applied between W and O is scanned and the current I due to the ion transfer at the W/O interface is measured as the function of E. The voltammogram has been recorded at the stationary W/O interface, micro W/O interface or dropping electrolyte solution (W or O) interface. The voltammogram recorded at the dropping electrolyte solution interface is called as a polarogram. The feature of the ion transfer at the W/O interface will be discussed by referring mainly to polarograms in the following, since polarograms are much simpler to analyse than voltammograms at the stationary and micro W/O interfaces.

Figure 6.1 shows polarograms for the transfer of a cation at the W/O interface schematically. The E is defined as the potential of W versus that of O. Hence, the transfer of a cation from W to O or an anion from O to W gives a positive current wave, and that of a cation from O to W or an anion from W to O gives a negative current wave. The positive or negative current wave attains a plateau called the positive or negative limiting current, I_{lp} or I_{ln}, when E is sufficiently positive or negative,

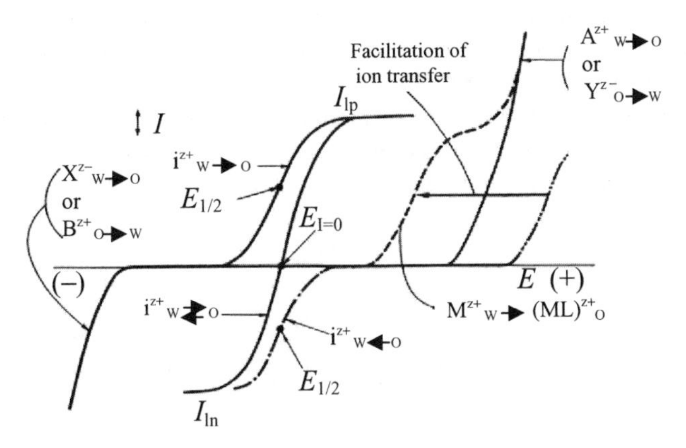

FIGURE 6.1. Schematic depiction of voltammograms. i^{z+}: objective cation; A^{z+}: cation of supporting electrolyte (SE) in W; B^{z+}: cation of SE in O; X^{z-}: anion of SE in W; Y^{z-}: anion of SE in O; M^{z+}: metal ion in W; L: ionophore in O.

respectively. The E where I is a half of I_{ln} or I_{lp} is defined as the half-wave potential, $E_{1/2}$.

The polarogram for a reversible ion transfer in the absence of both the ion-pair and complex formations can be expressed as

$$E = E_{1/2} + \frac{RT}{zF} \ln \frac{I_{lp} - I}{I - I_{ln}} \tag{10}$$

$$E_{1/2} = E^0 + \frac{RT}{2zF} \ln \frac{D_{i,W}}{D_{i,O}} \tag{11}$$

where E^0 is the standard potential for the transfer of i^z and $D_{i,S}$ the diffusion coefficient of i^z in S phase. The I_{lp} and I_{ln} are given as Equations (12) and (13), respectively, when polarograms are recorded by dropping W into O.

$$I_{lp} = KzFD_{i,W}{}^{1/2}m^{2/3}t^{1/6}a_{i,W} \tag{12}$$

$$I_{ln} = KzFD_{i,O}{}^{1/2}m^{2/3}t^{1/6}a_{i,O} \tag{13}$$

where K is the Ilkovič constant, m the flow rate of W, t the drop time of W and a_i the activity of i^z.

The E^0 depends on the potential of the reference electrode in O, and can be connected to $\Delta_O^W \phi_{tr,i}^0$ as Equation (14).

$$E^0 = \Delta_O^W \phi_{tr,i}^0 + \text{const.} \tag{14}$$

where const. includes the potential of the reference electrode in O and the junction potential.

Taking into account Equations (7), (11) and (14), it is clear that $E_{1/2}$ in the positive or negative current wave is more negative or more positive, respectively, when the ion transfer occurs more easily.

The polarogram in the presence of the ion-pair formation as Equation (2) or the complex formation as Equation (3) is expressed by substituting $E_{1/2}$ of Equation (15) or (16) instead of Equation (11) to Equation (10).

$$E_{1/2} = E^0 + \frac{RT}{2zF} \ln \frac{D_{i,W}}{D_{i,O}} + \frac{RT}{zF} \ln K_{ip,ij}a_{j,O} \tag{15}$$

$$E_{1/2} = E^0 + \frac{RT}{2zF} \ln \frac{D_{i,W}}{D_{i,O}} + \frac{RT}{zF} \ln K_{st,(iYp)}c_{Y,O}{}^p \tag{16}$$

The $E_{1/2}$ and $\Delta G_{tr,i}$ calculated from $E_{1/2}$ for various ions are summarized in References [7] and [8]. Fundamental factors useful for the measurement of *VCTIES* such as solvents available as O, supporting electrolytes in W or O, buffer solutions in W, reference and counter electrodes were described in Reference [7].

6.3. VOLTAMMOGRAM FOR ELECTRON TRANSFER AT THE W/O INTERFACE

Not only the charge carried by an ion but also that carried by an electron proceeds at the W/O interface. The electron transfer results from the redox reaction occurring at the interface when an oxidant, O1(W), or a reductant, R1(W), in W is brought into contact with a reductant, R2(O), or an oxidant, O2(O) in O.

$$n_2\text{O1(W)} + n_1\text{R2(O)} \rightleftharpoons n_2\text{R1(W)} + n_1\text{O2(O)} \tag{17}$$

where n_1 and n_2 are related to the redox couples in the individual phases,

$$\text{O1(W)} + n_1\text{e}^- \rightleftharpoons \text{R1(W)} \qquad \left(E^0_{\text{O1/R1}}\right) \tag{18}$$

$$\text{O2(O)} + n_2\text{e}^- \rightleftharpoons \text{R2(O)} \qquad \left(E^0_{\text{O2/R2}}\right) \tag{19}$$

$E^0_{\text{O/R}}$is the standard potential for reactions (18) and (19).

The electron transfer can also be observed as a voltammogram or a polarogram, recording I, which corresponds to the amount of electrons transferred as the function of E. Since voltammetry for the electron transfer will be discussed also in Chapter 8, only the outline of the equation for the polarogram will be introduced in this chapter.

The equation to express the polarogram for the reversible electron transfer was proposed by Samec [9], which can be rearranged [10] as

$$E = A + \frac{RT}{n_1 n_2 F} \ln \frac{(I_{\text{lR1}} - I)^{n_2}(I_{\text{lO2}} - I)^{n_1}}{(I_{\text{lO1}} - I)^{n_2}(I_{\text{lR2}} - I)^{n_1}} \tag{20}$$

$$A = E^0 + \frac{RT}{n_1 n_2 F} \ln \frac{D_{\text{O}_1}^{n_2} D_{\text{R}_2}^{n_1}}{D_{\text{R}_1}^{n_2} D_{\text{O}_2}^{n_1}} \tag{21}$$

where E^0 is the standard potential for the reaction of Equation (17) and D_{i} the diffusion coefficient of a species i. I_{li} is given as Equation (22) similarly to Equations (12) and (13) by using K, m, t and a_{i}.

$$I_{\text{li}} = KnFD_{\text{i,W}}{}^{1/2}m^{2/3}t^{1/6}a_{\text{i,W}} \tag{22}$$

where n is the number of electrons involved in the redox reaction of i.

The $E_{1/2}$ is given as Equation (23) by substituting $(I_{\text{lp}} + I_{\text{ln}})/2$ for I in Equation (20).

$$E_{1/2} = A + \frac{RT}{n_1 n_2 F} \ln \frac{(2I_{\text{lR1}} - I_{\text{lp}} - I_{\text{ln}})^{n_2}(2I_{\text{lO2}} - I_{\text{lp}} - I_{\text{ln}})^{n_1}}{(2I_{\text{lO1}} - I_{\text{lp}} - I_{\text{ln}})^{n_2}(2I_{\text{lR2}} - I_{\text{lp}} - I_{\text{ln}})^{n_1}} \tag{23}$$

where I_{lp} and I_{ln} are the positive and negative limiting currents. Here, I_{lp} corresponds to the smaller one between I_{lO1} and I_{lR2}, and I_{ln} to the smaller one between I_{lR1} and I_{lO2}. As an example, $E_{1/2}$ under the condition that only R1 and O2 are present in bulks of W and O, respectively, $I_{\text{lR1}}$is smaller thanI_{lO2} and $n_1 = n_2 = 1$ can be simplified as

Equation (24) from Equation (23) by substituting $I_{\mathrm{IO1}} = I_{\mathrm{IR2}} = 0$, $I_{\mathrm{lp}} = 0$ and $I_{\mathrm{ln}} = I_{\mathrm{IR1}}$.

$$E_{1/2} = A + \frac{RT}{F} \ln \frac{2I_{\mathrm{IO2}} - I_{\mathrm{IR1}}}{I_{\mathrm{IR1}}} \tag{24}$$

The E^0 involved in A in Equation (21) was confirmed to be the difference between $E^0_{\mathrm{O2/R2}}$ in O and $E^0_{\mathrm{O1/R1}}$ in W [9,10].

$$E^0 = E^0_{\mathrm{O2/R2}} - E^0_{\mathrm{O1/R1}} \tag{25}$$

6.4. ION TRANSFER COUPLED WITH ELECTRON TRANSFER AT THE W/O INTERFACE

When an ion transfer and an electron transfer occur simultaneously at the W/O interface, these two processes may be accelerated or restrained by one another. The coupling of the transfer reactions is considered to be very interesting from the viewpoints of the energy transformation at biological membranes, the active transport, selective transport of ions or electrons, etc.

Since both the ion and electron transfer reactions are electrochemical reactions which can be interpreted in terms of the current/Galvani potential difference relation as described in previous sections, *VCTIES* is expected to be very powerful for the quantitative understanding of the coupling.

6.4.1. Estimation of the Coupling Between the Ion and Electron Transfers on the Basis of VCTIES [11]

Polarograms illustrated schematically in Figure 6.2 were recorded for the subsequent discussion on the interrelation between the ion and electron transfer reactions.

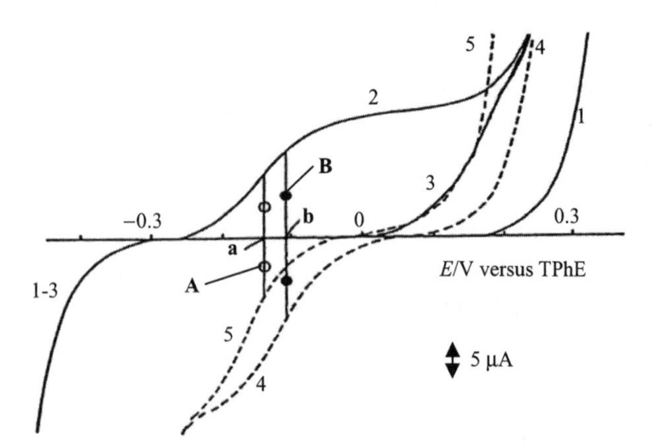

FIGURE 6.2. Polarograms for transfer of K^+ or Na^+ from W to NB facilitated by Val and for transfer of electron from W to NB. (1) residual current, (2) 10^{-4} M K_2SO_4 in W and 10^{-3} M Val in NB, (3) 10^{-4} M Na_2SO_4 in W and 10^{-3} M Val in NB, (4) 5×10^{-4} M $H_4Fe(CN)_6$ in W and 5×10^{-4} M TCNQ and 10^{-3} M $TCNQ^-$ in NB, (5) 5×10^{-4} M $H_4Fe(CN)_6$ in W and 5×10^{-4} M TCNQ and 10^{-2} M $TCNQ^-$ in NB. Supporting electrolytes: 1 M $MgSO_4$ in W, 0.05 M CV^+TPhB^- in NB. TPhE: see Ref. [7].

Curve 1 is the residual current recorded at the interface of W containing 1 M ($=\text{mol dm}^{-3}$) $MgSO_4$ as a supporting electrolyte (SE) and nitrobenzene (NB) containing 0.05 M crystalviolet tetraphenylborate (CV^+TPhB^-) as an SE. Curve 2 or 3 shows the polarogram for the transfer of K^+ or Na^+, respectively, facilitated by valinomycin (Val) recorded at the interface between W containing 10^{-4} M K_2SO_4 or Na_2SO_4 and NB containing 10^{-3} M Val. Curve 4 or 5 is the electron transfer polarogram observed with W containing 5×10^{-4} M $H_4Fe(CN)_6$ and NB containing 5×10^{-3} M 7,7,8,8-tetracyanoquinodimethane (TCNQ) and 10^{-3} or 10^{-2}M $TCNQ^-$.

Let us consider the interrelation between the ion transfer and the electron transfer at the W/NB interface which occurs spontaneously (i.e., without applying potential or current) when W containing 2×10^{-4} M K^+, 2×10^{-4} M Na^+ and 5×10^{-4} M $Fe(CN)_6{}^{4-}$ is put into contact with NB containing 5×10^{-3} M TCNQ, 10^{-2} $TCNQ^-$ and 10^{-3} M Val. The rate of the transfer of electrons that carry minus charges from W to NB because of the redox reaction between $Fe(CN)_6{}^{4-}$ and TCNQ should be equal to that of the transfer of cations from W to NB or that of anions from NB to W in order to hold the electroneutrality in both phases. In the present case, the transfer of K^+ from W to NB occurs since the transfer of K^+ is much easier than the transfer of Na^+ from W to NB or $TPhB^-$ from NB to W (see curves 2 and 3 in Figure 6.2). The E at the W/NB interface established by the coupling of two charge transfer reactions can be estimated based on the charge transfer polarograms, taking into account that the current corresponds to the rate of charge transfer. That is, under the present experimental condition, E is determined to be **a** in Figure 6.2 where the current indicated as **A** in polarogram 2 is equal to that in polarogram 5. When the concentration of $TCNQ^-$ is 10^{-3} M instead of 10^{-2} M, the rates of the ion and electron transfers increase to those corresponding to current **B**, and E (indicated by **b**) is more positive than **a**. When 2×10^{-4} M Na^+ is added to W instead of K^+, rates of the transfers Na^+ and electron are negligible regardless of the concentration of $TCNQ^-$, because the polarogram for the transfer of Na^+ lies at much more positive potentials than those for the transfer of K^+.

The above consideration based on *VCTIES* indicates that the coupling of the ion and electron transfers at the W/O interface is influenced not only by the difference between E^0 for the ion transfer and that for the electron transfer, but also by the concentration ratio of O1(W) to R2(O) and that of O2(O) to R1(W).

6.4.2. *Estimation of the Concentration of the Ion Transferred with the Electron Transfer After the Equilibrium [11]*

When W containing 2×10^{-4} M K^+ and 5×10^{-4} M $Fe(CN)_6{}^{4-}$ is put into contact with an equal volume of NB containing 5×10^{-3} M TCNQ, 10^{-2} $TCNQ^-$ and 10^{-3}M Val, the initial rate of the ion transfer and that of the electron transfer correspond to **A** in curves 2 and 5 in Figure 6.2 (which are reproduced as curves 1 and 1′ in Figure 6.3) as described previously. The transfer of K^+ from W to NB and that of equivalent number of electrons from NB to W proceed during the system is left standing. Consequently, concentrations of K^+ and $Fe(CN)_6{}^{4-}$ in W and that of TCNQ in NB decrease, and that of $Fe(CN)_6{}^{3-}$ in W and those of K^+ and $TCNQ^-$ in NB increase instead. Hence, the ion and electron transfer polarograms are transformed as curves 1 to 4 and curves 1′ to

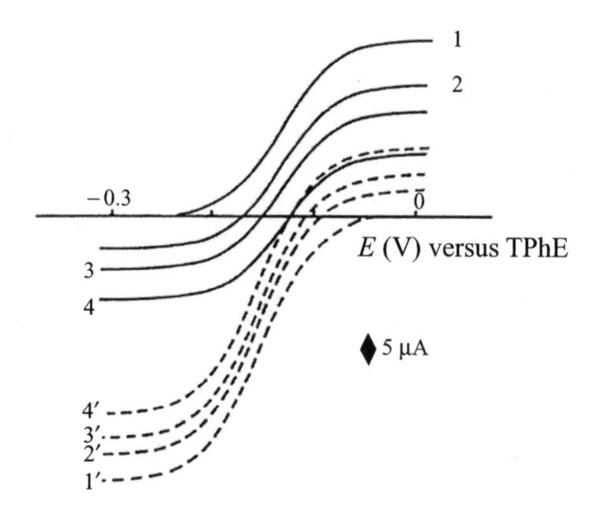

FIGURE 6.3. Time causes of ion transfer and electron transfer polarograms after W containing K^+ and $Fe(CN)_6{}^{4-}$ is brought into contact with NB containing TCNQ, $TCNQ^-$ and Val.

4′, respectively, with an increase of the standing time, if the ion and electron transfer polarograms could be measured individually.

As can be understood intuitively with the assistance of Figure 6.3, the rates of the ion and electron transfers decrease and E at current = 0 in the ion transfer polarogram, $E_{\text{i,I=0}}$, and that in the electron transfer polarogran, $E_{\text{el,I=0}}$, come close to each other with an increase of the standing time. Finally, when the equilibrium is attained, the apparent rates of the charge transfers become zero and $E_{\text{i,I=0}}$ coincides with $E_{\text{el,I=0}}$.

The $E_{\text{i,I=0}}$ in the polarogram for the ion transfer facilitated by an ionophore and $E_{\text{el,I=0}}$ in the polarogram for the electron transfer are given as Equations (26) and (27), respectively, by substituting 0 for I in Equations (10) with (16) and (20) and using (12), (13) and (22).

$$E_{\text{i,I=0}} = E^0 - \frac{RT}{2zF} \ln \frac{a_{\text{i,W}}}{a_{\text{i,O}}} - \frac{RT}{zF} \ln K_{\text{st(iY)}p} c_{\text{Y,O}}{}^p \tag{26}$$

$$E_{\text{el,I=0}} = E^0 + \frac{RT}{n_1 n_2 F} \ln \frac{a_{\text{O1}}^{n2} a_{\text{R2}}^{n1}}{a_{\text{R1}}^{n2} a_{\text{O2}}^{n1}} \tag{27}$$

The relation among $a_{\text{i,W}}$, $a_{\text{i,O}}$, a_{O1}, a_{R1}, a_{O2} and a_{R2} after the equilibrium is given under the condition that

$$E_{\text{i,I=0}} = E_{\text{el,I=0}} \tag{28}$$

6.4.3. *Biomimetic Redox Reactions at the W/O Interface Coupled with Ion Transfer*

Fundamental investigations on redox reactions of substances important for physiological processes have been carried out extensively for the better understanding of vital phenomena. Most of the works have been done in homogeneous solutions, however, though the reaction in living bodies has been believed to proceed at the solution/membrane interface.

In the following, the outline of the redox reaction between O_2 in W and decamethylferrocene in O catalysed by flavin mononucleotide (FMN) in W evolving CO_2, which was elucidated by applying *VCTIES*, will be introduced [12,13]. The selective ion transfer controlled by the redox reaction will also be discussed. The respiration mimetic reactions such as the redox reactions between β-nicotinamide adenine dinucleotide and ascorbic acid or oxygen in W and quinone derivatives or hydroquinone derivatives, respectively, were also investigated with the aid of *VCTIES* [14,15], though they are not introduced here.

6.4.3.1. Spontaneous Evolution of CO_2 Coupled with Selective Ion Transfer Observed at the W/NB Interface [13]. When W containing FMN, pyruvic acid and O_2 was brought into contact with NB containing Na^+, K^+, Val and bis(1,2,3,4,5-pentamethylcyclopentadienyl)iron (DMFC), Na^+ was transferred from NB to W though K^+ remained in NB. The evolution of CO_2 was also observed in W. Here, Val was added to stabilize Na^+ and K^+ in NB. Pyruvic acid is a well-known intermediate of glycolysis pathway.

As an example of the respiration mimetic reaction, when the W/NB system as Equation (29) had been stood for 5 hours with bubbling O_2 into W, the spontaneous transfer of 1.9×10^{-5} mol of Na^+ from NB to W, the evolution of 9.8×10^{-6} mol of CO_2 in W and the decrease of pyruvic acid in W to be 5.3×10^{-4} M were found. The transfer of K^+ was negligible as less than 10^{-7} mol.

$$\begin{array}{l|l} 2.5\times10^{-4}\ \text{M FMN}, & 0.01\ \text{M DMFC}, \\ 1.0\times10^{-3}\ \text{M pyruvic acid}, & 1.0\times10^{-3}\ \text{M Na}^{+}\text{TFPB}^{-} \\ \text{pH buffer}(1.6\times10^{-2}\ \text{M Li}_2\text{HPO}_4+ & 1.0\times10^{-3}\ \text{M K}^{+}\text{TFPB}^{-} \\ 4\times10^{-3}\ \text{M LiH}_2\text{PO}_4;\ \text{pH 7.0}), & 4.0\times10^{-3}\ \text{M valinomycin} \\ \quad(\text{W: 20 ml}) & \quad(\text{NB: 20 ml}) \end{array} \qquad (29)$$

where $TFPB^-$ denotes tetrakis[3,5-bis(trifluoromethyl)phenyl]borate.

Although 1.8×10^{-5} mol of Na^+ was transferred from NB to W, even in the absence of pyruvic acid, the transfer was not found in the absence of FMN or DMFC. In the absence of one of the reagents among Na^+TFPB^-, DMFC, FMN and pyruvic acid, CO_2 was scarcely generated (less than 5×10^{-7} mol). When N_2 was bubbled into W instead of O_2, Na^+ transferred was much less than that observed with O_2 bubbling, and CO_2 formed was negligible.

These results indicate that (a) the selective transfer of Na^+ from NB to W occurs even when W contains only FMN (in the absence of O_2 and pyruvic acid), (b) the existence of FMN, O_2 and pyruvic acid in W and DMFC and Na^+ in NB is necessary for the spontaneous evolution of CO_2, and (c) although the coexistence of O_2 enhances the transfer of Na^+, the transfer is independent of pyruvic acid.

6.4.3.2. Evolution of CO_2 by Controlled Potential Difference Electrolysis at the W/NB Interface. The evolution of CO_2 was also observed when controlled potential difference electrolysis (CPDE) was carried out by applying a definite potential difference at the W/NB interface of the cell of Equation (30), which was similar to that of Equation (29), but in the absence of Na^+ and K^+ in NB. Li_2SO_4 (0.1 M) and $TPenA^+TFPB^-$ (0.05 M; $TPenA^+$, tetrapentyl ammonium ion) were added to W and NB, respectively, as SE.

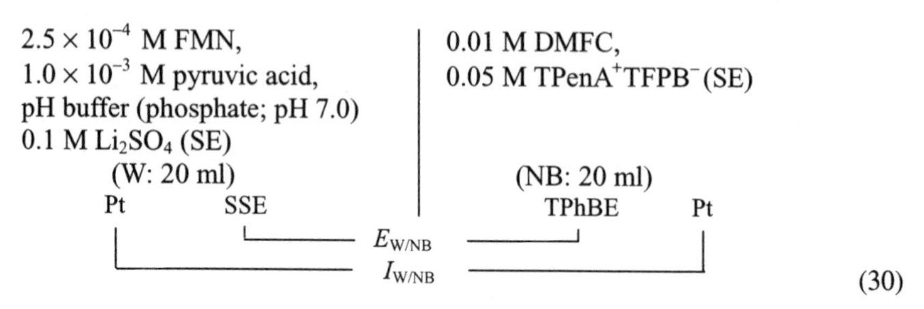

(30)

The electrolysis was attained by applying the potential difference as the potential of silver/silver chloride electrode (SSE) in W versus $TPhB^-$-ion selective electrode (TPhBE) in NB.

After the electrolysis for 5 hours at −0.15 V with the bubbling of O_2 into W, the amount of CO_2 produced was found to be 1.6×10^{-5} mol, and 3.1×10^{-5} mol of 1 electron oxidation product of DMFC ($DMFC^+$) was found in NB. However, when one of the reagents among FMN, O_2, pyruvic acid and DMFC was not added, or the applied potential difference was −0.55 or −0.35 V instead of −0.15 V, the evolution of CO_2 was negligible. The amount of $DMFC^+$ produced after the electrolysis was not affected by pyruvic acid, but was smaller when the electrolysis was carried out in the absence of O_2 as 9.2×10^{-6} mol. When W in the absence of pyruvic acid was saturated with O_2, 9.20×10^{-6} mol of H_2O_2 was found in W after the electrolysis.

6.4.3.3. Estimation of Process Involved in the Evolution Reaction of CO_2 with the Aid of VCTIES. Curve 1 in Figure 6.4 realizes the polarogram at the interface between

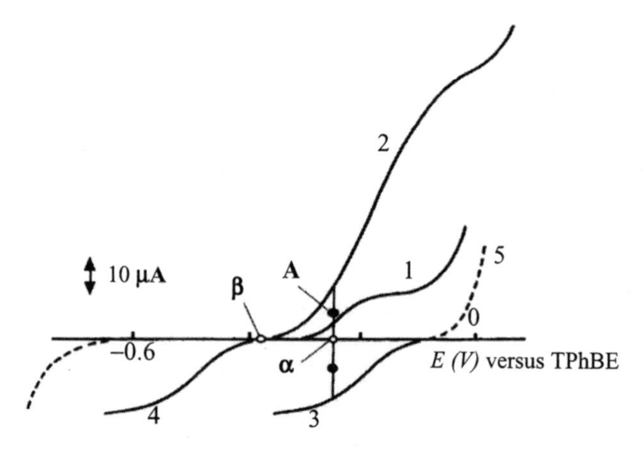

FIGURE 6.4. Polarograms for electron and ion transfers at the W/NB interface. (1) electron transfer observed with deaerated W containing 2.5×10^{-4} M FMN, 0.02 M phosphate buffer and deaerated NB containing DMFC, (2) the same as (1), but O_2 was bubbled into W, (3) ion transfer observed with W containing only SE and NB containing 5×10^{-4} M Na^+TPhB^- and 4×10^{-3} M Val, (4) the same as (3), but K^+ was added instead of Na^+, (5) only supporting electrolytes (0.1 M Li_2SO_4 in W and 0.05 M $TPenA^+TFPB^-$).

W containing 2×10^{-4} M FMN, 0.02 M phosphate buffer (pH 7.0) and 0.1 M Li_2SO_4 (SE), and NB containing 0.01 M DMFC and 0.05 M $TPenA^+TFPB^-$ (SE). Prior to polarographic measurement, both W and NB were carefully deaerated by bubbling high-purity N_2 gas. The limiting current of the positive current wave in the polarogram was proportional to the square root of the height of aqueous-solution reservoir and to the concentration of FMN in the range between 5×10^{-5} to 10^{-3} M. These facts and the result after CPDE described in Section 6.4.3.2 suggest that the wave is due to the reversible electron transfer reaction controlled by the diffusion of FMN in W as

$$\mathrm{FMN(W) + DMFC(NB) + H^+(W) \rightleftharpoons FMNH^\bullet\,(W) + DMFC^+(NB)} \tag{31}$$

The logarithmic analysis based on the theoretical equation for a reversible electron transfer at the W/O interface [see Equation (20)] supported that the current was caused by 1 electron transfer at the W/NB interface.

Polarogram 2 in Figure 6.4 was recorded at the same interface as curve 1, but by bubbling O_2 into W instead of N_2. Although a positive current wave appeared in the potential range identical with that in curve 1, the limiting current was about six times larger than that in curve 1 with the deaerated solution. The large current can be explained by considering the reduction of O_2 catalysed by the FMN/FMNH$^\bullet$ reaction as Equation (32) [12], i.e., the reduction product, FMNH$^\bullet$, at the interface is oxidized by O_2 in the vicinity of the interface regenerating FMN that can be reduced again through the reaction of Equation (31).

$$\mathrm{2FMNH^\bullet\,(W) + O2(W) \rightleftharpoons 2FMN(W) + H_2O_2(W)} \tag{32}$$

Summarizing the results obtained by CPDE and polarography, the reaction process for the electrolytic evolution of CO_2 was estimated to be as follows: The first step was 1 electron transfer from DMFC in NB to FMN in W as Equation (31). The second step was the catalytic reduction of O_2 by FMNH$^\bullet$ as Equation (32). The final step was the oxidation of pyruvic acid by the reduction product of O_2, H_2O_2, in W as Equation (33) that is well-known as an oxidative decarboxylation of α-keto acids [16].

$$\mathrm{CH_3(CO)COOH(W) + H_2O_2(W) \rightleftharpoons H_3COOH(W) + CO_2 + H_2O} \tag{33}$$

6.4.3.4. Polarogram for Transfer of Na^+ or K^+ from NB to W. Curves 3 and 4 in Figure 6.4 are polarograms for transfers of Na^+ and K^+ from NB containing 5×10^{-4} M Na^+TFPB^- and K^+TFPB^-, respectively, in addition to 4×10^{-3} M Val to W. The characteristics of the negative current waves suggested that the transfer processes were controlled by the diffusion of Na^+ or K^+ in NB.

6.4.3.5. Voltammetric Evaluation of Coupling of Ion Transfer with Electron Transfer at the W/NB Interface. The selective transfer of Na^+ from NB to W in the cell of Equation (29) can be understood by referring to the ion transfer and electron transfer polarograms shown in Figure 6.4.

When simultaneous transfers of ions and electrons occur spontaneously at one W/NB interface, the rate of the transfer of electrons (which carry minus charges) from NB to W should be equal to that of the transfer of cations (which carry plus charges) from

NB to W in order to maintain the electroneutrality in both phases. In other words, the positive current due to the electron transfer should be equal to the negative current due to the cation transfer in their absolute values, since the currents correspond to the rates of electron and ion transfers. Applying this condition to the present case, the potential difference established by the coupling of transfers of Na^+ and electron at the W/NB interface is expected to be **α** in Figure 6.4, where the magnitude of the current (indicated as **A**) in polarogram 2 for the electron transfer is equivalent to that in polarogram 3 for the transfer of Na^+.

When K^+ is added to NB instead of Na^+, the potential difference established by the coupling of transfers of electron and K^+ was expected to be **β** in Figure 6.4. The potential **β** is more negative than **α**, and the magnitude of the current at this potential is much smaller than **A**, since the polarogram for the transfer of K^+ lies at much more negative potentials than that for the transfer of Na^+.

The difference in magnitudes of the currents for ion transfers coupled with the electron transfer from DMFC in NB to FMN in W is responsible for that Na^+ transfers from NB to W spontaneously in the system of Equation (29), though K^+ does not.

6.5. PROCESS OF ION TRANSPORT THROUGH A MEMBRANE

A quantitative understanding of the membrane transport is very important for elucidating physiological reactions occurring at biomembranes such as nervous transmission, respiration and metabolism as well as the application of the membrane transports to analytical methods such as liquid membrane type ion sensors and membrane separations [17–19].

The group of the present author investigated the relation between the membrane potential (the potential difference between two aqueous solutions, W1 and W2, separated by a membrane) and the membrane current (the current flowing between W1 and W2), and obtained the voltammogram for the ion transfer through a membrane (VITTM). The membrane used was a liquid membrane, M. Then, the VITTM was compared with the voltammograms at the W1/M and M/W2 interfaces recorded simultaneously with the VITTM, and the membrane transport process was elucidated [20,21] as follows.

An example of VITTM is realized as curve 1 in Figure 6.5. The cell system investigated for the measurement was as Equation (34) in which W1 and W2 (5 ml each) containing an SE ($MgSO_4$) were separated by NB containing an SE (CV^+TPhB^-) and an ionophore (dibenzo-18-crown-6, DB18C6). The NB solution worked as the M of interfacial area of 1 cm^2 and thickness of 1 cm. The cell used was similar to that illustrated in Figure 6.7a.

2.5×10^{-4} M K_2SO_4, 1 M $MgSO_4$ (W1)	0.02 M DB18C6, 0.1 M CV^+TPhB^- (M)	2 M $MgSO_4$ (W2)

CE1 RE1 — $E_{W1/M}$ — RE3 RE4 — $E_{M/W2}$ — RE2 CE2

RE1 ——— $E_{W1\text{-}W2}$ ——— RE2

CE1 ——— $I_{W1\text{-}W2}$ ——— CE2 (34)

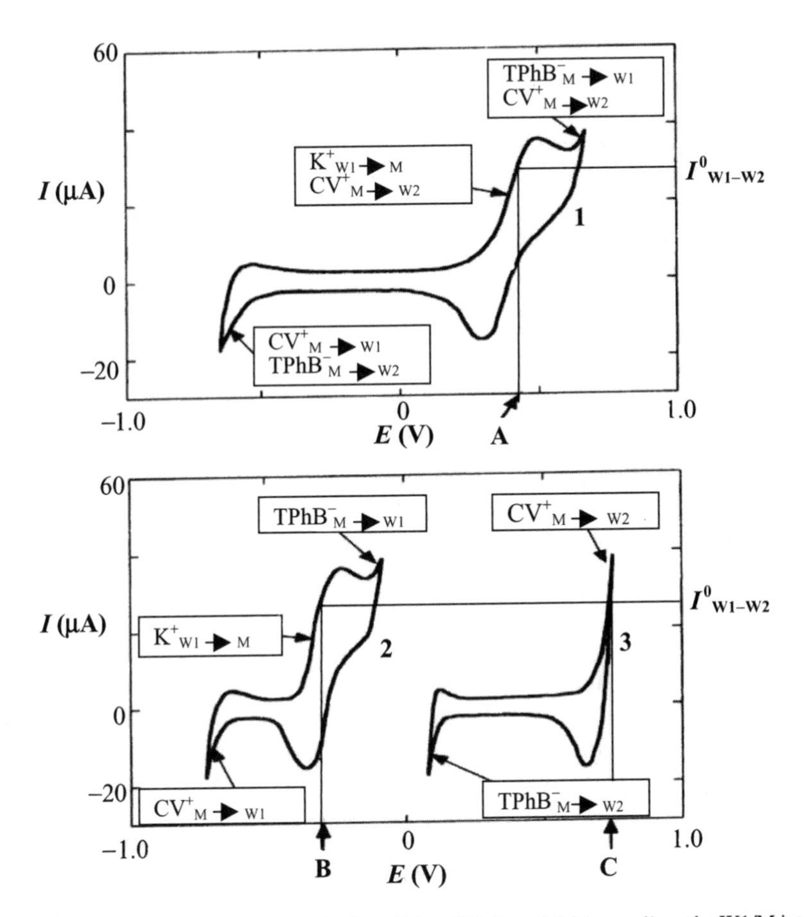

FIGURE 6.5. Voltammograms for ion transfer from W1 to W2 through M (curve 1), at the W1/M interface (curve 2) and at the M/W2 interface (curve 3). See Equation (34) for compositions of W1, M and W2. Scan rate of E_{W1-W2}, 0.01 V · s^{-1}.

The VITTM was recorded by scanning the membrane potential, E_{W1-W2}, and measuring the membrane current, I_{W1-W2}. Here, E_{W1-W2} was applied using two SSE, RE1 and RE2, as the potential of RE1 versus RE2. During the recording of the VITTM, voltammograms for ion transfers at the W1/M and M/W2 interfaces (curves 2 and 3) were also recorded by monitoring the potential differences at the W1/M and M/W2 interfaces, $E_{W1/M}$ and $E_{M/W2}$, as the function of I_{W1-W2}. Here, $E_{W1/M}$ and $E_{M/W2}$ were measured as the potential of RE1 versus RE3 and that of RE2 versus RE4, where RE3 and RE4 were TPhBE placed in M near to the W1/M and M/W2 interfaces, respectively. Consulting with the voltammetric work on the transfer of K^+ at the W/NB interface, the positive peak, the final rise and the final descent in curve 2 are attributable to the transfer of K^+ from W1 to M facilitated by DB18C6, $TPhB^-$ from M to W1 and CV^+ from M to W1, respectively. Here, the final rise and the final descent mean large positive and negative currents, respectively, limiting the potential window. The negative peak is due to the transfer of K^+, which has moved into M during the positive scan from M to W1.

The final rise and the final descent in curve 3 correspond to the transfer of CV^+ from M to W2 and that of $TPhB^-$ from M to W2, respectively.

Comparing curve 1 with curves 2 and 3, it is obvious that (a) the potential window in curve 1 is about twice that in curve 2 or 3, (b) the potential regions where the positive and the negative peaks appear in curve 1 are different from those in curve 2, and (c) the slopes of the positive peak, negative peak, final rise and final descent in curve 1 are much smaller than those in curves 2 and 3.

In order to elucidate the reaction involved in curve 1, the relation among E_{W1-W2} in curve 1, $E_{W1/M}$ in curve 2 and $E_{M/W2}$ in curve 3 was investigated at a definite membrane current, I^0_{W1-W2}, taking into account that currents flowing across both the W1/M and M/W2 interfaces must be the same and equal to the membrane current. The relation

$$E_{W1-W2} = E_{W1/M} + E_{M/W2} + I_{W1-W2}{}^0 R \tag{35}$$

was found to be held at any $I_{W1-W2}{}^0$, where R is the resistance between RE1 and RE2.

When W1, W2 and M contain sufficient concentrations of ions, the contribution of $I_{W1-W2}{}^0 R$ to E_{W1-W2} is not significant, and Equation (36) can be approximated by

$$E_{W1-W2} = E_{W1/M} + E_{M/W2} \tag{36}$$

Equation (36) suggests that the membrane potential in the presence of sufficient electrolytes in W1, W2 and M is primarily determined by the potential differences at two interfaces that depend on ion transfer reactions at the interfaces, although the potential differences at interfaces are not apparently taken into account in theoretical equations such as Nernst–Planck, Henderson and Goldman–Hodgkin–Katz equations which have been often adopted in the discussion of the membrane potential [19,22–25].

The characteristics of VITTM in Figure 6.5 can easily be understood taking into account the relation of Equation (36) as follows. Since the membrane potential where the positive current wave appears (E_{W1-W2} indicated by **A** in curve 1) is the sum of $E_{W1/M}$ indicated by **B** in curve 2 and $E_{M/W2}$ indicated by **C** in curve 3, the positive current wave in curve 1 is considered to be caused by the coupled reactions of both the transfer of K^+ from W1 to M facilitated by DB18C6 (the positive current wave in curve 2) and that of CV^+ from M to W2 (the final rise in curve 3). Hence, the potential region for the positive wave in curve 1 differs from that in curve 2.

On the basis of the similar analysis, the negative current wave in curve 1 is considered to consist of the transfer of K^+ from M to W1 (the negative current wave in curve 2) and that of CV^+ from W2 to M (the negative current wave in curve 3). The final rise in curve 1 involves the transfer of $TPhB^-$ from M to W1 (the final rise in curve 2) and that of CV^+ from M to W2 (the final rise in curve 3), and the final descent in curve 1 involves the transfer of CV^+ from M to W1 (the final descent in curve 2) and that of $TPhB^-$ from M to W2 (the final descent in curve 3). These coupled reactions are responsible for the wide potential window.

One of the reasons for the small slope of the positive current wave in curve 1 is the membrane resistance R in Equation (35). However, the slope is still smaller than the slope of the positive current wave in curve 2 even after it has been corrected for R. The small slope after the correction can be explained by considering that the slope of the positive current wave in curve 1 is composed of the slope of the positive current wave in curve 2 and that of the final rise in curve 3. The small slopes of the final rise and the final descent in curve 1 are also attributable to R and the coupled reactions at two W/M interfaces.

As described above, the ion transfer through a membrane is controlled practically by the complementary ion transfer reactions at two W/M interfaces when the M contained sufficient electrolytes. This idea was successfully applied to explanations of the following subjects concerning with membrane phenomena [20,21,26]. (1) Influence of ion transfer reaction at one W/M interface on that at another W/M interface under an applied membrane potential; (2) Ion transfers through an M in the presence of the objective ion in W1, M and/or W2; (3) Ion separation by electrolysis under an applied membrane potential; (4) Ion transfer through a thin supported liquid membrane. The idea was also demonstrated to be very useful for the elucidation of ion or electron transport process through a bilayer lipid membrane (BLM), which is much thinner than a liquid membrane [21,26].

6.6. NEW TYPES OF MEMBRANE REACTIONS MIMICKING BIOLOGICAL PROCESSES [27,28]

Generally speaking, the membrane transport of a charge (ion or electron) means the transfer of a charge from W1 to W2 across M, as shown in Figure 6.6a. This membrane transport is realized in the presence of a potential gradient between W1 and W2 perpendicular to the W1/M or W2/M interface, and hence is called "perpendicular transport" hereafter. However, the reaction that a charge is incorporated from W1 into M at one site (site A) of the W1/M interface and released from M to W1 at another site (site B) of the same interface after transfer in M can also be regarded as a membrane transport

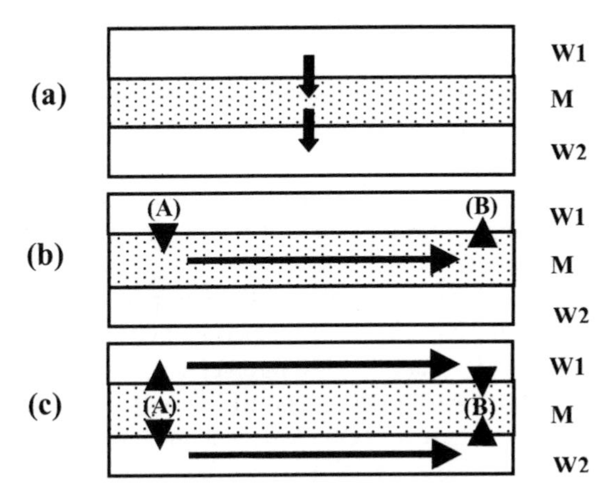

FIGURE 6.6. Three types of membrane transport: (a) perpendicular transport, (b) parallel transport of type I and (c) parallel transport of type II.

[see Figure 6.6b]. This transport is realized under a potential gradient between sites A and B in W1 parallel to the W1/M interface [27], and hence is called "parallel transport of type I" hereafter. Another variety of "parallel transport" is the reaction that a charge is released from M to W1 or W2 at one site (site A) of the W1/M or W2/M interface and incorporated from W1 or W2 to M at another site (site B) of the W1/M or W2/M interface after transfer in W1 or W2 [see Figure 6.6c]. This variety is called "parallel transport of type II" hereafter [28].

The parallel transports are considered to occur often at the interface between an aqueous solution and a heterogeneous biomembrane with various domains [29–31]. Therefore, the electrochemical elucidation of the mechanisms of parallel transports is expected to be important for better understanding of biomembrane phenomena as well as for the design of novel analytical methods or other chemical techniques mimicking the phenomena, though such investigations have been very few so far.

In the following, the fundamental feature of parallel transports of types I and II elucidated with the aid of *VCTIES* is introduced briefly.

6.6.1. The Cell Used to Observe "Perpendicular Transport" and "Parallel Transports of Types I and II"

The cell used for the investigation of perpendicular transport is illustrated in Figure 6.7a [20,21]. The M (1 cm thick) was composed of DCE (about 1 ml) containing an SE (e.g., 0.05 M CV^+TPhB^-). In order to stabilize M between W1 (5 ml) and W2 (5 ml) by means of the difference in specific gravities, concentrations of $MgSO_4$ added as SE in W1 and W2 were made to be 1 and 2 M, respectively. Two SSE, RE_{W1} and RE_{W2}, were set in W1 and W2. Two platinum wires were set in W1 and W2 as counter electrodes, CE_{W1} and CE_{W2}. In order to monitor potential differences at the W1/M and W2/M interfaces, two TPhBE, $RE_{M,1}$ and $RE_{M,2}$, were set in M near to the interfaces.

The cell used for the investigation of parallel transport of type I is illustrated in Figure 6.7b [27]. The M (0.7 cm thick) was composed of DCE (10 ml) containing a rather concentrated SE (e.g., 0.05 M CV^+TPhB^-, CV^+; crystal violet ion), W1 (10 ml) was distilled water without any electrolytes and W2 (10 ml) contained 2 M $MgSO_4$. Two SSE, $RE_{A,W1}$ and $RE_{B,W1}$, were set at sites A and B in W1 in the vicinity of the W1/M interface. The distance between two sites was 12 cm. Two platinum wires were set at sites A and B in W1 as counter electrodes, $CE_{A,W1}$ and $CE_{B,W1}$. In order to monitor potential differences at the W1/M interfaces of sites A and B, two TPhBE, $RE_{A,M}$ and $RE_{B,M}$, were set in M near to the interfaces.

The cell used for the investigation of parallel transport of type II is illustrated in Figure 6.7c [28]. The M (0.7 cm thick) was composed of DCE (10 ml) containing dilute SE (e.g., 10^{-4} M CV^+TPhB^-), and W1 (10 ml) and W2 (10 ml) contained 1 and 2 M $MgSO_4$, respectively. Two TPhBE, $RE_{A,M}$ and $RE_{B,M}$, were set at sites A and B in M. The distance between two sites was 12 cm. Two platinum wires were set at sites A and B in M as counter electrodes, $CE_{A,M}$ and $CE_{B,M}$. In order to monitor potential differences at the W1/M or W2/M interfaces of sites A and B, four SSE, $RE_{A,W1}$ and $RE_{B,W1}$ or $RE_{A,W2}$ and $RE_{B,W2}$, were set at sites A and B in W1 or W2 in the vicinity of the W1/M or W2/M interfaces.

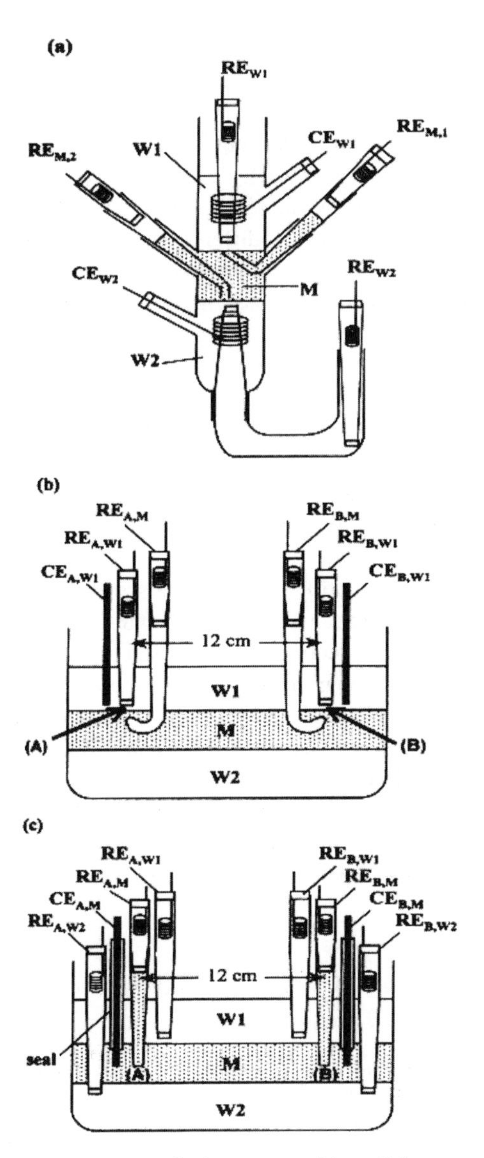

FIGURE 6.7. Electrolytic cells for (a) perpendicular transport, (b) parallel transport of type I and (c) parallel transport of type II. RE_{W1}, RE_{W2}, $RE_{A,W1}$, $RE_{B,W1}$, $RE_{A,W2}$ and $RE_{B,W2}$ are silver/silver chloride reference electrodes. $RE_{M,1}$, $RE_{M,2}$, $RE_{A,M}$ and $RE_{B,M}$ are $TPhB^-$ ion-selective reference electrodes. CE_{W1}, CE_{W2}, $CE_{A,W1}$, $CE_{B,W1}$, $CE_{A,M}$ and $CE_{B,M}$ are platinum wire counter electrodes.

6.6.2. Characteristics of Voltammograms for "Perpendicular Transport" and "Parallel Transports of Types I and II"

The voltammogram for perpendicular transport was recorded by scanning the potential difference E_{W1-W2} between W1 and W2 and measuring the current I_{W1-W2} between W1 and W2 [see Figure 6.7a]. During the recording of the voltammogram, variations of potential differences $E_{W1/M}$ and $E_{M/W2}$ at the W1/M and W2/M interfaces were monitored as the function of I_{W1-W2}, and voltammograms for ion transfers at the W1/M and W2/M interfaces were obtained.

The voltammogram for parallel transport of type I was recorded by scanning the potential difference $E_{W1(A-B)}$ between $RE_{A,W1}$ and $RE_{B,W1}$ and measuring the current $I_{W1(A-B)}$ between sites A and B in W1 and W2 [see Figure 6.7b]. During the recording of the voltammogram, variations in potential differences $E_{W1/M,A}$ and $E_{M/W1,B}$ at the W1/M interface of sites A and B were monitored as the function of $I_{W1(A-B)}$, and voltammograms for the charge transfer at the W1/M interface of sites A and B were obtained.

The voltammogram for parallel transport of type II was recorded by scanning the potential difference $E_{M(A-B)}$ between $RE_{A,M}$ and $RE_{B,M}$, applied as the potential of $RE_{A,M}$ versus $RE_{B,M}$, and measuring the current between sites A and B in M, $I_{M(A-B)}$ [see Figure 6.7c]. During the recording of the voltammogram, variations of potential differences $E_{M/W1,A}$ and $E_{W1/M,B}$ at the W1/M interface of sites A and B or $E_{M/W2,A}$ and $E_{W2/M,B}$ at the W2/M interface of sites A and B were monitored as the function of $I_{M(A-B)}$, and voltammograms for the charge transfer at the W1/M or W2/M interface of sites A and B were obtained.

The voltammograms recorded for parallel transports of types I and II were very similar to those for perpendicular transport, and had characteristics which were described as (a), (b) and (c) in the previous section and expressed by Equations (35) and (36). The results indicate that the voltammograms were realized mainly by the composite of two interfacial ion transfer reactions, i.e., reactions at the W1/M interface of sites A and B for parallel transport of type I and reactions at the W1/M and W2/M interfaces of sites A and B for parallel transport of type II. In other words, parallel transport of type I (W1→M→W1 transport) and parallel transport of type II (M→W1→M or M→W2→M transport) could be realized when potential differences were applied between two sites in one aqueous phase of the system of Figure 6.7b and between two sites in the membrane of the system of Figure 6.7c, respectively.

6.6.3. Unique Reactions Realized Under the Condition for "Parallel Transport of Type I or II"

When $E_{W1(A-B)}$ or $E_{M(A-B)}$ was applied between sites A and B in W of the cell of Figure 6.7b or in M of the cell of Figure 6.7c, respectively, not only the ion transfer reactions at vicinities of sites A and B, but also a reaction at a site in a region (site C) between sites A and B was found to occur. The transfer at site C was explained by considering the gradient of potential differences at the W/M interface, $E_{W/M,C}$, generated when $E_{W1(A-B)}$ or $E_{M(A-B)}$ was applied. The gradient of $E_{W/M,C}$ was investigated by using two reference electrodes, $RE_{C,W}$ and $RE_{C,M}$, set at site C in W and M, respectively, and was found to change almost linearly with the distance from site A when $E_{W1(A-B)}$ or $E_{M(A-B)}$ was 1 V.

Example A: Ion transfer at the W1/M interface in a region between sites A and B under the potential difference applied between sites A and B in W1. This example was investigated with a membrane system of Figure 6.7b in which M contained 2×10^{-4} M picrate (Pic^-) in addition to SE, and W1 did not contain any electrolytes [27]. After the electrolysis for 3 hours by applying 1 V of $E_{W1(A-B)}$, the transfer of Pic^- from M to W1 was found to occur only in a region between site A and about 6 cm from site A where $E_{W/M,C}$ generated was appropriate for the transfer of Pic^-.

Example B: Ion transport from W2 to W1 through a domain under the potential difference applied between sites A and B in W1. This example was examined with a view to simulate the ion transport through a membrane in the presence of a domain by using the membrane system of Figure 6.7b, in which W1 did not contain any electrolytes, M contained SE and W2 contained 10^{-3} M $Mg(TPhB)_2$ and 2 M $MgSO_4$ [27]. The domain was formed in a region of 1 cm wide in M separated from other parts of M by using two porous tetrafluoroethylene resin films as physical boundaries. When 500 μl of 10^{-2} M bis(diphenylphosphinyl)methane (BDPPM), which complexes strongly with Mg^{2+}, was added into the domain in M and then stood for 3 hours by applying $E_{W1(A-B)}$ of 1 V between sites A and B in W1, Mg^{2+} was found to be transferred from W2 to W1 through the domain. The transfer occurred only when the domain was formed in a region between 7 cm from site A and site B. The transport of Mg^{2+} can be explained as follows. Mg^{2+} in W2 is extracted spontaneously with $TPhB^-$ into the domain, since BDPPM facilitates Mg^{2+} transfer owing to the strong complex formation, and the counter ion ($TPhB^-$) is highly hydrophobic. The Mg^{2+} moved into M can be transferred to W1 when the domain is formed in the region described above since $E_{W1/M,C}$ at the M/W1 interface is appropriate for the transfer of Mg^{2+} from M in the presence of BDPPM to W1.

Example C: Ion transport from W1 to a special region of W2 under the potential difference applied between sites A and B in M. This example was investigated by adopting the membrane system of Figure 6.7c of which W1 contained 2×10^{-4} M Pic^- in addition to 1 M $MgSO_4$, M contained 10^{-4} M $TPenA^+TPhB^-$ and W2 contained 2 M $MgSO_4$ [28]. After the electrolysis for 3 hours by applying $E_{M(A-B)}$ of 1 V between sites A and B in M, it was found that the transfer of Pic^- from W1 to W2 occurred in a narrow region around 3–6 cm from site A. The transfer was explained as follows. The standard potential for the transfer of Pic^- at the W/DCE interface coincides with $E_{W1/M,C}$ or $E_{W2/M,C}$ at site C of 4–5 cm from site A, which means that Pic^- in W1 transfers into M in a region between site A and the region around site C. On the other hand, Pic^- transfers from M to W2 in a region between around site C and site B, if Pic^- presents in M. In the present experiment, however, Pic^- did not exist in the region between sites C and B in M. Therefore, Pic^- transfers only in a small region around site C.

The result described in this section suggests that the ion transport from W1 to W2 at a special region of a membrane that resembles the transport at a biomembrane with an ion channel can be realized even in the absence of any channel proteins.

6.7. OSCILLATION OF MEMBRANE CURRENT OR MEMBRANE POTENTIAL [32]

The oscillation of membrane current or membrane potential is well known to occur in biomembranes of neurons and heart cells, and tremendous experimental and theoretical studies on oscillations in biomembranes as well as artificial membranes [32,33] have been carried out from the viewpoint of the biological importance.

The mechanisms of the oscillations in biomembranes have been explained based on the gating of membrane protein called "ion channel", and enormous efforts have been made to elucidate the gating process, mainly by reconstitution of channel proteins into bilayer membranes [34–36]. However, oscillations observed with artificial membranes such as thick liquid membranes, lipid-doped filter or BLM suggest that the oscillations occur even in the absence of any channel proteins. Even if the composition of an artificial membrane is different from that of a biomembrane, the oscillations at artificial membranes are expected to provide fundamental information useful for the elucidation of the oscillation processes in living membrane systems.

In the following, a unique oscillation of membrane current observed with a liquid membrane system by the present authors [32,37], which has characteristics similar to those of the oscillation at a biomembrane with so-called "sodium channel", will be introduced as an example, and the mechanisms for the oscillation will be clarified by using *VCTIES*, taking into consideration ion transfer reactions and adsorptions at two W/M interfaces in the membrane system. In this connection, various oscillations other than this example and the elucidation of their mechanisms were described elsewhere [32,37,38].

6.7.1. *Characteristics of Oscillation of Membrane Current Observed Under an Applied Membrane Potential*

Curve 1 in Figure 6.8 gives an example of the oscillation of membrane current observed with the liquid membrane system as Equation (37) by applying a constant E_{W1-W2} of -0.48 V and measuring the time course of the current through the M, I_{W1-W2},

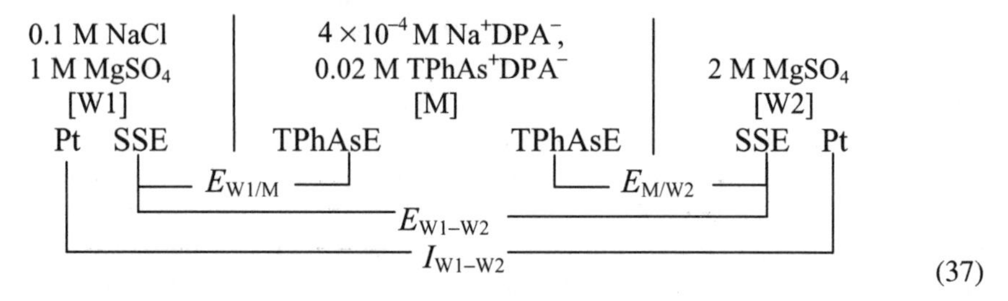

(37)

where $TPhAs^{+}$ and DPA^{-} denote tetraphenylarsonium ion and dipicrylaminate, respectively. The cell used was identical with that of Figure 6.7a except that the solvent

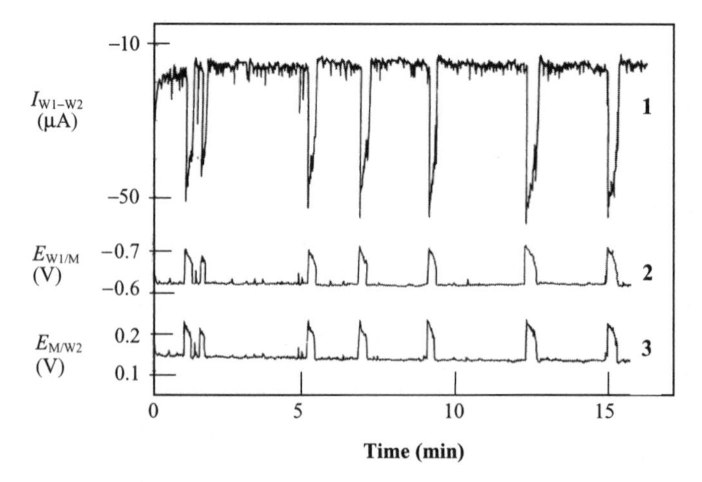

FIGURE 6.8. Time courses of membrane current (curve 1), I_{W1-W2}, and $E_{W1/M}$ (curve 2) and $E_{M/W2}$ (curve 3) observed by applying -0.48 V to the cell of Equation (37) as E_{W1-W2}.

of M was NB and $TPhAs^+$ ion-selective electrodes [7] (TPhAsE) were employed as reference electrodes in M.

The current oscillation lasted for more than 2 hours. Curves 2 and 3 show time courses of $E_{W1/M}$ and $E_{M/W2}$ observed simultaneously with curve 1. From these curves, it is obvious that the current oscillation occurs with the oscillations of $E_{W1/M}$ and $E_{M/W2}$, though the amplitudes are not large (about 0.08 V). Here, it is notable that the amplitude of the oscillation of $E_{W1/M}$ is the same as that of $E_{M/W2}$ because of the relation of Equation (36).

The range of applied E_{W1-W2} available for the current oscillation was between -0.40 and -0.55 V. When E_{W1-W2} was more negative in the range, the amplitude of the oscillation was larger and the period was longer as follows: Amplitude was 53, 38 or 22 $\mu A\ cm^{-2}$ and period was 2.1, 1.8 or 1.6 minutes when E_{W1-W2} was -0.53, -0.48 or -0.43 V, respectively.

The oscillation at $E_{W1-W2} = -0.48$ V was not observed within 2 hours, when the concentration of Na^+DPA^-, c_{NaDPA}, was less than 10^{-4} M or more than 2×10^{-3} M (concentration of $TPhAs^+DPA^-$, $c_{TPhAsDPA}$: 0.02 M) and when $c_{TPhAsDPA}$ was less than 0.01 M or more than 0.07 M (c_{NaDPA}: 4×10^{-4} M). In this connection, the dependence of the oscillation on the activity of dissociated Na^+, a_{Na^+}, and that of associated Na^+ with DPA^-, a_{NaDPA}, in M were examined by varying c_{NaDPA} or $c_{TPhAsDPA}$. The amplitude of the current oscillation strongly depended on a_{Na^+}, but little on a_{NaDPA}.

6.7.2. *Induction or Inhibition of the Current Oscillation*

Similarly to the induction or the inhibition of the current oscillation at a biomembrane with a "sodium channel", the current oscillation observed with the liquid membrane system of Equation (37) without any channel proteins can be induced by acetylcholine ion, Ach^+, or inhibited by such rather hydrophobic ions as alkylammonium ions and glutamate.

The current oscillation was induced by adding 2×10^{-4} M Ach^+ into W1 of the membrane system as Equation (37), to which E_{W1-W2} of -0.30 V had been applied instead of -0.48 V. Here, -0.30 V is not the E_{W1-W2} effective for the oscillation in the absence of Ach^+. The lifetime of the induced current oscillation was much shorter than that observed by applying -0.48 V to the system in the absence of Ach^+, and was from 20 to 40 minutes. The concentration range of Ach^+ effective for the induction was from 7×10^{-5} to 3×10^{-4} M. The lifetime was longer with higher concentration of Ach^+ as follows: 5–15 or 30–45 minutes when concentration was 7×10^{-5} or 3×10^{-4} M, respectively. The amplitude was larger with higher concentration of Ach^+ as follows: about 25 or 50 $\mu A\ cm^{-2}$ when concentration was 7×10^{-5} or 3×10^{-4} M, respectively.

The current oscillation observed by applying $E_{W1-W2} = -0.48$ V was inhibited when tetraethylammonium ion, TEA^+ or glutamate was added to W1 to be more than 10^{-3} M.

6.7.3. *Voltammograms for the Ion Transfer Through M and at the W/M Interfaces*

The voltammograms were investigated in order to understand the processes involved in the current oscillation.

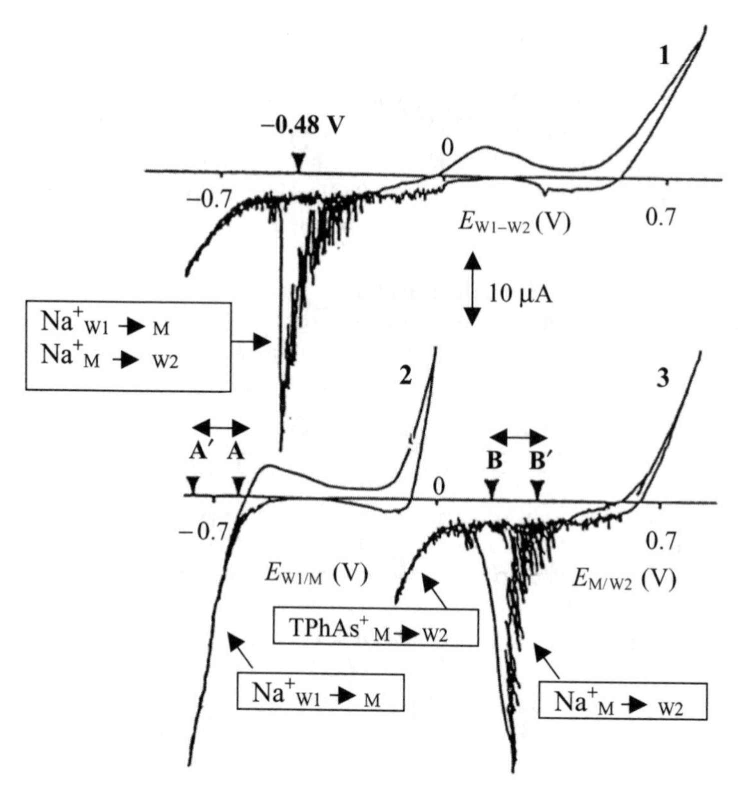

FIGURE 6.9. Voltammograms for ion transfer through a membrane (curve 1), at the W1/M interface (curve 2) and at the M/W2 interface (curve 3) recorded with the cell of Equation (37).

Curve 1 in Figure 6.9 is the voltammogram recorded with the cell of Equation (37) by scanning E_{W1-W2} and measuring I_{W1-W2}. Curves 2 and 3 are voltammograms at W1/M and M/W2 interfaces recorded simultaneously with curve 1, measuring $E_{W1/M}$ and $E_{M/W2}$ as the function of I_{W1-W2}.

An extremely large current peak appeared in voltammograms 1 and 3. Consulting with voltammograms at the W/NB interface recorded under various conditions, the negative current peak in voltammogram 3 was attributed to the maximum because of the transfer of Na^+ from M to W2 enhanced by the stirring caused by the interfacial adsorption. The final rise and the final descent in voltammogram 2 or 3 were confirmed to be due to transfers of $TPhAs^+$ from M to W1 and Na^+ from W1 to M or DPA^- from M to W2 and $TPhAs^+$ from M to W2, respectively.

Taking into account the relation of Equation (36), the negative current peak in voltammogram 1 in Figure 6.9 is considered to be composed of the peak in voltammogram 3 because of the enhanced transfer of Na^+ from M to W2 and the final descent in voltammogram 2 owing to the transfer of Na^+ from W1 to M.

The maximum peaks in voltammograms 1 and 3 in Figure 6.9 were not observed when c_{NaDPA} was less than 10^{-4} M and $c_{TPhAsDPA}$ was 0.02 M. The dependence of the maximum peak on a_{Na^+} or a_{NaDPA} was examined under the condition that a_{Na^+} or a_{NaDPA}

was constant, respectively. The magnitude of the maximum peak depended strongly on a_{Na^+}, although it was practically independent of a_{NaDPA}.

The above-mentioned results on the voltammograms indicate that the condition available for the oscillation of the membrane current resembles closely that for the appearance of the maximum.

6.7.4. *Drop Time–Potential Curves at the W/NB Interface*

The drop time–potential curve was measured with an aqueous solution dropping electrode [7,39] in order to investigate the adsorption at the interface between W containing 1 M $MgSO_4$ and NB containing 4×10^{-4} M Na^+DPA^- and 0.02 M $TPhAs^+DPA^-$. In the curve, there existed a depressed part as compared with the curve in the absence of Na^+DPA^-, suggesting the interfacial adsorption of chemical species relative to Na^+. Assuming that W and NB in the polarographic measurement correspond to W2 and M, respectively, in the membrane system of Equation (37), the potential range for the depression resembles that available for both the transfer of Na^+ from M to W2 and the appearance of the maximum peak at the M/W2 interface. Here, the potential range for the depression lies at more positive potentials than the point of zero charge (pzc) of the interface at around 0.2 V, and hence W is polarized to be positive in the potential range.

Taking into account that both the voltammetric maximum and the depression in the drop time–potential curve were affected by the ion pair formation equilibrium of Na^+DPA^- in M, it is concluded that Na^+ which has been transferred from NB to W may be adsorbed at the interface from the side of W inducing the adsorption of DPA^- as a counter ion from the side of NB. At the interface, the adsorbed Na^+ may exist as an ion pair, which is denoted as $Na^+DPA^-_{ip}$ hereafter.

6.7.5. *Mechanism of the Current Oscillation*

The mechanism will be discussed by referring to voltammograms in Figure 6.9. The oscillation of the membrane current is brought about by the mutually dependent ion transfer reactions at the W1/M and M/W2 interfaces. When E_{W1-W2} in the range effective for the maximum in curve 1 of Figure 6.9 is applied to a membrane system of Equation (37), $E_{M/W2}$ settles at a potential **B** available for both the transfer of Na^+ from M to W2 and the adsorption of $Na^+DPA^-_{ip}$ at the M/W2 interface, and $E_{W1/M}$ at a potential **A** in the final descent due to the transfer of Na^+ from W1 to M. Here, it should be reminded that the relation of Equation (36) holds among E_{W1-W2}, $E_{W1/M}$ and $E_{M/W2}$, and magnitudes of currents flowing across the W1/M and M/W2 interfaces, $I_{W1/M}$ and $I_{M/W2}$, are of the same magnitude.

At $E_{M/W2}$ of **B**, Na^+ transfers from M to W2, and Na^+ transferred to W2 is adsorbed at the M/W2 interface as $Na^+DPA^-_{ip}$. The adsorption generates the stirring of the solution in the vicinity of the interface because the adsorption induces the change of the interfacial tension at the M/W2 interface. Since the stirring enhances the transfer of Na^+ at the interface, $I_{M/W2}$ grows up as the maximum current. At the same time, $I_{W1/M}$ grows up along the final descent of the voltammogram at the W1/M interface (see curve 1 in Figure 6.9), causing the negative shift of $E_{W1/M}$ to **A′**. The negative shift of $E_{W1/M}$ results in the positive shift of $E_{M/W2}$ to **B′** because of the relation of Equation (36). The adsorption at **B′** may be stronger than that at the original $E_{M/W2}$ (**B**) since **B′** is a potential remoter from pzc than **B**, and hence the polarization of the M/W2 interface is more significant at **B′** than that at **B**. The duration of the adsorption brings about the saturation of $Na^+DPA^-_{ip}$ at the interface and the reduction of the stirring. Because of the reduced stirring and the consumption of Na^+ in M near to the interface owing to the enhanced transfer of Na^+, $I_{M/W2}$ decreases. Simultaneously, $I_{W1/M}$ decreases accompanying the positive shift of $E_{W1/M}$ to around **A** and the negative shift of $E_{M/W2}$ to around **B** [cf. Equation (36)]. At the $E_{M/W2}$ around **B** which is nearer to pzc than **B′**, $Na^+DPA^-_{ip}$, which has been adsorbed at the M/W2 interface, is desorbed into M, because the adsorption at this $E_{M/W2}$ (around **B**) is weaker than that at the positive $E_{M/W2}$ (around **B′**) and $Na^+DPA^-_{ip}$ is hydrophobic rather than hydrophilic. The desorption followed by the dissociation of $Na^+DPA^-_{ip}$ restores the activity of Na^+ in M in the vicinity of the interface. Hence, the current increases again, causing the positive shift of $E_{M/W2}$. These processes may repeat to realize the oscillation of the membrane current.

6.7.6. Mechanism of Induction or Inhibition of the Current Oscillation

It is noteworthy that the current oscillation was induced and inhibited by the addition of Ach^+ and tetraalkylammonium ion or glutamate, respectively, into W1 although the W1/M interface is not directly related to the voltammetric maximum essential for the oscillation.

The induction of the oscillation of the membrane current by the addition of Ach^+ can be understood with the aid of ion transfer voltammograms at the W1/M and M/W2 interfaces (Figure 6.10). Curves 1 and 2 are schematic illustrations of voltammograms 2 and 3 in Figure 6.9, and curve 3 is the voltammogram for the transfer of Ach^+ from W1 to M.

When −0.30 V is applied as E_{W1-W2} to the cell of Equation (37) in the absence of Ach^+, $E_{W1/M}$ and $E_{M/W2}$ settle at potentials indicated as **C** and **D**, respectively, and satisfy the relation of Equation (36). The potential **D** is not effective for the adsorption

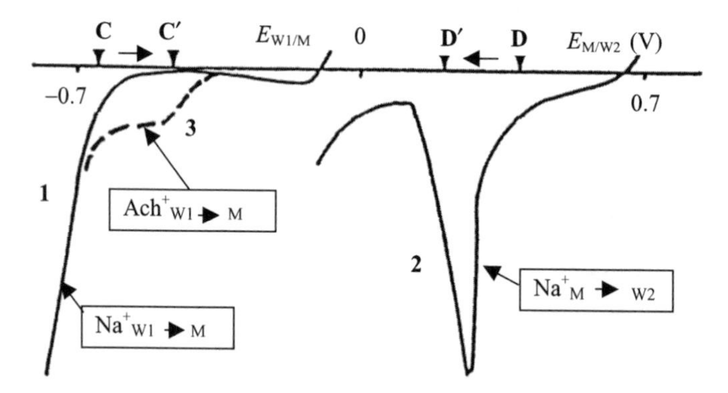

FIGURE 6.10. Voltammetric explanation for the induction of the oscillation by acetylcholine, Ach^+, under $E_{W1-W2} = -0.30$ V. Curves 1 and 2: schematic illustrations of curves 2 and 3 in Figure 6.9. Curve 3: voltammogram recorded at the W1/M interface in the presence of Ach^+ in W1.

of $Na^+DPA^-_{ip}$, and hence for the maximum current. When Ach^+ is added into W1 to be 2×10^{-4} M (cf. curve 3), $E_{W1/M}$ shifts to a less negative potential **C′**, because $I_{W1/M}$ is undertaken by the transfer of Ach^+ from W1 to M. At the same time, according to the relation of Equation (36), $E_{M/W2}$ shifts to a potential **D′** less positive than **D** and effective for the transfer of Na^+ and the adsorption of $Na^+DPA^-_{ip}$. At **D′**, $I_{M/W2}$ grows up as the maximum current, since the transfer of Na^+ from M to W2 is enhanced by the stirring caused by the adsorption. With the growth of $I_{M/W2}$, $I_{W1/M}$ increases as a result of the transfer of Na^+ as well as Ach^+ from W1 to M. The processes similar to those for the oscillation of the membrane current in the absence of Ach^+ mentioned previously may succeed, and the oscillation induced by Ach^+ may be realized.

The lifetime of the induced current oscillation is considered to be determined by the concentration of Ach^+, and the oscillation may terminate when the concentration of Ach^+ in W1 is decreased owing to its transfer from W1 to M, which explains the result that the lifetime was shorter with lower concentrations of Ach^+. The higher concentration limit of Ach^+ effective for the induction can be understood by considering as follows. When Ach^+ of concentration higher than 3×10^{-4}M is added to W1, the positive shift of $E_{W1/M}$ as a result of the transfer of Ach^+ is too large and causes a negative shift of $E_{M/W2}$ to a potential less positive than the range for the maximum current. Therefore, the oscillation cannot be induced when the concentration of Ach^+ is higher than 3×10^{-4} M.

The inhibition of the oscillation by tetraalkylammonium ion or glutamate can be explained referring to voltammograms in Figure 6.11. In the following, the inhibition by

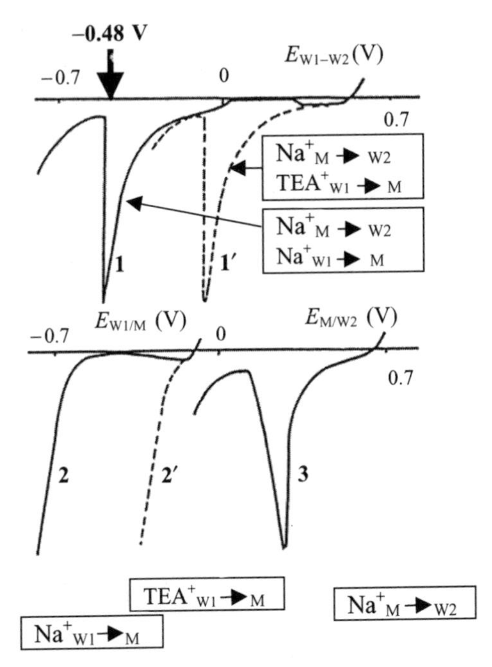

FIGURE 6.11. Voltammetric explanation for the inhibition of the oscillation by tetraethyl ammonium ion, TEA^+, under $E_{W1-W2} = -0.30$ V. Curves 1 and 2: schematic illustrations of curves 2 and 3 in Figure 6.9. Curves 1′ and 2′: voltammograms the same as curves 1 and 2, but recorded in the presence of TEA^+ in W1.

0.01 M TEA^+ will be adopted as an example. Curves 1, 2 and 3 in the figure schematically illustrate voltammograms 1, 2 and 3 in Figure 6.9, respectively. Curves 1′ and 2′ are voltammograms observed under the same condition as that for curves 1 and 2, but adding 0.01 M TEA^+ to W1.

Since the final descent in the voltammogram at the W1/M interface in the presence of 0.01 M TEA^+ in W1 (curve 2′) lies at much more positive potentials than those in the absence of TEA^+ (curve 2), the maximum in the voltammogram for the transfer of Na^+ through the M in the presence of TEA^+ (curve 1′) appears at more positive potentials than that in the absence of TEA^+ (curve 1) for the sake of the relation of Equation (36). Therefore, the E_{W1-W2} value necessary to observe the current oscillation in the absence of TEA^+ (e.g., −0.48 V) is too negative to continue the oscillation after the addition of TEA^+.

6.8. CONCLUSION

In this chapter, the ion or electron transfer and their coupling at the W/O interface were discussed on the basis of results obtained by *VCTIES* after a brief introduction of the fundamental features of *VCTIES*. Applying the method and concept of *VCTIES*, the charge transport through a liquid membrane in the presence of sufficient electrolytes was demonstrated to be controlled by mainly charge transfers at two aqueous/membrane interfaces in the membrane system, which means that the quantitative understanding of charge transfers at two interfaces is inevitable for the elucidation of the membrane transport process. This fundamental fact was successfully applied to the interpretation of new types of membrane transports, parallel transport of type I or II, and the elucidation of the mechanism of the oscillation of the membrane current.

The respiration mimetic redox reactions at the W/O interface and membrane phenomena introduced here might endow some views useful for better understanding of physiological phenomena at biomembranes.

Finally, it should be stressed that the voltammetric concept and method are very helpful for the elucidation of electrochemical reactions at biomembranes as well as artificial membranes [26,40–42].

REFERENCES

1. H.H. Girault, *Modern Aspects of Electrochemistry*, Eds., J.O'M. Bockris, B.E. Conway and R.E. White, Plenum Press, New York, 1993, Vol. 25, p. 1.
2. V. Marecek and Z. Samec (Eds.), Electrochim. Acta **40**, 2839 (1995).
3. A.G. Volkov and D.W. Deamer (Eds.), *Liquid–Liquid Interfaces, Theory and Methods*, CRC Press, Boca Raton, FL, 1996, Chapters 1, 2, 8, 12, 14, 17 and 18.
4. A.G. Volkov, D.W. Deamer, D.L. Tanelian and V.S. Markin, *Liquid Interfaces in Chemistry and Biology*, John Wiley & Sons, New York, 1998, Chapters 4, 5 and 8.
5. A.G. Volkov (Ed.), *Liquid Interfaces in Chemical, Biological, and Pharmaceutical Applications*, Marcel Dekker, New York, 2001, Chapters 5, 15, 17, 20, 22, 25, 26 and 29.
6. A.G. Volkov (Ed.), *Interfacial Catalysis*, Marcel Dekker, New York, 2003, Chapters 2, 4, 19 and 20.
7. S. Kihara, M. Suzuki, K. Maeda, K. Ogura, S. Umetani, M. Matsui and Z. Yoshida, Anal. Chem. **58**, 2954 (1986).

8. H.H.J. Girault and D.J. Schiffrin, *Electroanalytical Chemistry*, Ed., A.J. Bard, Marcel Dekker, New York, 1989, Vol. 15, p. 1.
9. Z. Samec, J. Electroanal. Chem. **103**, 11 (1979).
10. S. Kihara, Z. Yoshida, M. Suzuki, K. Maeda, K. Ogura and M. Matsui, J. Electroanal. Chem. **271**, 107 (1989).
11. K. Maeda, S. Kihara, M. Suzuki and M. Matsui, J. Electroanal. Chem. **303**, 171 (1991).
12. M. Suzuki, M. Matsui and S. Kihara, J. Electroanal. Chem. **438**, 147 (1997).
13. K. Maeda, M. Nishihara, H. Ohde and S. Kihara, Anal. Sci. **14**, 85 (1998).
14. H. Ohde, K. Maeda, Y. Yoshida and S. Kihara, Electrochim. Acta **44**, 23 (1998).
15. H. Ohde, K. Maeda, Y. Yoshida and K. Kihara, J. Electroanal. Chem. **483**, 108 (2000).
16. J. March, *Advanced Organic Chemistry*, John Wiley & Sons, New York, 1992, p. 1175.
17. H.T. Tien, *Bilayer Lipid Membranes*, Marcel Dekker, New York, 1974, Chapters 1–5.
18. R.B. Gennis, *Biomembranes*, Springer-Verlag, Tokyo, 1990, Chapter 7.
19. J. Koryta, *Ions, electrodes, and membranes*, John Wiley & Sons, New York, 1982, Chapter 3.
20. O. Shirai, S. Kihara, M. Suzuki, K. Ogura and M. Matsui, Anal. Sci. **7**(Suppl.), 607 (1991).
21. O. Shirai, S. Kihara, Y. Yoshida and M. Matsui, J. Electroanal. Chem. **389**, 61 (1995).
22. G.E. Goldman, J. Gen. Physiol. **25**, 37 (1943).
23. A.L. Hodgkin and B. Katz, J. Physiol. **108**, 37 (1949).
24. P. Henderson, Z. Physik. Chem. **59**, 118 (1907).
25. N. Laksminarayanaiah, *Equations of Membrane Biophysics*, Academic Press, Florida, 1984, Chapters 1–7.
26. O. Shirai, S. Kihara, Y. Yoshida, K. Maeda and M. Matsui, Bull. Chem. Soc. Jpn. **69**, 3151 (1995).
27. N. Kurauchi, Y. Yoshida, N. Ichieda, H. Ohde, O. Shirai, K. Maeda and S. Kihara, J. Electroanal. Chem. **496**, 118 (2001).
28. N. Kurauchi, Y. Yoshida, N. Ichieda, M. Kasuno, K. Banu, K. Maeda and S. Kihara, J. Electroanal. Chem. **526**, 101 (2002).
29. P. Mitchell, Nature **191**, 144 (1961), Eur. J. Biochem. **95**, 1 (1979).
30. L.S. Yaguzhinsky, L.I. Boguslavsky and A.D. Ismailov, Biochim. Biophys. Acta **368**, 22 (1974).
31. R.G. Gennis, *Biomembranes: Molecular Structure and Function*, Springer-Verlag, Tokyo, 1990, Chapters 5 and 6.
32. S. Kihara and K. Maeda, Prog. Surf. Sci. **47**, 1 (1994).
33. R. Larter, Chem. Rev. **90**, 355 (1990).
34. B. Sakmann and E. Neher, *Single-Channel Recording*, Plenum Press, New York, 1983, Chapter 6.
35. C. Miller, *Ion Channel Reconstitution*, Plenum Press, New York, 1986, Chapters 8–10.
36. S. Numa, The Harvley Lectures **83**, 121 (1989).
37. S. Kihara, K. Maeda, O. Shirai, Y. Yoshida and M. Matsui, *Bioelectroanalysis 2*, Ed., E. Pungor, Akademiai Kaido, Budapest, 1993, p. 331.
38. K. Maeda, W. Hyogo, S. Nagami and S Kihara, Bull. Chem. Soc. Jpn. **70**, 1505 (1997).
39. Z. Yoshida and S. Kihara, J. Electroanal. Chem. **227**, 171 (1987).
40. N. Ichieda, O. Shirai, M. Kasuno, K. Banu, A. Uehara, Y. Yoshida and S. Kihara, J. Electroanal. Chem. **542**, 97 (2003).
41. H. Shiba, K. Maeda, N. Ichieda, M. Kasuno, Y. Yoshida, O. Shirai and S. Kihara, J. Electroanal. Chem. **556**, 1 (2003).
42. N. Ichieda, O. Shirai, Y. Yoshida, M. Torimura and S. Kihara, Electroanalysis, **16**, 779 (2004).

7

Electrochemical Instability at Liquid/Liquid Interfaces

Takashi Kakiuchi
Kyoto University, kyoto 615-8510, Japan

7.1. INTRODUCTION

When a liquid/liquid two-phase system contains ionic surfactants, the interface can exhibit a variety of intriguing features, such as spontaneous emulsification, kicking movements of the interface and oscillations of the interfacial tension and of phase-boundary potential [1–6]. These interfacial turbulences have been known for a long time in surface chemistry. A more recent interest in the interfacial turbulences originates in solvent extraction, in which the adsorption of ligands and metal–ligand complexes at the interface often results in the interfacial turbulences [7,8]. Such observations prompted theoretical studies of the surface instability, among which a seminal work by Sterling and Scriven resorting to the linear stability theory [9] has been an important basis of the theoretical studies of the Marangoni instability in liquid extraction and related systems[10].

On the other hand, the chaotic oscillation of electric current associated with the transfer of ions was known already at the beginning of the electrochemical studies of the electrified liquid/liquid interfaces [11,12]. To avoid such interfacial turbulences in recording current–potential curves, surfactant was added to the system in analogy with the suppression of the polarographic maxima by adding surface-active substances [13]. The interfacial turbulence in the voltammetry of ion transfer has been rather considered to be a nuisance than made use of as a phenomenon carrying rich information of the interfacial properties. As a matter of fact, the chaotic current oscillation appears in voltammograms with some regularity [14]. The current oscillation occurs at the potential in the vicinity of the standard ion- transfer potential of the surface-active ions. Moreover, the potential where the interface becomes unstable is often limited to a certain region around the standard ion transfer potential. The origin of such interfacial turbulence in voltammetry of ion transfer across the liquid/liquid interface can be understood in terms of the concept of the electrochemical instability we proposed recently [15]. The interplay of the adsorption and partition of surface-active ions can bring the liquid/liquid interface into a thermodynamically unstable state where the second derivative of the intefacial tension with respect to the phase-boundary potential becomes positive.

In the interfacial turbulences in solvent extraction, the phase-boundary potential has not been considered to be a major player, probably because no net electric current flows across the interface in conventional extraction experiments. It seems, however, that the electrochemical instability established through the study of voltammetry of ion transfer can explain salient features of the turbulences in such electrically neutral conditions, using the phase-boundary potential as the key parameter, as is the case of the theoretical understanding of the ion-pair extraction from the viewpoint of electrochemistry of ion transfer where the phase-boundary potential plays a pivotal role [16,17].

In the following, we describe first a theoretical basis of the electrochemical instability and then introduce some characteristic features of experimental behaviour of the electrochemical instability.

7.2. THEORETICAL BACKGROUND

7.2.1. *Potential-Dependent Adsorption of Ions at the Liquid/Liquid Interface*

We consider the adsorption and partition of an ionic surfactant i in two immiscible electrolyte solutions, O and W, in contact, as a function of the phase-boundary potential between O and W, $\Delta_{\mathrm{O}}^{\mathrm{W}}\phi = \phi^{\mathrm{W}} - \phi^{\mathrm{O}}$, where ϕ^{W} and ϕ^{O} are the inner potentials in W and O. When a surface-active substance partitions between O and W, the partition of the surfactant and the two adsorption processes from both sides of the interface are not independent of each other, as these are all affected by $\Delta_{\mathrm{O}}^{\mathrm{W}}\phi$ [18]. The dependence of the partition of an ionic substance on $\Delta_{\mathrm{O}}^{\mathrm{W}}\phi$ is expressed by the Nernst equation,

$$\Delta_{\mathrm{O}}^{\mathrm{W}}\phi = \Delta_{\mathrm{O}}^{\mathrm{W}}\phi_i^{\ominus} + \frac{RT}{z_i F} \ln \frac{c_i^{\mathrm{O}}}{c_i^{\mathrm{W}}} \tag{1}$$

where $\Delta_{\mathrm{O}}^{\mathrm{W}}\phi_i^{\theta}$ is the standard ion-transfer potential, z_i is the signed unit of electronic charge on ion i and c_i^{α} is the concentration of i in the phase $\alpha(\alpha = \mathrm{O}$ or W).

The dependence of the adsorption on $\Delta_{\mathrm{O}}^{\mathrm{W}}\phi$ is represented in the adsorption Gibbs energy. A simple model for this dependence is the linear variation of the adsorption Gibbs energy of the ionic surfactant on $\Delta_{\mathrm{O}}^{\mathrm{W}}\phi$ [18]. For the adsorption from O, we may write [18]

$$\Delta G_{\mathrm{ads},i}^{\mathrm{O},0} = \Delta G_{\mathrm{ads},i}^{\ominus} + z_i F \beta_i \left[\Delta_{\mathrm{O}}^{\mathrm{W}}\phi - \Delta_{\mathrm{O}}^{\mathrm{W}}\phi_i^{\ominus}\right] \tag{2}$$

where $\Delta G_{\mathrm{ads},i}^{\mathrm{O},0}$ is the standard Gibbs energy of adsorption from O , β_i is the constant that does not depend on $\Delta_{\mathrm{O}}^{\mathrm{W}}\phi$ and $\Delta G_{\mathrm{ads},i}^{\ominus}$ is the standard adsorption Gibbs energy at $\Delta_{\mathrm{O}}^{\mathrm{W}}\phi_i^{\ominus}$. We have chosen $\Delta_{\mathrm{O}}^{\mathrm{W}}\phi_i^{\ominus}$ as the reference point of the potential-dependent $\Delta G_{\mathrm{ads},i}^{\mathrm{O},0}$ and $\Delta G_{\mathrm{ads},i}^{\mathrm{W},0}$ for convenience [18], because at $\Delta_{\mathrm{O}}^{\mathrm{W}}\phi = \Delta_{\mathrm{O}}^{\mathrm{W}}\phi_i^{\ominus}$

$$\Delta G_{\mathrm{ads},i}^{\mathrm{O},0} = \Delta G_{\mathrm{ads},i}^{\mathrm{W},0} = \Delta G_{\mathrm{ads},i}^{\ominus} \tag{3}$$

Similarly, for the adsorption from W,

$$\Delta G^{\mathrm{W},0}_{\mathrm{ads},i} = \Delta G^{\ominus}_{\mathrm{ads},i} - z_i F(1-\beta_i)\left[\Delta^{\mathrm{W}}_{\mathrm{O}}\phi - \Delta^{\mathrm{W}}_{\mathrm{O}}\phi^{\ominus}_i\right] \tag{4}$$

where $\Delta G^{\mathrm{W},0}_{\mathrm{ads},i}$ is the standard Gibbs energy of adsorption from W. The sign of the second term on the rhs of Equation (2) has been chosen so that the adsorption of anions from O becomes favourable when $\Delta^{\mathrm{W}}_{\mathrm{O}}\phi$ is positive. A fraction of the phase-boundary potential referred to the standard ion-transfer potential thus works as the electrochemical driving force for the adsorption from each phase, which, incidentally, resembles the Butler–Volmer equation for the electrode reaction.

Using Equation (4), we can calculate the dependence of the adsorption on $\Delta^{\mathrm{W}}_{\mathrm{O}}\phi$ for a particular adsorption isotherm. When the adsorption and partition processes are in a thermodynamic equilibrium, the adsorption of i is described by an adsorption isotherm in the from

$$B^{0,\ominus}_i \xi_i m_i / V^{\mathrm{W}} = f(\theta_i) \tag{5}$$

where $B^{0,\ominus}_i$ is the adsorption coefficient of i at $\Delta^{\mathrm{W}}_{\mathrm{O}}\phi^{\ominus}_i$, m_i is the total amount of the surfactant i in the system, θ is the surface coverage of i and V_{w} is the volume of the phase W. We have neglected the contribution of the adsorption of i to m_i. The potential-dependent term in Equation (5)is ξ_i, defined by

$$\xi_i = \frac{e^{(1-\beta_i)y_i}}{1 + r\, e^{y_i}} \tag{6}$$

where r is the ratio of the volume of O (V^{O}) to V^{W}, that is, $r = V^{\mathrm{O}}/V^{\mathrm{W}}$ and

$$y_i = (z_i F/RT)\left(\Delta^{\mathrm{W}}_{\mathrm{O}}\phi - \Delta^{\mathrm{W}}_{\mathrm{O}}\phi^{\ominus}_i\right) \tag{7}$$

Given a functional form of $f(\theta)$, we can calculate the dependence of the adsorbed amount of the ionic surfactant, Γ_i, as a function of $\Delta^{\mathrm{W}}_{\mathrm{O}}\phi$.

Figure 7.1a shows the dependence of θ_i on y_i at $\beta_i = 0.5$ for two values of the interaction parameter a, 1.0 (curve 1) and 2.0 (curve 2), in the Frumkin isotherm [19]:

$$f(\theta_i) = \frac{\theta_i}{1-\theta_i}\exp(-2a\theta_i) \tag{8}$$

It is seen that the adsorption reaches a maximum right at $\Delta^{\mathrm{W}}_{\mathrm{O}}\phi = \Delta^{\mathrm{W}}_{\mathrm{O}}\phi^{\ominus}_i$. The location of the adsorption maximum on the y_i-axis depends on the value of β_i that may vary between 0 and 1. An important point is that the adsorption decreases at both sufficiently positive and sufficiently negative values of y_i, whatever the value of β_i is [18]. This characteristic dependence of adsorption in the phase-boundary potential is the origin of the electrochemical instability (vide infra).

According to the thermodynamics of the interface, the adsorption induces the variation in the interfacial tension, γ, a decrease (increase) in the case of the positive (negative) adsorption. In the case of the Frumkin isotherm, the decrease in γ is given by [19]

$$\Delta\gamma = \gamma_0 - \gamma = RT\Gamma_{\mathrm{m}}[\ln(1-\theta) + a\theta^2] \tag{9}$$

where Γ_{m} is the maximum adsorption. Equation (9) represents the variation of γ as a

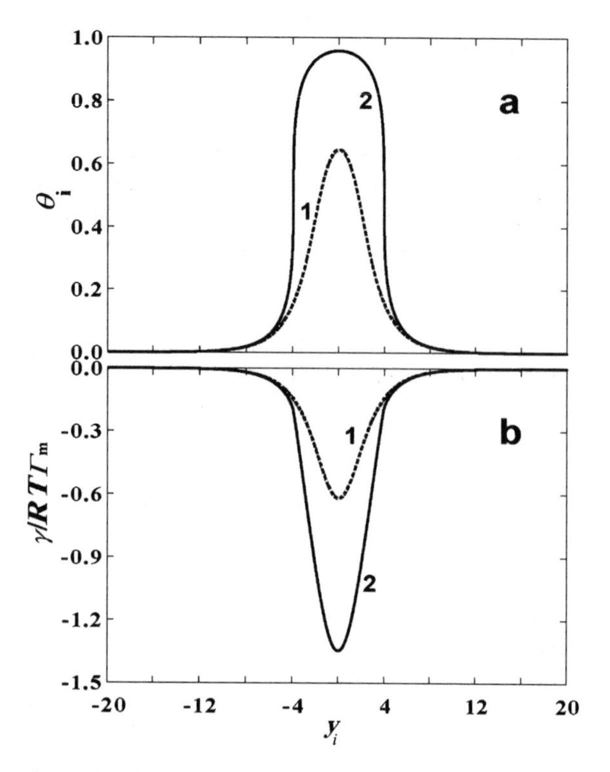

FIGURE 7.1. Dependence of surface coverage on phase-boundary potential; (*a*) and corresponding depression of interfacial tension (*b*) for absorption modelled by Frumkin isotherm at two values of interaction parameters ($a = 1.0$ (curve 1) and 2.0 (curve 2). Adapted from Figure 1 in Ref. [15].

function of $\Delta_{\mathrm{O}}^{\mathrm{W}}\phi$ through the potential-dependent term $\theta = \Gamma_i/\Gamma_{\mathrm{m}}$. Figure 7.1b shows the decrease in γ corresponding to the adsorption shown in Figure 7.1a. The bell-shaped adsorption thus naturally leads to the dip in the γ vs. y_i curves.

7.2.2. Electrocapillary Curves Based on the Gouy–Chapman Theory

At the interface between O and W, the presence of the electrical double layers on both sides of the interface also causes the variation of γ with $\Delta_{\mathrm{O}}^{\mathrm{W}}\phi$. In the absence of the specific adsorption of ions at the interface, the Gouy–Chapman theory satisfactorily describes the double-layer structure at the interface between two immiscible electrolyte soultions [20,21]. For the diffuse part of the double layer for a z:z electrolyte of concentration c^{W} in the phase W whose permittivity is ε^{W}, the Gouy–Chapman theory [22,23] gives an expression

$$\phi^{\mathrm{W}} - \phi_2^{\mathrm{W}} = \frac{2RT}{zF}\sinh^{-1}\left[\frac{q^{\mathrm{W}}}{(8RT\varepsilon^{\mathrm{W}}c^{\mathrm{W}})^{1/2}}\right] \tag{10}$$

where ϕ_2^{W} in the potential at the outer Helmholtz plane in W and q^{W} is the excess surface charge density in W. Similarly, for the diffuse part of the double layer in O,

$$\phi_2^{\mathrm{O}} = \frac{2RT}{zF} \sinh^{-1}\left[\frac{q^{\mathrm{W}}}{(8RT\varepsilon^{\mathrm{O}}c^{\mathrm{O}})^{1/2}}\right] \tag{11}$$

where ϕ_2^{O} is the potential at the outer Helmholtz plane in O.

The total phase-boundary potential is composed of three parts:

$$\Delta_{\mathrm{W}}^{\mathrm{O}}\phi = (\phi^{\mathrm{W}} - \phi_2^{\mathrm{W}}) + (\phi_2^{\mathrm{W}} - \phi_2^{\mathrm{O}}) + (\phi_2^{\mathrm{O}} - \phi^{\mathrm{O}}) \tag{12}$$

Once we know the value of the inner potential difference, $\phi_2^{\mathrm{W}} - \phi_2^{\mathrm{O}}$, we can correlate $\Delta_{\mathrm{W}}^{\mathrm{O}}\phi$ with q^{W} using Equations (10) and (11) and then calculate the variation of γ with $\Delta_{\mathrm{W}}^{\mathrm{O}}\phi$ due to the presence of the electrical double layers. In the present model, we simply neglect $\phi_2^{\mathrm{W}} - \phi_2^{\mathrm{O}}$, which is known to be small and shows no strong dependence on $\Delta_{\mathrm{W}}^{\mathrm{O}}\phi$ [24].

For further simplification, we calculate γ for the case when $\varepsilon^{\mathrm{W}}c^{\mathrm{W}} = \varepsilon^{\mathrm{O}}c^{\mathrm{O}} = \varepsilon c$. Then we obtain

$$q^{\mathrm{W}} = (8RT\epsilon c)^{1/2} \sinh\left(\frac{zF}{4RT}\Delta_{\mathrm{O}}^{\mathrm{W}}\phi\right) \tag{13}$$

As the electrocapillary equation at constant pressure and temperature has the form

$$-\mathrm{d}\gamma = q^{\mathrm{W}}\mathrm{d}\Delta_{\mathrm{O}}^{\mathrm{W}}\phi + \sum_j \Gamma j \mathrm{d}\mu_j \tag{14}$$

integrating Equation (13) from the potential of zero charge, $\Delta_{\mathrm{O}}^{\mathrm{W}}\phi_{\mathrm{pzc}}$, to $\Delta_{\mathrm{O}}^{\mathrm{W}}\phi$ at constant temperature, pressure and composition gives

$$\gamma_0 - \gamma = \left(\frac{4RT}{zF}\right)(8RT\varepsilon c)^{1/2}\cosh\left(\frac{zF}{4RT}\Delta_{\mathrm{O}}^{\mathrm{W}}\phi\right) \tag{15}$$

Under the assumptions employed, the parabolic shape of electrocapillary curves in this model is expressed as the hyperbolic cosine of $\Delta_{\mathrm{O}}^{\mathrm{W}}\phi$. Electrocapillary curves calculated as 25°C and $z = 1$ using Equation (15) are shown for $\varepsilon c/\varepsilon_0 = 8$ and 80, where ε_0 is the vacuum permittivity, as dashed lines in Figures 7.2a and 7.2b, respectively. It is seen that the curvature of the electrocapillary curves becomes greater with increasing value of εc.

7.2.3. *Electrocapillary Curves in the Presence of the Adsorption of Ionic Surfactant*

When ions specifically adsorb at the interface, the excess surface charge density is divided into three parts, the charges in the two diffuse parts of the double layer, $q^{\mathrm{d,O}}$ and $q^{\mathrm{d,W}}$, and the charge due to the specifically adsorbed ions, q^{σ} [25]. The electroneutrality condition of the entire interfacial region is

$$q^{\mathrm{d,O}} + q^{\sigma} + q^{\mathrm{d,W}} = 0 \tag{16}$$

For showing the presence of the instability window, we here employ a simlified assumption that the variation in γ due to the specific adsorption of ions (Equation (9)) and that

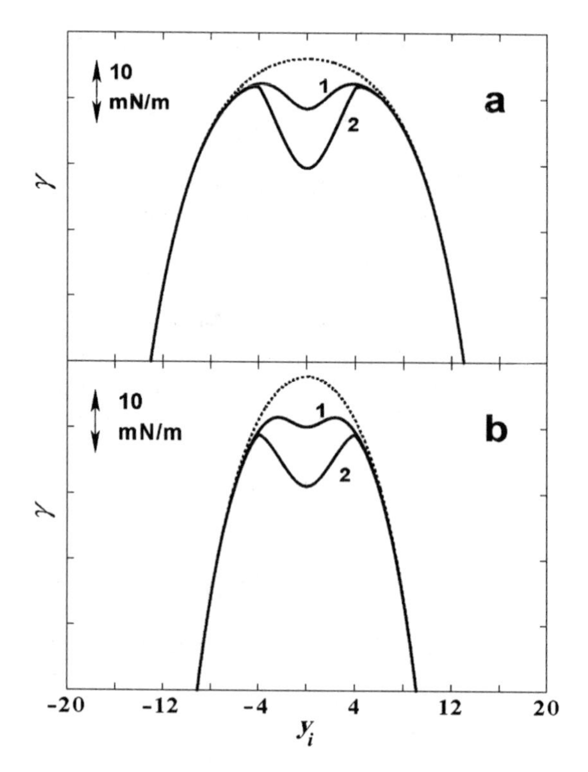

FIGURE 7.2. Electrocapillary curves at two values of εc, 8 (a) and 80 (b), when $\Delta_{\mathrm{O}}^{\mathrm{W}}\phi_{\mathrm{pzc}} = \Delta_{\mathrm{O}}^{\mathrm{W}}\phi^{\ominus}i$ in the absence (dashed lines) and presence (curves 1 and 2) of surfactant adsorption corresponding to the conditions in Figure 7.1. Adapted from Figure 2 in Ref. [15].

due to the presence of the diffuse part of the double layer (Equation (15)) are additive, though, strictly speaking, the specific adsorption of surface-active ions at a given value of $\Delta_{\mathrm{O}}^{\mathrm{W}}\phi$ necessarily alters $q^{\mathrm{d,W}}$ through Equation (16). This approximation may be justified when the system contains sufficient amounts of supporting electrolytes in both phases, while the concentration of surfactant ions is small.

Curves 1 ($a = 1.0$) and 2 ($a = 2.0$) in Figure 7.2a and those in Figure 7.2b were obtained by adding the curves 1 and 2 in Figure 7.1b to the curves in the absence of the specific adsorption (dashed lines in Figures 7.2a and 7.2b) for the special case when $\Delta_{\mathrm{O}}^{\mathrm{W}}\phi_i^{\ominus}$ coincides with the electrocapillary maximum, $\Delta_{\mathrm{O}}^{\mathrm{W}}\phi_{\mathrm{pzc}}$. Resultant electrocapillary curves exhibit dips and the curvature of the electrocapillary curves becomes positive in the middle of the dips both in Figures 7.2a and 7.2b.

As the second derivative of γ with respect to $\Delta_{\mathrm{W}}^{\mathrm{O}}\phi$ is the double layer capacitance C_{dl}, that is,

$$C_{\mathrm{dl}} = \left(\frac{\partial q^{\mathrm{W}}}{\partial \Delta_{\mathrm{O}}^{\mathrm{W}}\phi}\right)_{T,P,\mu_j} = -\left(\frac{\partial^2 \gamma}{\partial \Delta_{\mathrm{O}}^{\mathrm{W}}\phi^2}\right)_{T,P,\mu_j} \tag{17}$$

the positive curvature in the middle of the dips in Figure 7.2 means that C_{dl} is negative.

7.2.4. Electrochemical Instability

The positive curvature is not realized in actual systems from the thermodynamic requirement for the stability of a system, that is, the second derivative of the excess Gibbs energy, $\Delta G^{\sigma} = \gamma$, with respect to $\Delta_{\mathrm{O}}^{\mathrm{W}}\phi$ variable be negative, at T and P constant:

$$C_{\mathrm{dl}} = -\left(\frac{\partial^2 \gamma}{\partial \Delta_{\mathrm{O}}^{\mathrm{W}}\phi^2}\right)_{T,P,\mu_j} = -\left(\frac{\partial^2 \Delta G^{\sigma}}{\partial \Delta_{\mathrm{O}}^{\mathrm{W}}\phi^2}\right)_{T,P,\mu_j} > 0 \tag{18}$$

We call the instability of the interface caused by the violation of this stability condition the electrochemical instability. This connotes the stability intrinsic to the electrochemical interfaces where the phase-boundary potential contributes to the energy state of the interface.

The relationships between γ, q^{W} and C_{dl} are shown in Figure 7.3 as a function of y_i. The solid line in Figure 7.3a is the same as curve 1 in Figure 7.2a, and the dashed line and solid line in Figure 7.3b are the corresponding slope and curvature shown in the unit of charge density and capacitance, respectively. In this particular case, the capacitance becomes negative when $|y| < 1.6$, that is, there is an instability window of about 82 mV at room temperature for a 1:1 eletrolyte. In actual systems, the system escapes from the instability by dissipating energy through emulsification or other Marangoni-type movements of the interface [14,26]. There are two remarkable features in this instability; first, the region where the interface becomes unstable exists as a potential window, i.e., the instability window, on the axis of the phase-boundary potential, and, second, the interface becomes unstable even when the interfacial tension has a large positive value.

7.2.5. Width and Location of the Instability Window

The width of the instability window depends on three factors, the magnitude of $\Delta\gamma_i$ due to the surfactant adsorption, the curvature of an electrocapillary curve, and the relative location of $\Delta_{\mathrm{O}}^{\mathrm{W}}\phi_i^{\ominus}$ with respect to $\Delta_{\mathrm{O}}^{\mathrm{W}}\phi_{\mathrm{pzc}}$. The quantity $\Delta\gamma_i$ is a function of $B_i^{0,\ominus}m_i$, i.e., $\Delta G_{\mathrm{ads},i}^{\ominus}$ and the amount of ionic surfactant, and the form of the adsorption isotherm, in addition to r. The effect of the relative location of $\Delta_{\mathrm{O}}^{\mathrm{W}}\phi_{\mathrm{pzc}}$ and $\Delta_{\mathrm{O}}^{\mathrm{W}}\phi_i^{\ominus}$ on the instability window is illustrated in Figure 7.4 at two different values of $\varepsilon c/\varepsilon_0$, 0.8 (0.01 mol $\cdot$ dm^{-3} in water: Figure 7.4a) and 80 (1.0 mol $\cdot$ dm^{-3} in water: Figure 7.4b) for four values of $B_i^{0,\ominus}m_i$: 10 (curves 1), 1 (curves 2), 0.1 (curves 3) and 0.01 (curves 4), together with the curve with no adsorption (base). When $\Delta_{\mathrm{O}}^{\mathrm{W}}\phi_i^{\ominus} - \Delta_{\mathrm{O}}^{\mathrm{W}}\phi_{\mathrm{pzc}} = 4RT/F$ (a and b) and the supporting electrolyte concentration is low, the instability window exists unless the concentration of the surfactant is low or its surface activity is very weak (curve 4 in Figure 7.4a). In contrast, when the supporting electrolyte concentration is high (Figure 7.4b), no positive curvature develops in the electrocapillary curves, even though the ion is surface-active. The shape of the curve resembles archetypal electrocapillary curves of mercury electrodes in the presence of the specific adsorption of anions [25].

The interface is stable in the entire range of the applied potential or, more realistically, the interfacial tension easily becomes zero with applying the potential, possibly resulting in the interfacial instability of another kind at $\gamma \simeq 0$, where the electrocapillary emulsification [27] may take place.

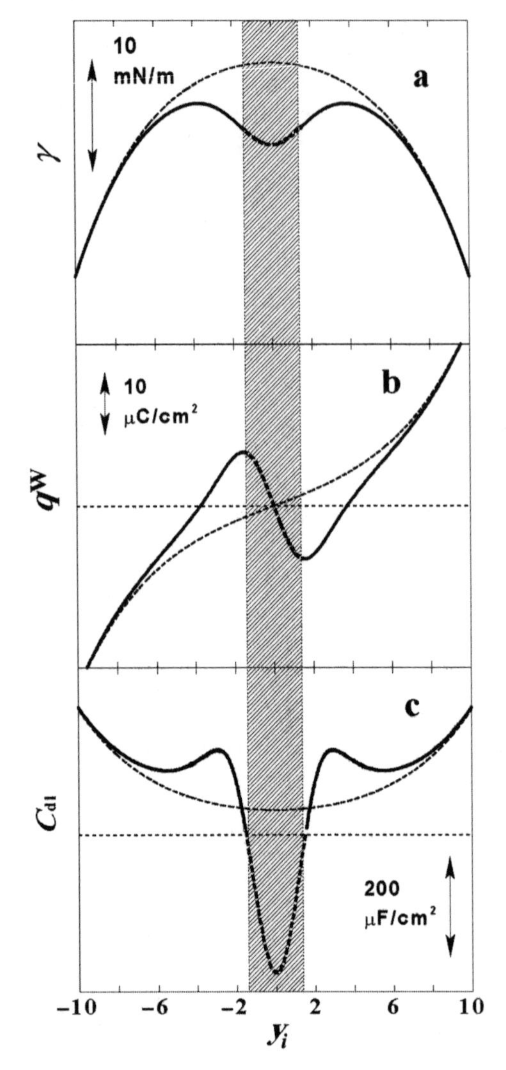

FIGURE 7.3. Illustration of the instability window in electrocapillary curve (a), excess surface charge density vs. potential curve (b) and double-layer capacitance vs. potential curve (c) in the absence (dashed lines) and the presence (solid lines) of the adsorption and partition of an ionic surfactant. Parameters used for calculation are the same as those in Figure 7.1. Shaded region shows the potential range where the system is thermodynamically unstable. Horizontal dashed lines in the middle of b and c represent the lines of zero surface charge and of zero capacitance, respectively. See text for parameters used for the calculation. Adapted in Figure 1 in Ref. [14].

It is seen that an ionic surfactant that has $\Delta_{\mathrm{O}}^{\mathrm{W}}\phi_i^{\ominus}$ close to $\Delta_{\mathrm{O}}^{\mathrm{W}}\phi_{\mathrm{pzc}}$ is the most powerful in destabilizing the system. Even if the surface activity is strong, the electrochemical destabilization would be small for surfactants whose $\Delta_{\mathrm{O}}^{\mathrm{W}}\phi_i^{\ominus}$ is distant from $\Delta_{\mathrm{O}}^{\mathrm{W}}\phi_{\mathrm{pzc}}$.

The location of $\Delta_{\mathrm{O}}^{\mathrm{W}}\phi_{\mathrm{pzc}}$ is usually in the middle of the potential window for a polarized interface between the immiscible electrolyte solutions [21]. It is expected therefore

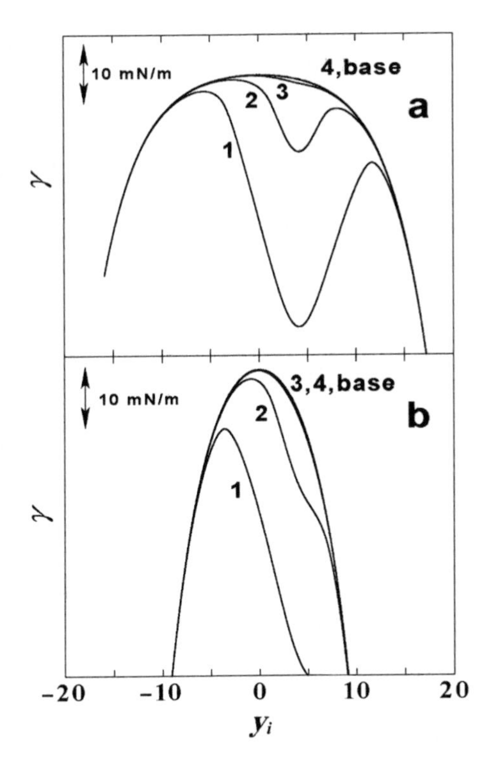

FIGURE 7.4. Effect of the location of $\Delta_O^W \phi_i^\ominus$ with respect to $\Delta_O^W \phi_{pzc}$ at four different values of $m_i B^{0,\ominus}$, 10 (curves 1), 1 (curves 2), 0.1 (curves 3) and 0.01 (curves 4) for two concentrations of supporting electrolytes, 0.01 mol · dm^{-3} (a) and 1.0 mol · dm^{-3} (b). Adapted from Figure 5 in Ref. [15].

that ionic surfactants having the values of $\Delta_O^W \phi_i^\ominus$ in the middle of the potential window are the most powerful for spontaneous emulsification, provided that the phase-boundary potential is optimized, that is, located close to $\Delta_O^W \phi_{pzc}$. In spontaneous emulsification, the distribution potential would be the most important factor in determining the phase-boundary potential. The type of the counterions is then crucial in determining the degree of the emulsification through their $\Delta_O^W \phi_i^\ominus$ values.

The concentration of the supporting electrolyte can also affect the width of the window, as can be seen in Figures 7.4a and 7.4b (see also Figures 7.2a and 7.2b). The value of 8 for $\varepsilon c/\varepsilon_0$ correspond to 0.1 mol· dm^{-3}1:1 electrolyte in water and 0.28 dm^{-3}1 : 1 electrolyte in nitrobenzene. The higher the concentration of the supporting electrolyte, the stronger the curvature of the electrocapillary curve, leading to shallower dents. The width of the instability window therefore becomes narrower with increasing concentration of the supporting electrolytes; the system becomes more stable.

7.2.6. *Electrochemical Instability Under Current Flow*

The concept of the electrochemical instability is based on the thermodynamic arguments. In most of experimental studies of electrochemical instability, however, voltammetry has been conveniently employed [14,26,28]. The concept of the electrochemical instability also applies to this voltammetric or other conditions where the net flow of ions across the interface exists.

The ion transfer across the liquid/liquid interface is fast in comparison with the mass transfer at least at the planar interface in conventional timescale of longer than millisecond; the ion transfer can be treated as an electrochemically reversible process [24,29]. The adsorption also seems to be fast under the same condition [14]. Thus it is a good approximation to assume the Nernst equation for the surface concentrations $c_i^{\mathrm{O}}(x=0)$ and $c_i^{\mathrm{W}}(x=0)$:

$$\Delta_{\mathrm{O}}^{\mathrm{W}}\phi = \Delta_{\mathrm{O}}^{\mathrm{W}}\phi_i^0 + \frac{RT}{z_i F}\ln\frac{c_i^{\mathrm{O}}(x=0)}{c_i^{\mathrm{W}}(x=0)}, \tag{19}$$

and a certain adsorption isotherm, for example, the Langmuir isotherm, also for the surface concentration:

$$B_i^{\alpha} c_i^{\alpha}(x=0) = \frac{\theta_i}{1-\theta_i} \tag{20}$$

Here, x is the distance along the axis normal to the interface at which x is taken to be zero and is directing towards the bulk of the W phase.

The mass balance of ion i is expressed as

$$D_i^{\mathrm{W}} \frac{\partial c_i^{\mathrm{W}}}{\partial x}\bigg|_{x=0} - \frac{\partial \Gamma_i}{\partial t} = D_i^{\mathrm{O}} \frac{\partial c_i^{\mathrm{O}}}{\partial x}\bigg|_{x=0} \tag{21}$$

where D_i^{α} is the diffusion coefficient of i in $\alpha(\alpha = \mathrm{W}$ or O). It can be shown that this continuity equation is basically the same as the mass balance in the equilibrium partition of i in the two phases [18]. In this case the ratio of the diffusion layer thickness, $\delta_i^{\mathrm{O}}/\delta_i^{\mathrm{W}}$, is the counterpart of $V^{\mathrm{O}}/V^{\mathrm{W}}$ in the partition. It is expected therefore that the arguments developed above basically hold also under the current flow. We can quantitatively estimate the time-dependent thickness of the diffusion layer by solving the diffusion equation under proper initial and boundary conditions including Equations (19)–(21).

Figure 7.5 shows the variation of the surface coverage of i with E on the forward (solid line) and reverse (dashed line) scans of E for two sets of parameters: $\beta_i = 0.5$, $B_i^0 = 1$, $\Gamma_{\mathrm{m}}\sqrt{z_i F v/RT}/(\sqrt{D^{\mathrm{W}}}{}^{\mathrm{b}}c_i^{\mathrm{W}}) = 10$ (curve 1) and $\beta_i = 0.5$, $B_i^0 = 100$, $\Gamma_{\mathrm{m}}\sqrt{z_i F v/RT}/(\sqrt{D^{\mathrm{W}}}{}^{\mathrm{b}}c_i^{\mathrm{W}}) = 1$ (curve 2), where B_i^0 is B_i^{α} at $E = E^0$, Γ_{m} is the maximum adsorption of i, and ${}^{\mathrm{b}}c^{\mathrm{W}}$ is the bulk concentration of i in W. In this model, $\Delta_{\mathrm{O}}^{\mathrm{W}}\phi_i^0$ is set to 0.

The adsorption reaches a maximum in the vicinity of the interface for both cases. When the surface activity is not strong ($B_i^0 = 1$), the adsorption maximum is located about 50 mV more positive with respect to $\Delta_{\mathrm{O}}^{\mathrm{W}}\phi_i^0$, though the symmetry factor β_i is 0.5. In the reverse scan, the adsorption in weaker. This predicts that the instability window is narrower in this case in the reverse scan, as exemplified by the horizontal dotted line which schematically indicates a hypothetical threshold level for the electrochemical instability.

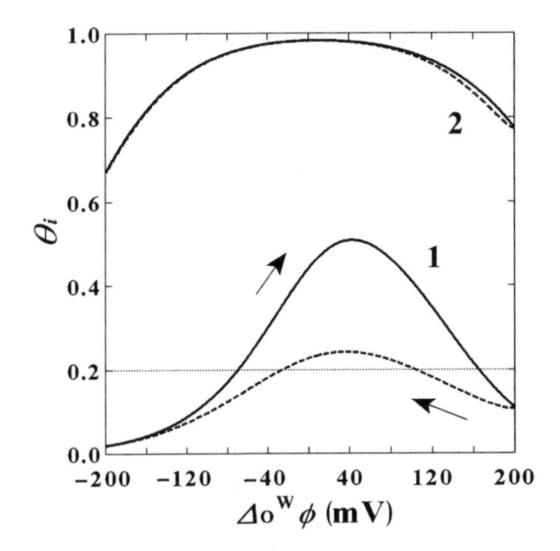

FIGURE 7.5. Dependence of adsorption on E for forward (solid line) and reverse (dashed line) scans calculated assuming that $\beta_i = 0.5$, $B_i^0 = 1$, $\Gamma_m \sqrt{z_i F v / RT} / (\sqrt{D^W}\, b_{c_i}^W) = 10$ (curve 1) and $\beta_i = 0.5$, $B_i^0 = 100$, $\Gamma_m \sqrt{z_i F v / RT} / (\sqrt{D^W}\, b_{c_i}^W) = 1$ (curve 2). Horizontal dashed line illustrates a threshold level for stability of the interface.

When the surface activity is stronger, the surface coverage is high throughout the values of $\Delta_O^W \phi$ calculated (curve 2). The peak becomes dull and there is no big difference between the surface coverages in the forward and reverse scans. Nevertheless, by and large, the results of numerical calculation in Figure 7.5 indicate that the properties of the electrochemical instability under the current flow are essentially the same as those in the thermodynamic equilibrium.

7.3. EXPERIMENTAL FEATURES OF ELECTROCHEMICAL INSTABILITY IN ELECTROCHEMICAL TRANSFER OF IONIC SURFACTANT ACROSS THE LIQUID/LIQUID INTERFACE

7.3.1. *Voltammetry of Anionic Surfactant Transfer*

Figure 7.6a shows cyclic voltammograms for the transfer of octanesulfonate ion, OS^-, at several scan rates at the DCE/W interface when the bulk concentration of OS^- is 0.5 mmol · dm^{-3}. The potential of the Ag/AgCl electrode in W with respect to the potential of the Ag/AgCl electrode in the reference W phase, Wr, is denoted as E. When the scan rate is higher, the voltammograms look like ordinary voltammograms for the diffusion-limited transfer of ions. A flat DCE/W interface was made at the orifice of a glass tube [14]. At slower scan rates, however, the voltammograms occasionally display irregular currents as illustrated by the voltammogram at 10 mV · s^{-1}. This irregular current might look like a noise irrelevant to the ion-transfer process. When the concentration of OS^- is raised, the irregular current appears more reproducibly and the extent of the irregularity also increases dramatically, as exemplified by the

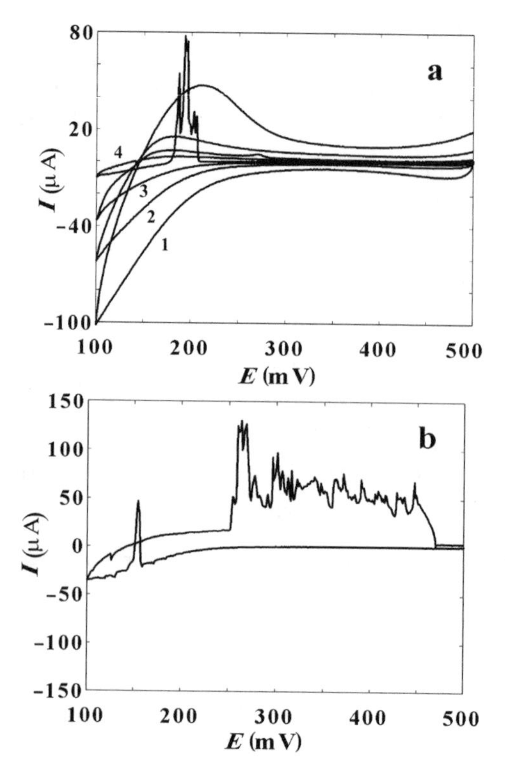

FIGURE 7.6. Cyclic voltammograms for the transfer of octanesulfonate (OS^-) across the DCE/W interface. Compositions of the solutions: W: 50 mmol · dm^{-3} LiCl + 0.5 mmol · dm^{-3} OS^-; DCE: 20 mmol · dm^{-3} TPnATPB; Wr: 5 mmol · dm^{-3} TPnACl + 10 mmol · dm^{-3} $MgCl_2$. Scan rates: 10, 50, 200 and 1000 mV · s^{-1}. b: 2 mmol · dm^{-3} OS^- in W, Scan rate: 10 mV · s^{-1}.

voltammogram in Figure 7.6b recorded at 2 mmol · dm^{-3} OS^- in W at the scan rate of 10 mV · s^{-1}. This irregularity suggests the presence of the electrochemical instability at the interface.

A similar irregularity in voltammograms of ion transfer is exhibited more clearly in the transfer of alkanesulfonates with longer alkyl chains. The solid line in Figure 7.7 shows a voltammogram for the transfer of dodecanesulfonate at the DCE/W interface. The dotted line is a voltammogram recorded after a drop of a DCE solution of sorbitan monooleate, known as a stabilizer against the hydrodynamic turbulences at the liquid|liquid interface, was added in the vicinity of the interface. This voltammogram has a form expected for the diffusion-limited transfer of ions, from which we can determine the mid-point potential E_{mid} of the transfer of dodecanesulfonate to be 320 mV. Comparing the two voltammograms, we can see that the current starts to increase when the potential reaches around E_{mid}. This is in line with one of the predictions of the electrochemical instability that the instability window is located around the standard ion-transfer potential of the transferring ion. Detailed experiments obtained by systematically changing the alkyl chain lengths of alkanesulfonates and alkyl sulfates unequivocally demonstrate that the centre of the instability window is

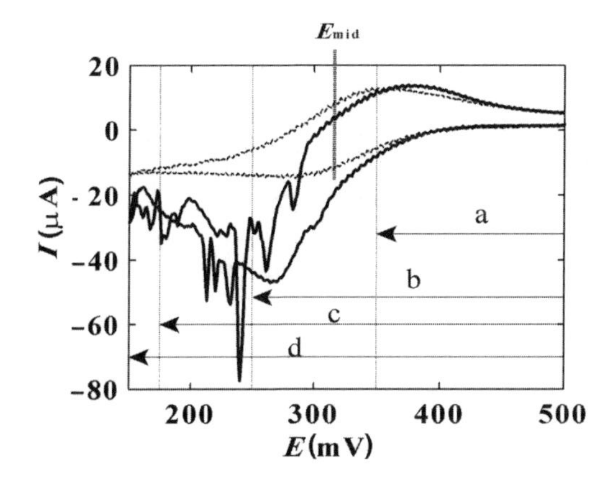

FIGURE 7.7. Cyclic voltammograms of the transfer of dodecanesulfonate ions at 0.5 mmol · dm^{-3} in W in the absence (solid line) and in the presence (dotted line) of sorbitan monooleate. Scan rate: 50 mV · s^{-1}. The location of the mid-point potential is shown as the thick vertical line. Arrows a, b, c and d indicates the potentials stepped in chronoamperometric measurements (Figure 7.8).

located around E_{mid} of the transferring ion, irrespective of the location of E_{mid} itself [14].

7.3.2. Potential-Step Chronoamperometry

Although voltammetry is conveniently used to study the instability, it is not ideal to examine whether the region of E where the interface is unstable is window-like or not, as the hydrodynamic movement of the solution phase due to the interfacial turbulence, once induced, tends to last for some time. The potential-step experiments would then be more suitable because we can make the potential jump over the instability window. Figure 7.8 shows current-transients when E is stepped from 500 mV to four different potentials, 350 (panel a), 250 (panel b), 175 (panel c) and 150 (panel d) mV. Dotted lines are the results when the DCE phase contains sorbitan monooleate for stabilizing the interface. Three solid lines are three independent runs. When E is 350 mV, that is 30 mV more positive to the mid-point potential, 320 mV, four traces are completely overlapped with each other (panel a). When E is stepped beyond the mid-point potential, three traces reproducibly exhibit irregular currents with some spikes (panel b). This irregularity becomes less pronounced but persists at E at 175 mV (panel c). However, when E is stepped to 150 mV, all traces again overlap almost perfectly. This clearly demonstrates that the instability region is window-like and the interface recovers the stability in the limiting current region.

This is very important, because the interface is stable in the presence of the diffusion-limited current, that is, the maximum current available, flowing across the interface. In other words, the flow of current or ions across the interface is not directly responsible for the interfacial turbulences, which in fact makes the strong contrast of the electrochemical instability with the instability associated with interfacial chemical reactions of the type treated within the framework of the linear stability theory [9,10,30].

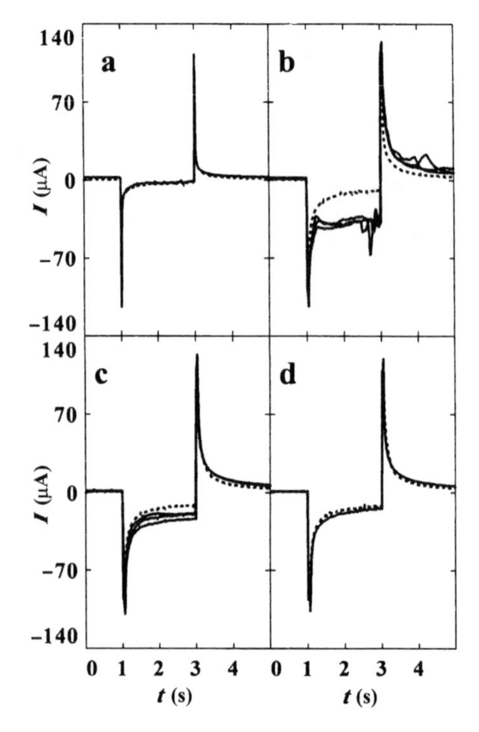

FIGURE 7.8. Potential-step chronoamperometry measurements for the transfer of dodecanesulfonate ions at four different potentials stepped from 500 mV to 350 (a), 250 (b), 175 (c) and 150 (d) mV. Other conditions are the same as those in Figure 7.7. Adapted from Figure 7.6 in Ref. [14].

7.3.3. Electrocapillarity Measurements

Figure 7.9 shows the electrocapillary curves recorded using the video-image processing of a pendant drop of the DCE solution in the W phase in the absence (curve 1) and presence (curve 2) of 1 mmol $\cdot$ dm^{-3} dodecanesulfonate in W. The presence of dodecanesulfonate ions does decrease γ. The values of γ are, however, never lower than 10 mN $\cdot$ m^{-1} over the entire potential range examined. During the recording of the image, the kicking motion of the drop was reproducibly observed in the middle of the potential. Simultaneously, the turbulence occurred inside the drop in the same potential region, indicating that the W/O type emulsion particles are formed in the DCE phase.

Curve 2 provides good evidence for the emulsification at large positive values of γ, which confirms the prediction from the electrochemical instability.

7.4. CONCLUSIONS

The concept of the electrochemical instability has been illustrated using a simple model for the coupling of the adsorption and partition of surface-active ions. Three main features of the electrochemical instability, that is, (1) the location of the instability in the values of the phase-boundary potential, (2) the existence of a window-like instability

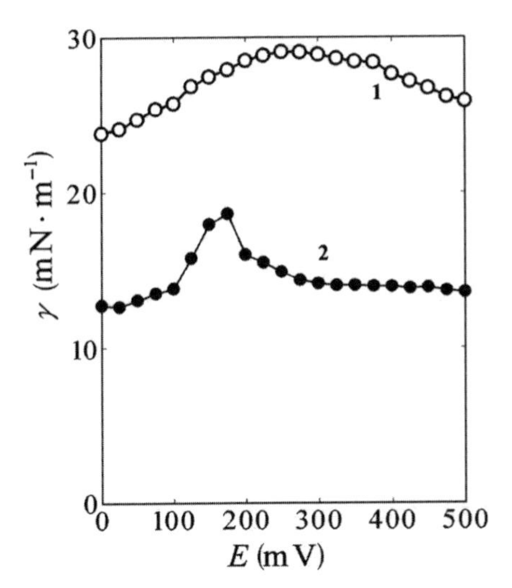

FIGURE 7.9. Electrocapillary curves in the absence (curve 1) and presence (curve 2) of 1 mmol · dm^{-3} in W. Supporting electrolytes: 0.1 mol · dm^{-3} LiCl (W) and 0.1 mol · dm^{-3} tetrapentylammonium tetraphenylborate (DCE). The aqueous solution on the reference side contains 5 mmol · dm^{-3} tetrapentylammonium chloride and 10 mmol · dm^{-3} $MgCl_2$.

zone in the phase-boundary potential and (3) the instability at a finite positive value of the interfacial tension, have all been confirmed experimentally for the transfer of anionic surfactant across the 1,2-dichloroethane/water interface. Similar electrochemical instability has been experimentally verified, recently, also in the transfer of cationic surfactants [28] and the facilitated transfer of metal ions with surface-active ligand at the liquid/liquid interfaces [31]. The electrochemical instability is a counterpart of the phase transitions associated with the pV, ST and $n\mu$ terms. In this respect, this newly introduced concept associated with the electrical energy of the interface would widen our views on the thermodynamic stability of not only macroscopic liquid/liquid two-phase systems but also a variety of other systems where the presence of the interface is decisive in determining their physicochemical properties.

ACKNOWLEDGEMENTS

The author thanks Minako Chiba, Naoya Nishi and Takuya Kasahara for the use of their unpublished data shown as Figures 7.6, 7.7 and 7.9.

REFERENCES

1. J. Gad, Arch. Anat. Physiol. 181 (1878).
2. G. Quincke, Ann. Phys. Chem. **35**, 561 (1888).
3. J.W. McBain and T.M. Woo, Proc. R. Soc. A **163**, 182 (1937).
4. A. Kaminski and J.W. McBain, Proc. R. Soc. A **198**, 447 (1949).

5. M. Dupeyrat and J. Michel, Experientia. Suppl. **18**, 269 (1971).
6. K.Yoshikawa and Y. Matsubara, J. Am. Chem. Soc. **105**, 5967 (1983).
7. J.B. Lewis and H.R.C. Pratt, Nature **171**, 1155 (1953).
8. J.B. Lewis, Chem. Eng. Sci. **8**, 295 (1958).
9. C.V. Sterling and L.E. Scriven, AIChEJ. **5**, 514 (1959).
10. P. Colinet, J.C. Legros and M.G. Velarde, *Nonlinear Dynamics of Surface-Tension-Driven Instabilities*, Wiley-VCH, Berlin, 2001.
11. D. Homolka and V. Mareček, J. Electroanal. Chem. **112**, 91 (1980).
12. P. Vanysek, J. Electroanal. Chem. **121**, 149 (1981).
13. J. Heyrovsky and J. Kuta, *Principles of Polarography*, Academic Press, New York, 1966, Chap. 19.
14. T. Kakiuchi, N. Nishi, T. Kasahara and M. Chiba, ChemPhysChem. **4**, 179 (2003).
15. T. Kakiuchi, J. Electroanal. Chem. **536**, 63 (2002).
16. T. Kakutani, Y. Nishiwaki and M. Senda, Bunseki Kagaku **33**, E175 (1984).
17. Y. Yoshida, M. Matsui, O. Shirai, K. Maeda and S.Kihara, Anal. Chim. Acta. **373**, 213 (1998).
18. T. Kakiuchi, J. Elertroanal. Chem. **496**, 137 (2001).
19. A.N. Frumkin, Z. Phys. Chem. **116**, 466 (1925).
20. M. Gros, S. Gromb and C. Gavach, J. Electroanal. Chem. **89**, 29 (1978).
21. T. Kakiuchi and M. Senda, Bull. Chem. Soc. Jpn. **56**, 1753 (1983).
22. G. Gouy, Ann. Chim. Phys. [9], **7**, 129 (1917).
23. D.L. Chapman, Philos. Mag. **25**, 475 (1913).
24. Z. Samec and T. Kakiuchi, in *Advances in Electrochemistry and Electrochemical Science*, Eds. H. Gerischer and C.W. Tobias, VCH, Weinheim, 1995, Vol. 4, p. 297.
25. D.C. Grahame, Chem. Rev. **41**, 441 (1947).
26. T. Kakiuchi, M. Chiba, N. Sezaki and N. Nakagawa, Electrochem. Commun. **4**, 701 (2002).
27. A. Watanabe, K. Higashituji and K. Nishizawa, J. Colloid Interface Sci. **64**, 278 (1978).
28. T. Kasahara, N. Nishi, M. Yamamoto and T. Kakiuchi, Langmuir **20**, 875 (2004).
29. N. Nishi, K. Izawa, M. Yamamoto and T. Kakiuchi, J. Phys. Chem. B **105**, 8162 (2001).
30. M.A. Mendes-Tatsis and E.S.P. de Ortiz, Chem. Eng. Sci. **51**, 3755 (1996) .
31. T. Kakiuchi, J. Electroanal. Chem. **569**, 287 (2004).

8

Electron Transfer at Liquid/Liquid Interfaces

Toshiyuki Osakai and Hiroki Hotta
Kobe University, Nada, Kobe 657-8501, Japan

8.1. INTRODUCTION

The study of electron transfer (ET) at the polarized oil (O)/water (W) (or liquid/ liquid) interface is useful for understanding not only certain catalytic reactions in two-phase systems (e.g., liquid membranes, microemulsions, micelles, etc.) but also energy conversion processes occurring at biomembranes. In 1979, Samec *et al.* [1,2] reported, as the first example, an ET between ferrocene (Fc) in nitrobenzene (NB) and $Fe(CN)_6^{3-}$ in W:

$$Fc(NB) + Fe(CN)_6^{3-}(W) \rightleftharpoons Fc^{+}(NB) + Fe(CN)_6^{4-}(W) \tag{1}$$

A well-defined voltammetric wave with positive- and negative-current peaks due to the forward and backward ET processes was observed. However, the possibility was suggested [3] that the reaction mechanism for this ET system is not simple, but involves ion transfer (IT) of the product, i.e., Fc^{+} (see the "IT mechanism" described below).

In 1988, Schiffrin's group [4] claimed that the use of lutetium biphthalocyanine ($LuPc_2$) complex more hydrophobic than Fc enabled observation and investigation of a "true", or heterogeneous, ET across the O/W interface. A series of subsequent papers reported ET systems by other hydrophobic metal complexes including tin diphthalocyanine [5], iron and ruthenium metalloporphyrin complexes with pyridine [6] and Fc derivatives [7]. These experimental studies then stimulated theoretical studies on the ET kinetics [8–12].

On the other hand, Kihara's group reported interesting ET systems for biological molecules including L-ascorbic acid [13], flavin mononucleotide (FMN) [14] and β-nicotinamide adenine dinucleotide (NADH) [15]. While these ET systems are very important from a biological viewpoint, their reaction mechanisms are often complicated by the coupling of ET and proton or ion transfer.

A recent introduction of scanning electrochemical microscopy (SECM) to this field [16–23] has revitalized the study of ET at the O/W interface. In contrast to the conventional, four-electrode cyclic voltammetry at externally polarized O/W interfaces, the SECM measurements not necessarily require supporting electrolytes, and thus can be carried out over a wide range of driving forces without the limitation of the potential window. This advantage of SECM allowed for an experimental verification of the Marcus theory in the driving-force dependence of the ET rate constant [18,21].

This chapter is focused on the reaction mechanisms of ET at O/W interfaces, and features recent developments of theories and methodologies. The reader is also referred to some excellent reviews [24–27] on the similar subject. This chapter does not cover photo-induced ET reactions, which are described in detail in the review of Fermń and Lahtinen [26].

8.2. ELECTRON TRANSFER AT THE POLARIZABLE O/W INTERFACE

8.2.1. *Principles*

Let us consider an ET reaction between a hydrophilic redox couple (O1/R1) in the W phase and a hydrophobic redox couple (O2/R2) in the O phase:

$$\mathrm{R2(O) + O1(W) \rightleftharpoons O2(O) + R1(W)} \tag{2}$$

in which n electrons are involved. If this ET system is in equilibrium, the Galvani potential difference ($\Delta_{\mathrm{O}}^{\mathrm{W}}\phi$) of the O/W interface is given by the Nernst equation:

$$\Delta_{\mathrm{O}}^{\mathrm{W}}\phi = \Delta_{\mathrm{O}}^{\mathrm{W}}\phi_{\mathrm{ET}}^{\circ'} + \frac{RT}{nF}\ln\left(\frac{[\mathrm{R1}]_{\mathrm{W}}[\mathrm{O2}]_{\mathrm{O}}}{[\mathrm{O1}]_{\mathrm{W}}[\mathrm{R2}]_{\mathrm{O}}}\right) \tag{3}$$

where the brackets $[\,]_{\mathrm{W}}$ and $[\,]_{\mathrm{O}}$ stand for the equilibrium concentrations of redox species in the W and O phases, respectively; R, T and F have their usual meanings; and $\Delta_{\mathrm{O}}^{\mathrm{W}}\phi_{\mathrm{ET}}^{\circ'}$ is the formal potential which is determined by

$$\Delta_{\mathrm{O}}^{\mathrm{W}}\phi_{\mathrm{ET}}^{\circ'} = \Delta_{\mathrm{O}}^{\mathrm{W}}\phi_{\mathrm{ET}}^{\circ} + \frac{RT}{nF}\ln\left(\frac{\gamma_{\mathrm{R1}}^{\mathrm{W}}\gamma_{\mathrm{O2}}^{\mathrm{O}}}{\gamma_{\mathrm{O1}}^{\mathrm{W}}\gamma_{\mathrm{R2}}^{\mathrm{O}}}\right) \tag{4}$$

Here γ represents the activity coefficient of a redox species in the indicated phase, and $\Delta_{\mathrm{O}}^{\mathrm{W}}\phi_{\mathrm{ET}}^{\circ}$ is the standard redox potential for the ET system, being given by the difference between the standard potentials of the respective redox couples that are expressed on a same potential scale:

$$\Delta_{\mathrm{O}}^{\mathrm{W}}\phi_{\mathrm{ET}}^{\circ} = E_{\mathrm{O2/R2}}^{\circ} - E_{\mathrm{O1/R1}}^{\circ} \tag{5}$$

If we choose a proper set of redox couples whose $\Delta_{\mathrm{O}}^{\mathrm{W}}\phi_{\mathrm{ET}}^{\circ}$ lies within the potential window, we can observe the voltammetric wave due to the ET shown in Equation (2). Figure 8.1 shows the formal potentials, $E^{\circ'}(\approx E^{\circ})$, of various redox species in W, NB and 1,2-dichloroethane (DCE) versus normal hydrogen electrode (NHE) [28]. In this figure are shown the data presented in the previous review [26] and other data [6,29–32].

The ET reaction at the polarizable O/W interface thus described is formally similar to that at a metal electrode surface. This allows us to employ usual electrochemical

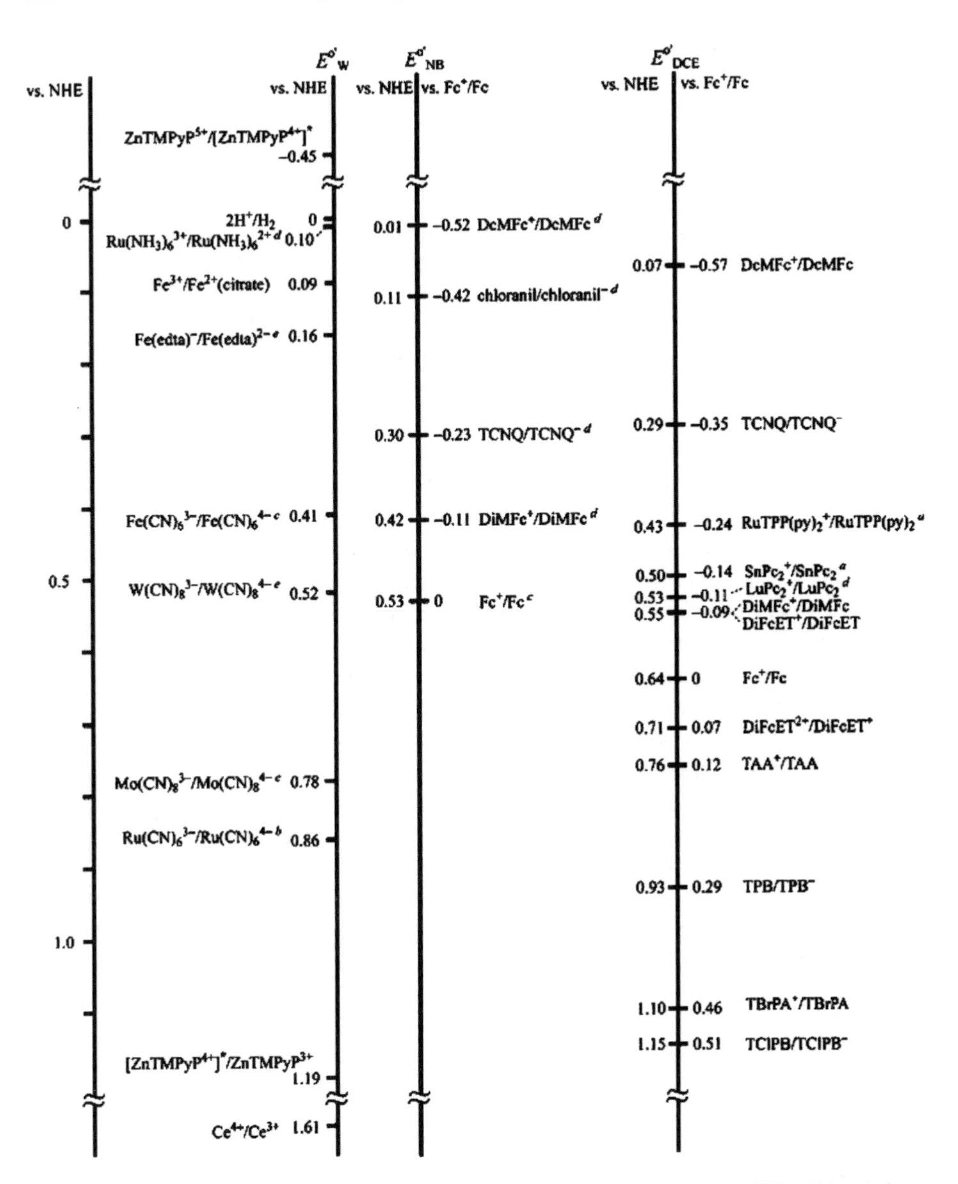

FIGURE 8.1. Formal potentials of redox species in W, NB and DCE versus NHE. Abbreviations: ZnTMPyP = zinc tetra-*N*-methyl-4-pyridium porphyrin, DcMFc = decamethylferrocene, TCNQ = 7,7,8,8-tetracyanoquinodimethane, DiMFc = dimethylferrocene, Fc = ferrocene, $RuTPP(py)_2$ = bis(pyridine) *meso*-tetraphenylporphyrinato ruthenium(II), $SnPc_2$ = tin(IV) diphthalocyanine, $LuPc_2$ = lutetium(III) diphthalocyanine, DiFcET = diferrocenylethane, TAA = tris(4-methoxyphenyl)amine, TPB = tetraphenylborate, TBrPA = tris(4-bromophenyl)amine, TClPB = tetrakis(4-chlorophenyl)borate. The data is from Ref. [26], except where otherwise noted: [a] Ref. [6], [b] Ref. [29], [c] Ref. [30], [d] Ref. [31], [e] Ref. [32]. Reprinted from Ref. [28], with permission from the Polarographic Society of Japan.

techniques including cyclic voltammetry (CV), polarography and a.c. impedance method for the study of ET at O/W interfaces.

8.2.2. *Voltammetric Behaviour*

Now let us assume that the redox couple in one phase (e.g., O1/R1 in W) exists in excess. In this case, the interfacial concentrations of O1 and R1 are kept constant even if the ET reaction proceeds. Then, the Nernst equation (3), which is valid for reversible ET systems, is reduced to $\Delta_O^W\phi = \text{const.} + (RT/nF)\ \ln([O2]_O/[R2]_O)$. This expression is the same as that for a typical electrode reaction. Thus, the W phase containing O1 and R1 in excess may be regarded as a metal electrode, and the ET reaction is limited by the diffusion of O2 in O from the bulk to the interface. The reversible cyclic voltammogram obtained under these conditions has a peak separation of (59/*n*) mV (at 25°C) in the same manner as that obtained for a typical electrode reaction.

Unless one of the redox couples exists in excess, however, the shape of a reversible wave in CV is changed by relative concentrations of the redox species. This was first claimed by Girault *et al.* [33]. Figure 8.2 represents reversible cyclic voltammograms (with $n = 1$) calculated for various concentration conditions. Panel (A) shows the case where the concentration of the redox couple in W, $[O1/R1]_W(=[O1]_W = [R1]_W)$, is changed at constant concentrations of the redox couple in O (i.e., $[O2]_O = 0$ mM; $[R2]_O = 1$ mM). When $[O1/R1]_W$ is high enough (i.e., 100 mM), the peak separation is 59 mV as described above. However, it should be noted that the peak separation becomes larger with decreasing $[O1/R1]_W$. Thus, for the ET at the O/W interface, the peak separation of a reversible wave is not necessarily (59/*n*) mV. Panel (B) in Figure 8.2 shows the case where $[R2]_O$ is changed at $[O1]_W = [R1]_W = 1$ mM and $[O2]_O = 0$ mM. The decrease of $[R2]_O$ from 100 mM to 1 mM leads to a shift of the midpoint potential from 71 to 101 mV. Thus, for the ET at the O/W interface, the midpoint potential does not necessarily coincide with the value of $\Delta_O^W\phi_{ET}^{o'}$, being generally influenced by the concentration conditions for redox species.

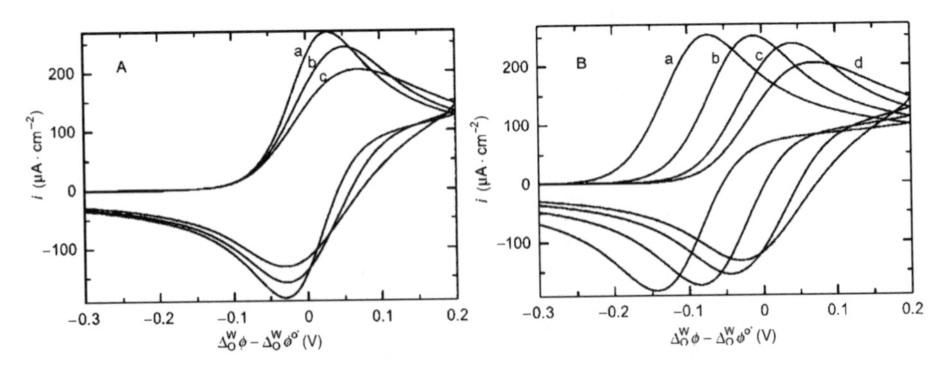

FIGURE 8.2. Reversible cyclic voltammograms of ET at the O/W interface under various concentration conditions. (A): $[O1]_W = [R1]_W =$ (a) 100, (b) 2, (c) 1 mM; $[O2]_O = 0$ mM; $[R2]_O = 1$ mM. (B): $[O1]_W = [R1]_W = 1$ mM; $[O2]_O = 0$ mM; $[R2]_O =$ (a) 100, (b) 10, (c) 2, (d) 1 mM. The diffusion coefficients of all redox species are 1×10^{-5} $cm^2 \cdot s^{-1}$. Sweep rate: 0.1 $V \cdot s^{-1}$. Reprinted from Ref. [28] with permission from the Polarographic Society of Japan.

8.3. NEW METHODOLOGIES

Conventional electrochemical techniques including CV [1–7], current-scan polarography [13–15,34,35] and a.c. impedance method [36,37] have so far been used for the study of ET at O/W interfaces. However, recent development in microtechnology brought to an introduction of SECM in this field [16–23]. Our group has also developed a few devices for characterizing ET reactions at O/W interfaces [38,39]. In this section, new methodologies recently introduced in this field are described.

8.3.1. SECM

In the SECM measurements, an ultramicroelectrode tip is used as a probe for ET occurring at an O/W interface. The electrode potential is controlled by a three-electrode potentiostat, whereas the potential drop across the O/W interface is usually determined by adding a common ion to both phases (except for the recent study [23] using an externally polarized interface). This sets the SECM measurements free from the restriction of the potential window. It should also be noted that in ordinary SECM measurements, all electrodes are in a single phase, so that it is possible to avoid the problems of *IR* drop and charging current. These advantages of SECM have been realized in kinetic studies of ET at O/W interfaces.

As another application of microtechnology, a microelectrochemical technique by laser trapping of a single oil droplet was developed by Nakatani *et al.* [40–42].

8.3.2. ECSOW System

Hotta *et al.* [38] have developed a new electrochemical device for studying ET at the O/W interface, in which the O and W phases are separated by an electron conductor (e.g., Pt). This system is named as electron-conductor separating oil–water (ECSOW) system. As shown in Figure 8.3, the EC phase that separates the O and W phases can be feasibly realized by connecting two platinum disk electrodes with an electric wire.

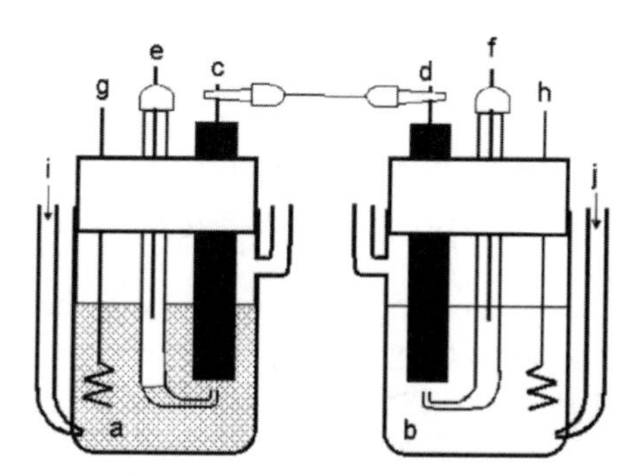

FIGURE 8.3. Electrolytic cell for the ECSOW system: (a) O phase; (b) W phase; (c), (d) platinum disk electrodes connected with an electric wire; (e), (f) reference electrodes with Luggin capillaries; (g), (h) platinum coil electrodes; (i), (j) N_2 gas inlet. Reprinted from Ref. [38], with permission from Elsevier Science.

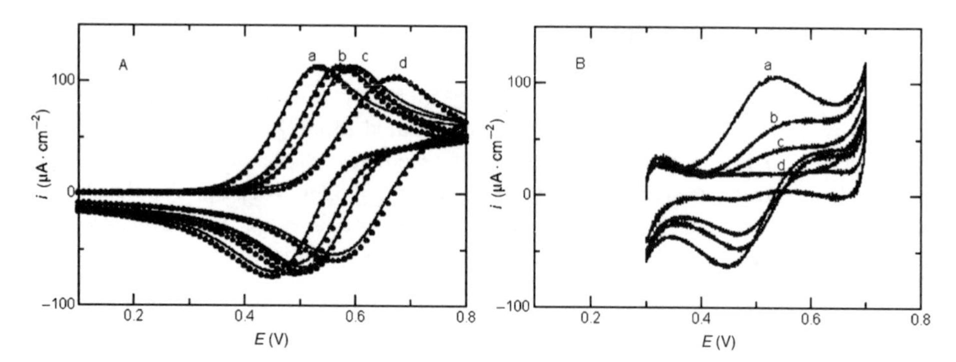

FIGURE 8.4. Cyclic voltammograms observed with (A) the ECSOW system and (B) the corresponding O/W interface in the presence of (a) 100, (b) 20, (c) 10 and (d) 1.0 mM Fc in NB and 0.5 mM $Fe(CN)_6^{3-}$ and 0.1 mM $Fe(CN)_6^{4-}$ in W. Sweep rate: 0.1 V · s^{-1}. The solid circles in panel (A) show the regression data. Reprinted from Ref. [38], with permission from Elsevier Science.

By controlling the Galvani potential difference between the O and W phases in a similar manner to the O/W interface, the ET across the EC phase can be observed voltammetrically. Panel (A) in Figure 8.4 shows, as an example, a set of cyclic voltammograms observed with the ESCOW system in the presence of various concentrations of Fc in NB and 0.5 mM $Fe(CN)_6^{3-}$ and 0.1 mM $Fe(CN)_6^{4-}$ in W. For comparison, cyclic voltammograms obtained at the corresponding O/W interfaces with the same reference electrodes are shown in panel (B). On the respective panels, curve (a) represents a voltammogram observed in the presence of an excess amount (i.e., 100 mM) of Fc in NB. For either the ECSOW system or the O/W interface, a well-defined wave has appeared at the same potential. This shows that the ECSOW system is thermodynamically equivalent with the corresponding O/W interface, as far as only a simple ET as shown in Equation (1) occurs either at the O/W interface or via the EC phase in the ECSOW system.

In Figure 8.4A, the change in the voltammogram with the decrease in $[Fc]_O$ is also shown. As seen in the figure, the midpoint potential shifted to more positive potentials. This change is totally in line with the theoretical prediction shown in Figure 8.2B for the reversible waves of ET at the O/W interface. The solid circles in Figure 8.4A show the regression data obtained by assuming that the ET across the EC phase occurs via two reversible ET reactions at the platinum electrode surfaces:

$$\mathrm{Fc(NB)} \rightleftharpoons \mathrm{Fc^+(NB)} + \mathrm{e^-} \tag{6a}$$

$$\mathrm{Fe(CN)_6^{3-}(W)} + \mathrm{e^-} \rightleftharpoons \mathrm{Fe(CN)_6^{4-}(W)} \tag{6b}$$

On the other hand, the voltammograms observed with the O/W interface showed quite different features. As shown in Figure 8.4B, the positive-current peak was depressed significantly by lowering $[Fc]_O$, suggesting the existence of a kinetically controlled process. Such a difference in the voltammetric behaviour between the ECSOW and O/W systems is due to the fact that no IT can take place in the ECSOW system. The voltammetric behaviour for the O/W interface shown in Figure 8.4B can be elucidated in terms of the IT mechanism described below.

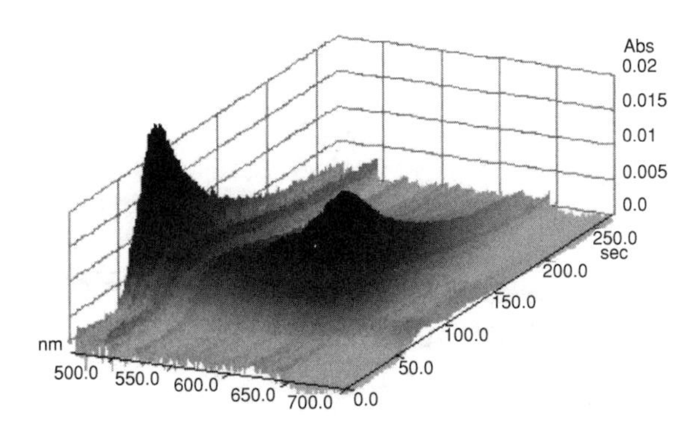

FIGURE 8.5. Spectrum change of the electrolyzed solution for the ET between Fc in NB and $Fe(CN)_6^{3-}$ in W, which was obtained with the microflow cell [39]. Applied potential: 0.15 V. Flow rate: 100 $\mu l \cdot min^{-1}$.

The ECSOW system has also been applied to a biomimetic redox system, i.e., the oxidation of L-ascorbic acid in W by chloranil added to NB [38]. A comparison of the cyclic voltammograms obtained with the ECSOW system and the O/W interface has provided important suggestions on the possible reaction mechanism at the O/W interface. Thus, the ECSOW system would offer important clues to clarify ET processes at O/W interfaces.

8.3.3. *Methods for Product Analysis*

Although product analysis seems essential for the clarification of complex ET processes involving biological molecules, only few attempts have so far been made. Ohde *et al.* [15,35] conducted bulk electrolysis to determine spectrophotometrically some redox products of interfacial ET reactions. Recently, Sawada *et al.* [39] have developed a microflow coulometric cell with a hydrophobic membrane-stabilized O/W interface. This microflow cell can accomplish complete electrolysis, and thus determination of the number of electrons for complex ET reactions at O/W interfaces. Also, its use for an on-line spectrophotometric detection of electrolysis products was made [43]. Figure 8.5 shows the spectrum change of the electrolyzed solution for the ET between Fc in NB and $Fe(CN)_6^{3-}$ in W. When relatively small potentials were applied to the microflow cell, Fc^+ could be detected in the electrolyzed solution. The characteristic absorbance peak at 620 nm showed an undoubted existence of Fc^+ in the W phase as the electrolysis product. This result would also support the IT mechanism. In situ UV-visible spectroscopy [44–46] also deserves attention for its usefulness in product analysis and clarification of reaction mechanisms.

8.4. REACTION MECHANISMS

Reaction mechanisms of ET at the O/W interfaces could be classified into two major categories: IT mechanism and ET mechanism. The former involves an IT of the ionic species produced by a homogeneous ET in one phase (usually the W phase). The latter corresponds to a heterogeneous, i.e., “true” ET across the O/W interface.

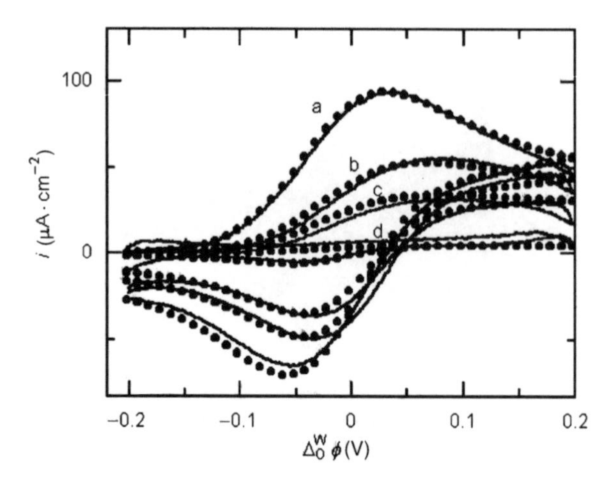

FIGURE 8.6. Base-line corrected cyclic voltammograms of the Fc–$Fe(CN)_6{}^{3-}$ system at the NB/W interface. The concentration conditions for (a)–(d) and sweep rate are as in Figure 8.4B. The solid circles show the regression data obtained by assuming the IT mechanism. Reprinted from Ref. [30], with permission from American Chemical Society.

8.4.1. *IT Mechanism*

Our recent study [30] has revealed that the reaction mechanism of the Fc(NB)–$Fe(CN)_6{}^{3-}$ (W) system is due to the IT mechanism as follows:

$$\mathrm{Fc(NB)} \rightleftharpoons \mathrm{Fc(W)} \tag{7a}$$

$$\mathrm{Fc(W)} + \mathrm{Fe(CN)_6{}^{3-}(W)} \rightleftharpoons \mathrm{Fc^+(W)} + \mathrm{Fe(CN)_6{}^{4-}(W)} \tag{7b}$$

$$\mathrm{Fc^+(W)} \rightleftharpoons \mathrm{Fc^+(NB)} \tag{7c}$$

In this mechanism, the ET (7b) takes place in a homogeneous media (i.e., in W), and the IT (7c) is responsible for the current flowing through the interface.

Figure 8.6 shows the base-line corrected cyclic voltammograms (solid lines) and the regression data obtained by assuming the IT mechanism (solid circles). The calculation of the voltammograms was performed by a normal explicit finite difference digital simulation technique with an exponentially expanding space grid method. As seen in the figure, a good fitting of the experimental values to the theoretical ones could be achieved. The partition coefficient of Fc ($K_D = [Fc]_O/[Fc]_W = 6000$) and the formal potential of the transfer of Fc^+ ($\Delta_O^W \phi_{Fc^+}^{o'} = -0.096$ V), which were used in the curve fitting, had been determined in advance by independent experiments. The forward rate constant of the homogeneous ET (7b), $k_1 = 2.0\times 10^7$ $M^{-1}\cdot s^{-1}$, was obtained as the sole fitting parameter. It should be noted that the same parameter set was successfully employed to achieve good fitting results in the voltammograms observed under any concentration conditions. Thus, cyclic voltammetry combined with a digital simulation technique has appeared to be a powerful method for the mechanistic study of ET at the O/W interface.

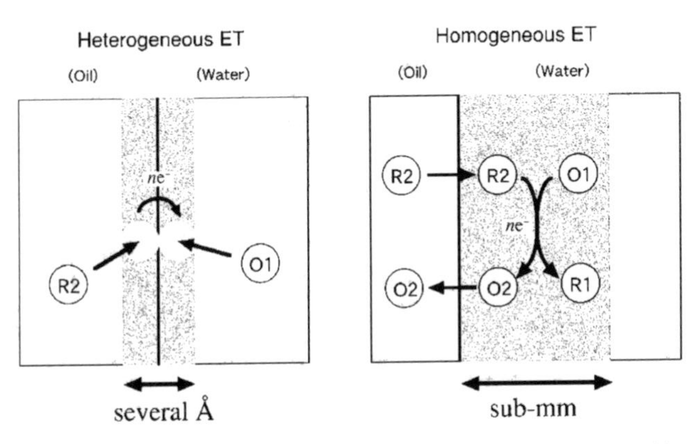

FIGURE 8.7. Comparison of the volume of reaction field between the heterogeneous and homogeneous ETs. The thickness of the reaction field for the homogeneous ET was estimated by a digital simulation analysis for the Fc (NB)–$Fe(CN)_6^{3-}$ (W) system [30]. Reprinted from Ref. [28], with permission from the Polarographic Society of Japan.

Then why does the ET for the Fc–$Fe(CN)_6^{3-}$ system occur not at the interface but in the W phase, in spite of the small partition of Fc into the W phase? One of the reasons is in the difference in the volume of reaction field between the ET and IT mechanisms. In Figure 8.7, the volume of the reaction field for the heterogeneous ET in the ET mechanism is compared with that for the homogeneous ET in the IT mechanism. In the former, the reaction field should be restricted to an interfacial layer as thin as several angstroms, whereas in the latter, it has been found, from the digital simulation analysis, that the reaction layer in the W phase grows up to ~200 μm in the course of the voltammetric sweep of a few seconds. Thus, the volume of the reaction field for the homogeneous ET would be $\sim 10^6$ times larger than that for the heterogeneous ET. Such a large difference in the volume of reaction field leads to the most crucial advantage of the IT mechanism, which would overcome its disadvantage, i.e., the small partition of Fc into W (cf. $K_D = 6000$).

It is evident from these considerations that the use of a less hydrophobic redox species in the O phase makes the homogeneous ET occur more favourably. Another example of the IT mechanism has been found in the ET between L-ascorbic acid in W and chloranil (with $K_D = 900$) in NB or DCE. This has been confirmed using potential-controlled polarography [47], potential modulated reflectance spectroscopy [46], microflow coulometry [39], ECSOW system [38] and digital simulation of cyclic voltammograms [48].

8.4.2. ET Mechanism

It is expected from the above considerations that the true ET can be obtained when a sufficiently hydrophobic redox species is used in the O phase. From this perspective, the previously reported ET systems involving highly hydrophobic complexes such as $LuPc_2$ [4–6] might be considered as heterogeneous ETs. However, the rate of the heterogeneous

ETs should suffer from a geometrical restriction, i.e., the existence of the O/W interface lying between the redox species in the respective phases.

8.4.2.1. Marcus Theory In the early 1990s, Marcus [8] presented a general equation for the second-order rate constant of a heterogeneous ET at the O/W interface:

$$k = Z \exp\left(\frac{-\Delta G^{\ddagger}}{RT}\right) \tag{8}$$

where Z is the pre-exponential frequency factor and $\Delta G^{\ddagger}$ is the standard Gibbs energy of the activation of the reaction, which is related to the standard Gibbs energy of ET (ΔG°) with

$$\Delta G^{\ddagger} = w^{\mathrm{r}} + \frac{\lambda}{4}\left(1 + \frac{\Delta G^{\circ} + w^{\mathrm{p}} - w^{\mathrm{r}}}{\lambda}\right)^2 \tag{9}$$

Here w^{r} and w^{p} are the work terms for bringing the reactants from d (the distance between two reactants) $= \infty$ and for removing the products to $d = \infty$, respectively. These work terms can be estimated based on a dielectric continuum model, from the size and charge of redox species, dielectric constants of solvents, and locations of the redox species. λ is given by the sum of the reorganization energy of "outer sphere", i.e., the solvents (λ_{o}) and that of "inner sphere", i.e., the intramolecular ligands (λ_{i}). The value of λ_{i} can be obtained experimentally from the normal vibrational analysis, whereas the value of λ_{o} can be estimated theoretically as the electrostatic energy, in a similar manner to the work terms. For the equations to calculate w^{r}, w^{p} and λ_{o}, see the original papers by Marcus [8].

Marcus proposed that the frequency factor Z for a sharp O/W interface is given by

$$Z = 2\pi(r_{\mathrm{A}} + r_{\mathrm{B}})\kappa\nu(\Delta R)^3 \tag{10}$$

where r_{A} and r_{B} are the molecular radii of the reactants in O and W, respectively; κ is the transmission coefficient ($\kappa = 1$ for a perfect adiabatic ET); ν is the frequency of molecular motion; and ΔR is the parameter appearing in an exponent for the dependence of the ET rate on separation distance R ($\propto \exp(-R/\Delta R)$). Marcus also estimated the frequency factor for the case in which both reactants are allowed to penetrate the interfacial region in such a way that their centres can effectively reach the liquid boundary:

$$Z \approx \pi(r_{\mathrm{A}} + r_{\mathrm{B}})^3\kappa\nu(\Delta R) \tag{11}$$

If we assume that $r_{\mathrm{A}} + r_{\mathrm{B}} = 10$ Å and $\Delta R = 1$ Å, it can be estimated that the Z value given by Equation (11) is approximately two orders of magnitude larger than the Z value given by Equation (10) for a sharp O/W interface.

In general, the phenomenological Butler–Volmer expression is assumed for the forward (k_{f}) and backward (k_{b}) rate constants of the ET reaction shown by Equation (2):

$$k_{\mathrm{f}} = k^{\circ} \exp\left[\frac{(1-\alpha)nF}{RT}\left(\Delta_{\mathrm{O}}^{\mathrm{W}}\phi - \Delta_{\mathrm{O}}^{\mathrm{W}}\phi_{\mathrm{ET}}^{\circ\prime}\right)\right] \tag{12a}$$

$$k_{\mathrm{b}} = k^{\circ} \exp\left[-\frac{\alpha nF}{RT}\left(\Delta_{\mathrm{O}}^{\mathrm{W}}\phi - \Delta_{\mathrm{O}}^{\mathrm{W}}\phi_{\mathrm{ET}}^{\circ\prime}\right)\right] \tag{12b}$$

where k° is the standard rate constant at $\Delta_O^W\phi = \Delta_O^W\phi_{ET}^{\circ\prime}$, and α the transfer coefficient. These kinetic parameters can be evaluated based on the Marcus theory.

Regarding the evaluation of k°, the value of $\Delta G^\ddagger$ is first calculated by setting $\Delta G^\circ = -nF(\Delta_O^W\phi - \Delta_O^W\phi_{ET}^\circ) = 0$ in Equation (9). Using this $\Delta G^\ddagger$ value and the Z value calculated with Equation (10) or (11), the value of k at $\Delta_O^W\phi = \Delta_O^W\phi_{ET}^\circ \approx \Delta_O^W\phi_{ET}^{\circ\prime}$, i.e., k° can be obtained.

For the evaluation of α, we substitute $\Delta G^\circ = -nF(\Delta_O^W\phi - \Delta_O^W\phi_{ET}^\circ)$ in Equation (9) to obtain

$$\Delta G^\ddagger = \Delta G^{\circ\ddagger} - \frac{1}{2}nF(\Delta_O^W\phi - \Delta_O^W\phi_{ET}^\circ) + \frac{1}{4\lambda}\left\{n^2F^2(\Delta_O^W\phi - \Delta_O^W\phi_{ET}^\circ)^2 - 2nF(\Delta_O^W\phi - \Delta_O^W\phi_{ET}^\circ)(w^p - w^r)\right\} \quad (13)$$

where $\Delta G^{\circ\ddagger}$ stands for $\Delta G^\ddagger$at $\Delta_O^W\phi = \Delta_O^W\phi_{ET}^\circ$. When $nF|\Delta_O^W\phi - \Delta_O^W\phi_{ET}^\circ|/2\lambda \ll 1$, Equation (13) is reduced to

$$\Delta G^\ddagger = \Delta G^{\circ\ddagger} - \frac{1}{2}\left(1 + \frac{w^p - w^r}{\lambda}\right)nF\left(\Delta_O^W\phi - \Delta_O^W\phi_{ET}^\circ\right) \quad (14)$$

Comparing this equation with the following relation in the classical Butler–Volmer formalism (note that $\Delta_O^W\phi_{ET}^\circ \approx \Delta_O^W\phi_{ET}^{\circ\prime}$),

$$\Delta G^\ddagger = \Delta G^{\circ\ddagger} - (1-\alpha)nF\left(\Delta_O^W\phi - \Delta_O^W\phi_{ET}^{\circ\prime}\right) \quad (15)$$

and using the values of w^r, w^p and λ, we can obtain the value of α.

8.4.2.2. Diffusion-Controlled Rate Constant Recently, we have calculated the diffusion-controlled (i.e., attainable maximum) rate constant of ET at an O/W interface [49]. Figure 8.8 shows models for diffusion-controlled bimolecular reactions (a) in homogeneous solution and (b) at an O/W interface.

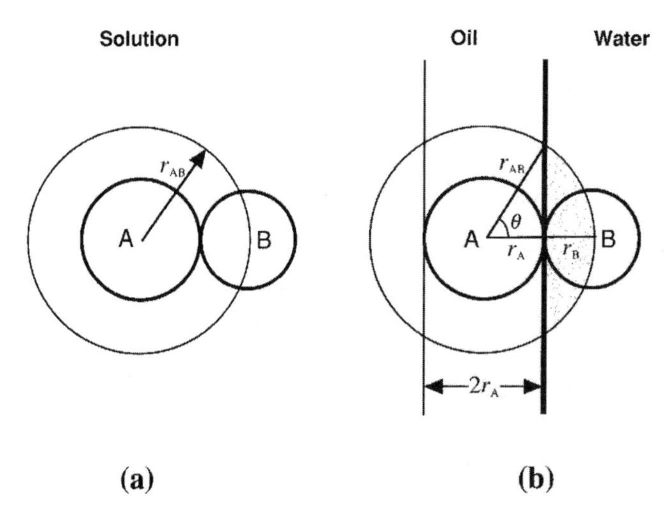

FIGURE 8.8. Models for diffusion-controlled bimolecular reactions (a) in homogeneous solution and (b) at an O/W interface. For details, see the text. Reprinted from ref [49], with permission from Elsevier Science.

In homogeneous solution, molecule B can diffuse from all directions to react with molecule A. In a diffusion-controlled reaction, when the contact of molecules A and B immediately yields the products, the molar flux of B, $j'_{\mathrm{B,hom}}$, across the spherical surface of the radius of r_{AB} $(= r_{\mathrm{A}} + r_{\mathrm{B}})$ around the centre of molecule A can be given by

$$j'_{\mathrm{B,hom}} = 4\pi r_{\mathrm{AB}} D_{\mathrm{AB}}[\mathrm{B}] \tag{16}$$

where D_{AB} is the relative diffusion coefficient that is assumed to be given by the sum of the diffusion coefficients of A and B (i.e., $D_{\mathrm{AB}} = D_{\mathrm{A}} + D_{\mathrm{B}}$) because of the simultaneous movement of both molecules. On this basis, Smoluchowski [50] and later Debye [51] provided an equation for the diffusion-controlled second-order rate constant as

$$k_{\mathrm{D,hom}} = 4\pi r_{\mathrm{AB}} D_{\mathrm{AB}} L \tag{17}$$

where L is Avogadro's number.

At an O/W interface, however, molecule B in the W phase can approach only from the W-phase side to molecule A staying in contact with the O-phase side of the interface. It is here assumed that molecule B reacts with molecule A just when it reaches the "reaction surface", i.e., the part of spherical surface of the radius of r_{AB} around the centre of A which bulges out to the W phase (see the shadow part in Figure 8.8b). The diffusion-controlled molar flux of B towards the reaction surface, $j'_{\mathrm{B,het}}$, would be obtained by analogy of Equation (16). However, the relative diffusion coefficient D_{AB} in Equation (16) should be replaced by the absolute diffusion coefficient of B in the W phase ($D_{\mathrm{B}}^{\mathrm{W}}$), because in this case, molecule A is regarded as staying at the interface for a reaction with B. Consequently, $j'_{\mathrm{B,het}}$ can be expressed by

$$j'_{\mathrm{B,het}} = 4\pi r_{\mathrm{AB}} \xi D_{\mathrm{B}}^{\mathrm{W}}[\mathrm{B}] \tag{18}$$

Note here that the additional coefficient ξ is the ratio of the reaction surface to the surface area of the sphere with the radius of r_{AB}, i.e.,

$$\xi = \frac{2\pi r_{\mathrm{AB}}{}^2(1 - \cos\theta)}{4\pi r_{\mathrm{AB}}{}^2} = \frac{1 - \cos\theta}{2} \tag{19}$$

where θ is the angle shown in Figure 8.8b and given by

$$\theta = \cos^{-1}\left(\frac{r_{\mathrm{A}}}{r_{\mathrm{AB}}}\right) = \cos^{-1}\left(\frac{r_{\mathrm{A}}}{r_{\mathrm{A}} + r_{\mathrm{B}}}\right) \tag{20}$$

Considering that $j'_{\mathrm{B,het}}$ given by Equation (18) is the flux of B towards one molecule of A, the total flux ($j_{\mathrm{B,het}}$) of B to reach all A molecules in the area of S may be given by $j_{\mathrm{B,het}} = N'(\mathrm{A}) j'_{\mathrm{B,het}}$, with $N'(\mathrm{A})$ being the number of A in S. It is further assumed that molecules A existing at the interface are fully isolated from each other so that diffusion layers of B are not overlapped. This assumption seems to be valid, so far as the concentrations of A are not much higher than that of B.

Since $j_{\mathrm{B,het}}$ is thus regarded as the diffusion-controlled rate, we obtain from the above relations

$$\text{Diffusion-Controlled Rate} = \frac{-\mathrm{d}n_{\mathrm{B}}{}'}{\mathrm{d}t} = j_{\mathrm{B,het}} = 4\pi r_{\mathrm{AB}} \xi D_{\mathrm{B}}^{\mathrm{W}}[\mathrm{B}] N'(\mathrm{A}) \tag{21}$$

where $n_{\mathrm{B}}{}'$ is the molar number of B existing in S.

On the other hand, the rate equation for a bimolecular reaction at an O/W interface is written as

$$\frac{1}{S}\left(-\frac{\mathrm{d}n_{\mathrm{B}}'}{\mathrm{d}t}\right) = k_{\mathrm{het}}[\mathrm{A}][\mathrm{B}] \tag{22}$$

where k_{het} is the second-order rate constant for a unit area.

If $N'(\mathrm{A})$ is assumed as the number of A existing in an interfacial layer of thickness of $2r_{\mathrm{A}}$, i.e., $N'(\mathrm{A}) = 2r_{\mathrm{A}}S[\mathrm{A}]L$, the combination of this equation and Equations (21) and (22) yields an expression of the diffusion-controlled rate constant:

$$k_{\mathrm{D,het}} = 8\pi r_{\mathrm{A}} r_{\mathrm{AB}} \xi D_{\mathrm{B}}^{\mathrm{W}} L \tag{23}$$

This equation has been derived for the case when the reaction rate is limited by the diffusion of B in W. If the diffusion of A in O is the rate-determining step, we can obtain the following equation:

$$k_{\mathrm{D,het}} = 8\pi r_{\mathrm{B}} r_{\mathrm{AB}} \zeta D_{\mathrm{A}}^{\mathrm{O}} L \tag{24}$$

where $D_{\mathrm{A}}^{\mathrm{O}}$ is the diffusion coefficient of A in O, and ζ is given by

$$\zeta = \frac{2\pi r_{\mathrm{AB}}{}^2(1-\cos\omega)}{4\pi r_{\mathrm{AB}}{}^2} = \frac{1-\cos\omega}{2} \tag{25}$$

with

$$\omega = \cos^{-1}\left(\frac{r_{\mathrm{B}}}{r_{\mathrm{AB}}}\right) = \cos^{-1}\left(\frac{r_{\mathrm{B}}}{r_{\mathrm{A}}+r_{\mathrm{B}}}\right) \tag{26}$$

Let us employ these equations to estimate $k_{\mathrm{D,het}}$ for a *hypothetical* ET between Fc (= A) in NB and $\mathrm{Fe(CN)_6}^{3-}$ (= B) in W. Using the parameters $r_{\mathrm{A}} = 3.8$ Å $r_{\mathrm{B}} = 4.4$ Å $D_{\mathrm{A}}^{\mathrm{O}} = 0.56 \times 10^{-5}\ \mathrm{cm^2 \cdot s^{-1}}$, $D_{\mathrm{B}}^{\mathrm{W}} = 0.90 \times 10^{-5}\ \mathrm{cm^2 \cdot s^{-1}}$, the $k_{\mathrm{D,het}}$ is evaluated to be $114\ \mathrm{M^{-1} \cdot cm \cdot s^{-1}}$ if the diffusion of B in W is the rate-determining step (using Equations (19), (20) and (23)). If the diffusion of A in NB limits the rate, $k_{\mathrm{D,het}}$ is likewise evaluated to be $71\ \mathrm{M^{-1} \cdot cm \cdot s^{-1}}$ (using Equations (24)–(26)). Since the latter value is smaller than the former one, $k_{\mathrm{D,het}}$ is regarded as $71\ \mathrm{M^{-1} \cdot cm \cdot s^{-1}}$ for this system.

The diffusion-controlled rate constant thus estimated may be related to the initial step of an ET reaction at an O/W interface:

$$\mathrm{A(O)} + \mathrm{B(W)} \underset{k_{\mathrm{uni}}}{\overset{\substack{\text{Diffusion}\\ k_{\mathrm{D,het}}}}{\rightleftharpoons}} \mathrm{A \ldots B} \overset{\substack{\text{ET}\\ k_{\mathrm{ET}}}}{\rightarrow} \text{products} \tag{27}$$

where A...B represents the encounter complex of A and B, which is formed at the interface; k_{uni} is the dissociation rate constant of A...B; and k_{ET} is the first-order heterogeneous rate constant of the intramolecular ET, which could be estimated using the Marcus theory. On the analogy of a bimolecular reaction in homogeneous solution [52,53], the steady-state approximation method is here used to obtain an expression of the overall rate constant (k):

$$k = \frac{k_{\mathrm{D,het}} k_{\mathrm{ET}}}{k_{\mathrm{ET}} + k_{\mathrm{uni}}} = \frac{k_{\mathrm{D,het}}}{1 + (k_{\mathrm{uni}}/k_{\mathrm{ET}})} \tag{28}$$

As is evident from this equation, $k_{\rm D,het}$ is the attainable maximum value of k; when $k_{\rm ET} \gg k_{\rm uni}$, the overall reaction becomes diffusion-controlled, i.e., $k = k_{\rm D,het}$. For slow reactions where $k_{\rm ET} \ll k_{\rm uni}$, the first step in the reaction scheme (27) is in equilibrium, i.e.,

$$k = K k_{\rm ET} \tag{29}$$

with $K = k_{\rm D,het}/k_{\rm uni}$. Further analysis based on the present scheme will be described elsewhere.

In the theory described above, as well as previous theoretical treatments of ET rate constants, the effect of the molecular-level diffusion process is dealt with by including it in the overall (i.e., observed) rate constant. However, a somewhat different approach to this problem has been advanced by Senda [54], who proposed a model that includes the bimolecular-reaction effect in the voltammetric theory of ET at the O/W interface.

8.4.2.3. Experimental Verification As described above, the $k_{\rm D,het}$ value for the *hypothetical* ET between Fc (NB) and $Fe(CN)_6{}^{3-}$ (W) is 71 $M^{-1} \cdot cm \cdot s^{-1}$. It is interesting to compare this value with the frequency factor Z in the Marcus' rate equation, though the Z factor in Equation (10) or (11) was not estimated as the number of molecular collisions. Using typical parameters adopted by Marcus [8], i.e., $\kappa\nu = 10^{12}$ s^{-1} and $\Delta R = 1$ Å the Z factor is calculated to be 310 $M^{-1} \cdot cm \cdot s^{-1}$ from Equation (10) for a "sharp" O/W interface. This value is comparable with, but larger than, the $k_{\rm D,het}$ value estimated. In our previous work [28], we estimated the rate constant for the Fc–$Fe(CN)_6{}^{3-}$ system on assuming that $Z = k_{\rm D,het}$, but this assumption was in error, in the context of the above-described reaction scheme (cf. Equation (27)).

Let us tentatively employ the Marcus theory to estimate the kinetic parameters (k° and α) for the Fc–$Fe(CN)_6{}^{3-}$ system. The values of $w^{\rm r}$ (= 3.4 kJ $\cdot$ mol^{-1}), $w^{\rm p}$(= −6.8 kJ $\cdot$ mol^{-1}) and $\lambda_{\rm o}$(= 78.8 kJ $\cdot$ mol^{-1}) are obtained from equations provided by Marcus [8] using static and optical dielectric constants of NB (35.7 and 2.4, respectively) and those of water (78.3 and 1.8). Using these work terms and $\lambda_{\rm i} = \lambda_{\rm i}({\rm Fc}) + \lambda_{\rm i}({\rm Fe(CN)_6}^{3-/4-}) = 0.6$ kJ $\cdot$ mol^{-1}[55] + 10.6 kJ $\cdot$ mol^{-1} [56], the activation energy is calculated as $\Delta G^{\ddagger} = 18.5$ kJ $\cdot$ mol^{-1} at $\Delta_{\rm O}^{\rm W}\phi = \Delta_{\rm O}^{\rm W}\phi_{\rm ET}^{\circ} \approx \Delta_{\rm O}^{\rm W}\phi_{\rm ET}^{\circ\prime}$. Substituting this value and $Z = 310$ $M^{-1} \cdot cm \cdot s^{-1}$ (for a sharp O/W interface) finally yields $k^\circ =$ 0.064 $M^{-1} \cdot cm \cdot s^{-1}$. The value of α can also be estimated as 0.56 using the above work terms and reorganization energies. In Figure 8.9, a simulated cyclic voltammogram based on these kinetic parameters is compared with the experimental voltammogram. As seen in the figure, the simulation curve is quite different, in both magnitude and shape, from the experimental one. It should particularly be noted that the α value of 1.0 obtained as a fitting parameter in the regression analysis is rather different from the theoretical value of 0.56. This would also suggest that the ET in the Fc–$Fe(CN)_6{}^{3-}$ system is not a "true" heterogenenous ET but is due to the IT mechanism.

There have been a few reports on the application of the Marcus theory to "true" ET systems. Cheng and Schiffrin [37] employed the a.c. impedance technique to determine the rate constants for some ET systems including the $LuPc_2$

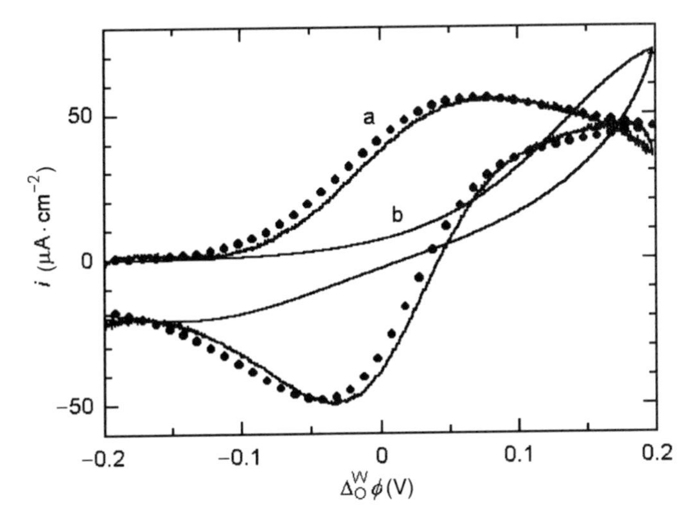

FIGURE 8.9. (a) Base-line corrected cyclic voltammogram observed at the NB/W interface in the presence of 20 mM Fc in NB and 0.5 mM $Fe(CN)_6^{3-}$ and 0.1 mM $Fe(CN)_6^{4-}$ in W. (b) Simulation curve obtained with $k^\circ = 0.064\ \mathrm{M^{-1} \cdot cm \cdot s^{-1}}$ and $\alpha = 0.56$. The solid circles show the regression data obtained by assuming the Butler–Volmer equation (Equations (12a) and (12b)); the fitting parameters are $k^\circ = 0.12\ \mathrm{M^{-1} \cdot cm \cdot s^{-1}}$ and $\alpha = 1.0$. Sweep rate: $0.1\ \mathrm{V \cdot s^{-1}}$.

(DCE)–$Fe(CN)_6^{3-}$ (W) system, and then claimed that the Marcus theory can be used to predict the rate constants. Recently, Bard's group [18,21] reported that the rate constants $k_{12}(= k)$ for the heterogeneous ETs between ZnPor^+ (the oxidized form of 5,10,15,20-tetraphenylporphyrinato zinc(II)) and various aqueous reductants, being determined by means of SECM, showed the driving-force (ΔG°) dependence in accordance with the Marcus theory. Figure 8.10 shows the plot of log k_{12} against $\Delta E_{1/2}(= -\Delta_{\rm O}^{\rm W}\phi + \Delta_{\rm O}^{\rm W}\phi_{\rm ET}^\circ = -\Delta G^\circ/nF)$ for three O/W systems [21]. It is expected from Equations (8) and (9) that log k_{12} shows a parabolic dependence on $\Delta E_{1/2}$ (or $-\Delta G^\circ$), if the work terms can be neglected. It has been claimed that the decrease in k_{12} at higher driving forces is an evidence of the Marcus inverted region. In this analysis, however, the differences in reorganization energy between aqueous redox couples are neglected, though their influences may not be significant. Further experimental verification is desirable.

The effect of adsorption of phospholipids [57–59] and non-ionic surfactants [60–62] on ET kinetics has been investigated. These studies would shed light on the ET mechanisms in biological systems.

8.5. CONCLUDING REMARKS

In the kinetic study of ET at O/W interfaces, the experiment seems to have not caught up with the theory. The experimental data is still insufficient for the verification of the Marcus theory. Further experimental approaches as well as theoretical advances are needed. Recently, novel interesting ET systems including metal nanoparticles [63,64]

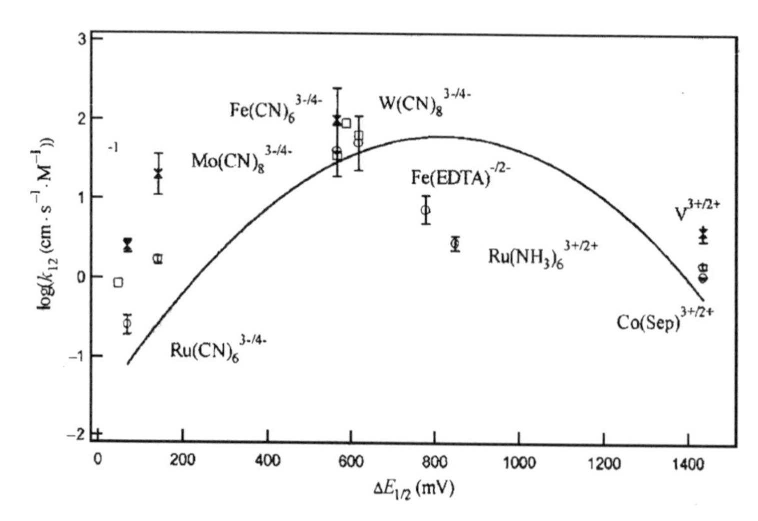

FIGURE 8.10. Dependence of the bimolecular rate constant, log k_{12}, on driving force, $\Delta E_{1/2}$, at the benzonitrile/W (○), benzene/W (□) and DCE/W (×) interfaces for heterogeneous ET between tip-generated organic phase $ZnPor^+$ and aqueous $Ru(CN)_6{}^{4-}$, $Mo(CN)_8{}^{4-}$, $Fe(CN)_6{}^{4-}$, $W(CN)_8{}^{4-}$, $Fe(EDTA)^{2-}$, $Ru(NH_3)_6{}^{2+}$, V^{2+} and $Co(Sep)^{2+}$ (cobalt sepulchrate). Data for $Fe(EDTA)^{2-}$ at the benzonitrile/W interface are reprinted from Ref. [18]. The solid line gives the Marcus prediction based on Equations (8) and (9). Reprinted from Ref. [21], with permission from American Chemical Society.

and redox proteins [65,66] have been developed. These ET systems seem important from technological and biological viewpoints. Fundamental studies, as described in this review, are expected to provide guidelines for design of such practically useful ET systems.

REFERENCES

1. Z. Samec, V. Marecek and J. Weber, J. Electroanal. Chem. **96**, 245 (1979).
2. Z. Samec, V. Marecek and J. Weber, J. Electroanal. Chem. **103**, 11 (1979).
3. J. Hanzlk, Z. Samec and J. Hovorka, J. Electroanal. Chem. **216**, 303 (1987).
4. G. Geblewicz and D.J. Schiffrin, J. Electroanal. Chem. **244**, 27 (1988).
5. V.J. Cunnane, D.J. Schiffrin, C. Beltran, G. Geblewicz and T. Solomon, J. Electroanal. Chem. **247**, 203 (1988).
6. Y. Cheng and D.J. Schiffrin, J. Electroanal. Chem. **314**, 153 (1991).
7. V.J. Cunnane, G. Geblewicz and D.J. Schiffrin, Electrochim. Acta **40**, 3005 (1995).
8. R.A. Marcus, J. Phys. Chem. **94**, 1050 (1990); **64**, 4152 (1990); **94**, 7742 (corrections) (1990); **95**, 2010 (1991); **99**, 5742 (corrections) (1995).
9. H.H. Girault, J. Electroanal. Chem. **388**, 93 (1995).
10. Y.I. Kharkats and A.M. Kuznetsov, *Liquid–Liquid Interfaces, Theory and Methods*, Eds. A.G. Volkov and D.W. Deamer, CRC Press, Boca Raton, FL, 1996, p. 139.
11. W. Schmickler, J. Electroanal. Chem. **428**, 123 (1997).
12. I. Benjamin and Y.I. Kharkats, Electrochim. Acta **44**, 133 (1998).
13. M. Suzuki, S. Umetani, M. Matsui and S. Kihara, J. Electroanal. Chem. **420**, 119 (1997).
14. M. Suzuki, M. Matsui and S. Kihara, J. Electroanal. Chem. **438**, 147 (1997).
15. H. Ohde, K. Maeda, Y. Yoshida and S. Kihara, Electrochim. Acta **44**, 23 (1998).

16. C. Wei, A.J. Bard and M.V. Mirkin, J. Phys. Chem. **99**, 16033 (1995).
17. T. Solomon and A.J. Bard, J. Phys. Chem. **99**, 17487 (1995).
18. A.L. Barker, P.R. Unwin, S. Amemiya, J. Zhou and A.J. Bard, J. Phys. Chem. **103**, 7260 (1999).
19. S. Amemiya, Z. Ding, J. Zhou and A.J. Bard, J. Electroanal. Chem. **483**, 7 (2000).
20. J. Zhang, A.L. Barker and P.R. Unwin, J. Electroanal. Chem. **483**, 95 (2000).
21. Z. Ding, B. M. Quinn and A.J. Bard, J. Phys. Chem. **105**, 6367 (2001).
22. J. Zhang and P.R. Unwin, Phys. Chem. Chem. Phys. **4**, 3820 (2002).
23. Z. Zhang, Yi Yuan, P. Sun, B. Su, J. Guo, Y. Shao and H.H. Girault, J. Phys. Chem. B **106**, 6713 (2002).
24. V.J. Cunnane and L. Murtomäki, *Liquid–Liquid Interfaces, Theory and Methods*, Eds. A.G. Volkov and D.W. Deamer, CRC Press, Boca Raton, FL, 1996, p. 401.
25. A.G. Volkov and D.W. Deamer, Prog. Colloid Polym. Sci. **103**, 21 (1997).
26. D.J. Fermñ and R. Lahtinen, *Liquid Interfaces in Chemical, Biological, and Pharmaceutical Applications*, Ed. A.G. Volkov, Marcel Dekker, New York, 2001, Chapter 8.
27. D.J. Fermñ, H. Jensen and H.H. Girault, *Encyclopedia of Electrochemistry, Vol. 2: Interfacial Kinetics and Mass Transport*, Eds. A.J. Bard, M. Stratmann and E.J. Calvo, Wiley-VCH, Weinheim, 2003, p. 360.
28. H. Hotta and T. Osakai, Rev. Polarogr. (Kyoto) **49**, 15 (2003) (in Japanese with English abstract, table and figure captions).
29. A.J. Bard, R. Parsons and J. Jordan (Eds.), *Standard Potentials in Aqueous Solution*, Marcel Dekker, New York, 1985.
30. H. Hotta, S. Ichikawa, T. Sugihara and T. Osakai, J. Phys. Chem. B **107**, 9717 (2003).
31. H. Hotta and T. Osakai, unpublished results, 2003.
32. K. Kano, Rev. Polarogr. (Kyoto) **48**, 29 (2002).
33. A.A. Stewart, J.A. Campbell, H.H. Girault and M. Eddowes, Ber. Bunsenges. Phys. Chem. **94**, 83 (1990).
34. S. Kihara, M. Suzuki, K. Maeda, K. Ogura, M. Matsui and Z. Yoshida, J. Electroanal. Chem. **271**, 107 (1989).
35. H. Ohde, K. Maeda, Y. Yoshida and S. Kihara, J. Electroanal. Chem. **483**, 108 (2000).
36. Qi-Z. Chen, K. Iwamoto and M. Senô, Electrochim. Acta **36**, 291 (1991).
37. Y. Cheng and D.J. Schiffrin, J. Chem. Soc., Faraday Trans. **89**, 199 (1993).
38. H. Hotta, N. Akagi, T. Sugihara, S. Ichikawa and T. Osakai, Electrochem. Commun. **4**, 472 (2002).
39. S. Sawada, M. Taguma, T. Kimoto, H. Hotta and T. Osakai, Anal. Chem. **74**, 1177 (2002).
40. K. Nakatani, T. Uchida, H. Misawa, N. Kitamura and H. Masuhara, J. Electroanal. Chem. **367**, 109 (1994).
41. K. Nakatani, K. Chikama, H.-B. Kim and N. Kitamura, Chem. Phys. Lett. **237**, 133 (1995).
42. K. Chikama, K. Nakatani and N. Kitamura, Bull. Chem. Soc. Jpn. **71**, 1065 (1998).
43. T. Osakai, H. Hotta, H. Tomida and S. Sawada, unpublished results, 2003.
44. Z. Ding, P.F. Brevet and H.H. Girault, Chem. Commun. 2059 (1997).
45. Z. Ding, D.J. Fermñ, P.-F. Brevet and H.H. Girault, J. Electroanal. Chem. **458**, 139 (1998).
46. T. Osakai, H. Jensen, H. Nagatani, D.J. Fermñ and H.H. Girault, J. Electroanal. Chem. **510**, 43 (2001).
47. T. Osakai, N. Akagi, H. Hotta, J. Ding and S. Sawada, J. Electroanal. Chem. **490**, 85 (2000).
48. T. Sugihara, H. Hotta and T. Osakai, Bunseki Kagaku **52**, 665 (2003) (in Japanese with English abstract, table, and figure captions).
49. T. Osakai, H. Hotta, T. Sugihara, and K. Nakatani, J. Electroanal, Chem. **571**, 201 (2004).
50. M. von Smoluchowski, Z. Phys. Chem. **92**, 129 (1917).
51. P. Debye, Trans. Electrochem. Soc. **82**, 265 (1942).
52. D. Rehm and A. Weller, Israel J. Chem. **8**, 259 (1970).
53. J.I. Steinfeld, J.S. Francisco and W.L. Hase, *Chemical Kinetics and Dynamics*, Prentice-Hall, Englewood Cliffs, NJ, 1989, Chapter 4.
54. M. Senda, Rev. Polarogr. (Kyoto) **49**, 31 (2003).
55. T. Gennett, D.F. Milner and M.J. Weaver, J. Phys. Chem. **89**, 2787 (1985).
56. P. Delahay, Chem. Phys. Lett. **99**, 87 (1983).
57. Y. Cheng and D.J. Schiffrin, J. Chem. Soc., Faraday Trans. **90**, 2517 (1994).
58. M. Tsionsky, A.J. Bard and M.V. Mirkin, J. Am. Chem. Soc. **119**, 10785 (1997).
59. M.-H. Delville, M. Tsionsky and A.J. Bard, Langmuir **14**, 2774 (1998).
60. J. Zhang and P.R. Unwin, J. Electroanal. Chem. **494**, 47 (2000).

61. J. Zhang, C.J. Slevin, L. Murtomäki, K. Kontturi, D.E. Williams and P.R. Unwin, Langmuir **17**, 821 (2001).
62. D.G. Georganopoulou, J. Strutwolf, C.M. Pereira, F. Silva, P.R. Unwin and D.E. Williams, Langmuir **17**, 8348 (2001).
63. C. Johans, R. Lahtinen, K. Kontturi and D.J. Schiffrin, J. Electroanal. Chem. **488**, 99 (2000).
64. C. Johans, K. Kontturi and D.J. Schiffrin, J. Electroanal. Chem. **526**, 29 (2002).
65. D.G. Georganopoulou, D.J. Caruana, J. Strutwolf and D.E. Williams, Faraday Discuss. **116**, 109 (2000).
66. G.C. Lillie, S.M. Holmes and R.A.W. Dryfe, J. Phys. Chem. B **106**, 12101 (2002).

9

Mass Transfer and Reaction Rate in the Nano-Region of Microdroplet/Solution Interfaces

Kiyoharu Nakatani and Takayuki Negishi
University of Tsukuba, Tennoudai, Tsukuba 305-8571, Japan

9.1. INTRODUCTION

The chemical reaction between a hydrophobic solute in an oil phase and a hydrophilic solute in a water phase proceeds in the vicinity of the oil/water interface. A cation and an anion dissolved in the water phase are extracted into the oil phase as a neutral species. A solute possessing both hydrophobic and hydrophilic groups is adsorbed at the oil/water interface. Because various chemical and physical processes proceed in the nano-region of the oil/water interfaces, characteristic chemical reactions, different from those in a homogeneous solution, can occur in the oil/water interface systems such as oil/water or water/oil emulsions, biological cells and so forth. In the heterogeneous systems, oil/water interfaces play significant and important roles, so kinetic analyses of the interfacial processes are necessary for understanding the microscopic mechanisms and controlling the chemical reactions. The overall chemical reaction in an oil/water system is highly dependent on the mass transfer in the oil and water phases (external mass transfer), chemical reaction, ion-pair formation, complexation, permeation and adsorption/desorption at the interface and so forth. When chemical and physical processes across an oil/water interface are induced by electron or ion transfer, kinetic analyses of these processes will be successfully performed in a polarized liquid/liquid interface system using electrochemical and spectroscopic methods [1–8]. By using a four-electrode technique, the Galvani potential difference between the oil and water phases can be controlled, so that kinetic and thermodynamic analyses can be performed by inducing the charge transfer across the liquid/liquid boundary. Such studies are described in detail in other chapters. On the other hand, if interfacial chemical and physical processes are induced by a reaction or mass transfer of a neutral species, it is difficult to measure the reaction rate since these interfacial processes proceed during the construction of the oil/water system. Thus, considerable efforts have been devoted to the kinetic analysis of the interfacial chemical and physical processes using rapid two-phase mixing, a scanning electrochemical microscope, dynamic surface tension and single droplet techniques [9–22], as described in other chapters.

In single micrometre-sized droplet systems, external mass transfer is spherical diffusion under stationary conditions. From the droplet size (droplet radius, r) dependence of a reaction rate, the rate-determining step of an overall reaction in the microdroplet/solution system can be experimentally determined [18,23]. When the reaction rate is governed by an interfacial process or diffusion in the surrounding solution phase, the rate constant is directly proportional to r^{-1} or r^{-2}, respectively. If the rate-determining step of the reaction is a reaction in the homogeneous phase, the rate constant is independent of r. Based on the characteristic features in the single microdroplet systems, the chemical and physical processes in the nano-region of the oil/water interface can be kinetically analyzed. In this chapter, the mass transfer and chemical processes across oil/water interfaces are described on the basis of the single microdroplet measurements. Particularly, the kinetic analysis of the ion-pair extraction as a neutral species is discussed in detail.

9.2. MANIPULATION, ELECTROCHEMISTRY AND SPECTROSCOPY OF SINGLE MICRODROPLETS

The Brownian motion of microdroplets is vigorous in solution. The volume of a micrometre-sized droplet is $10^{-15} - 10^{-11}$ dm^3. Therefore, a manipulation technique is indispensable for single microdroplet measurements. For the microanalysis of a single microdroplet, size of the probe should be smaller than that of the microdroplet. A light beam and a microelectrode are frequently used as a probe, and the analyses of small domains are performed by absorption/fluorescence microspectroscopy [24–29] and microelectrochemical methods [17,30–32]. In this section, single microdroplet techniques for the kinetic analysis of physical and chemical processes across a microdroplet/solution interface are described.

Figure 9.1 shows the laser and microcapillary manipulation, absorption/fluorescence microspectroscopy and microelectrochemistry system used in our laboratory [33–35]. A 1064-nm light beam from a continuous-wave Nd^{3+}:YVO_4 laser (Spectra-Physics, Millenia IR) is introduced into an optical microscope (Olympus, BX-60) and focused onto a single microdroplet (beam spot size of 1–2 μm) using an objective lens (×60 or ×100). Laser trapping is one of the most important techniques for single microdroplet manipulation [36–39]. By radiation pressure from the laser beam, the single droplet can be non-destructively manipulated in the vicinity of the focused laser beam. Thus, the Brownian motion of the laser-trapped microdroplet is suppressed and the microdroplet can be freely positioned in an appropriate space. Based on a microcapillary injection and manipulation technique (Narishige, MN-151, MMW-200, IM-16, IM-300), on the other hand, a single microdroplet with a volume greater than 10^{-15} dm^3 can be injected and positioned in a solution using a glass microcapillary with a tip diameter smaller than 10 μm.

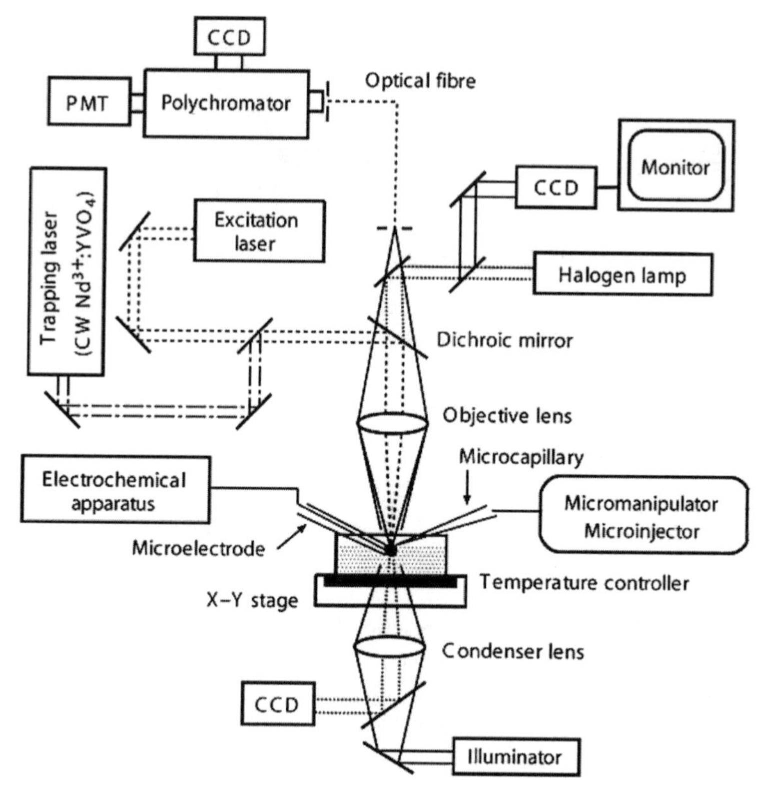

FIGURE 9.1. Block diagram of laser and microcapillary manipulation, absorption/fluorescence microspectroscopy and microelectrochemistry system.

For absorption microspectroscopy, a light beam from a halogen lamp is coaxially led into the microscope and focused onto a microdroplet (beam spot size of 2–3 μm). The transmitted light beam from the microdroplet center (I) is collected by a condenser lens and detected by a multichannel photodetector (Hamamatsu Photonics, PMA11). Using a transmitted light beam near the microdroplet as a reference (I_0), the absorption spectrum of the single microdroplet is recorded ($\log(I_0/I)$). For fluorescence microspectroscopy, a light beam from a continuous-wave diode laser (Neoark, TC20-40305-2F4.5) is used instead of the halogen lamp in the absorption microspectroscopy system. The fluorescence from a microdroplet is collected by the objective lens and detected by a polychromator (Solar TII, MS2001I)—multichannel photodetector (Andor Technology, DV401-BV) system. For microelectrochemical measurements, a gold wire (diameter of 10–25 μm) insulated in a glass capillary (tip diameter of 50–100 μm) is used as the working electrode. Pt wire and Ag/AgCl electrodes, used as the counter and reference electrodes, respectively, are placed in the surrounding solution phase. Various electrochemical measurements can be performed using an electrochemical analyser (BAS, BAS100B/W-low current module).

In actual experiments, a single microdroplet is manipulated by the microcapillary or laser manipulation technique and makes contact with the working microelectrode

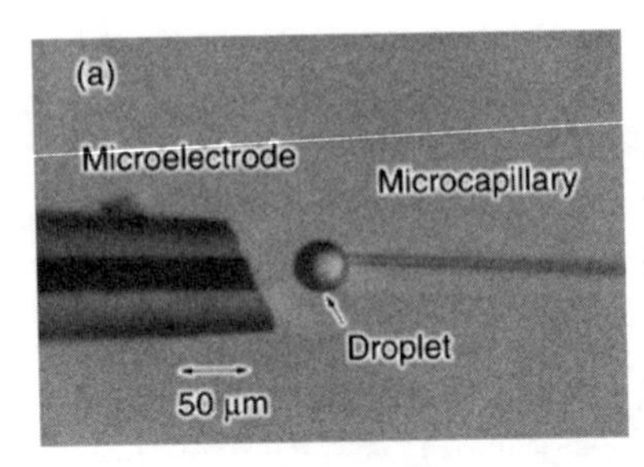

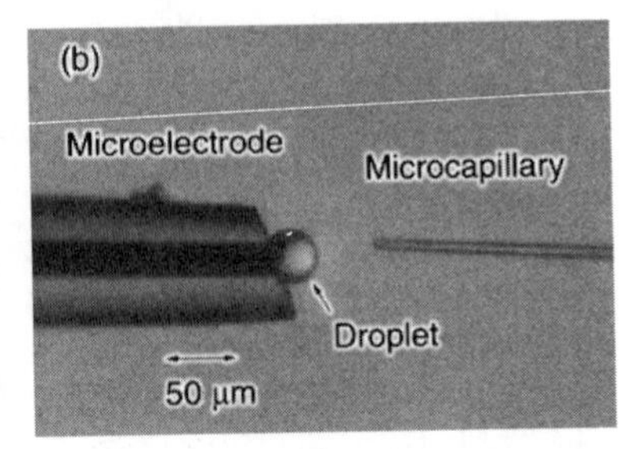

FIGURE 9.2. Photographs of a single microdroplet (a) injected into a solution and (b) contacted on a microelectrode by microcapillary manipulation technique.

(Figure 9.2). The electrochemical and spectroscopic measurements are then done using a single microdroplet.

9.3. MICRODROPLET SIZE EFFECT ON MASS TRANSFER AND REACTION RATE

The mass transfer and chemical reaction between a single microdroplet and the surrounding solution phase can be induced by the laser and microcapillary manipulation and microelectrochemistry system. An example of the electrochemically induced mass transfer across a single microdroplet/solution interface is shown in Figure 9.3. If

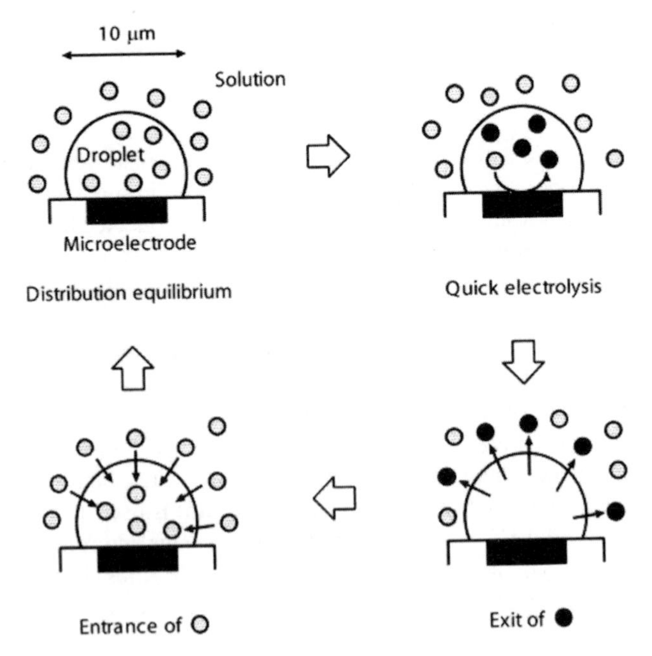

FIGURE 9.3. General scheme of electrochemically induced mass transfer across a single microdroplet/solution interface.

a solute distributed in a microdroplet is electrolyzed by a microelectrode, electrolysis of the solute is quickly completed owing to the micrometre-sized droplet. The diffusion time (t_d) of the solute in the microdroplet with $r = 5$ μm is calculated to be 25 ms using $r = (2Dt_d)^{1/2}$ (D, diffusion coefficient (5×10^{-6} $cm^2 \cdot s^{-1}$ as a typical value of a solute in solution)). In this time frame, the solute and the electrochemically produced compound have a non-equilibrated distribution between the droplet and the surrounding solution phase, so the exit of the produced compound from the microdroplet and entrance of the solute into the microdroplet are induced to attain an equilibrated distribution. The mass transfer rate across the microdroplet/solution interface can be determined by micro-electrochemical and microspectroscopic measurements of the single microdroplet. Such a relaxation method can be applied only in a single microdroplet/solution system. In micro-oil-droplet/water systems containing a ferrocene derivative as the neutral species, migration of an electrochemically produced ferrocenium cation has been reported to proceed quickly from the oil droplet into the water phase, depending on the Galvani potential difference between the oil and water phases [17,32,40]. Frequently, the exit rate of the produced compound is much faster than the entrance rate of the solute. In this section and the following sections, we describe the kinetic analyses of the interfacial mass transfer processes, focused on the entrance from the water phase into a single micro-oil-droplet.

For a simple extraction of a solute (X) from a water phase into a single spherical micro-oil-droplet, the time (t) dependence of the amount of the solute transferred across the microdroplet/water interface (m) is obtained by Equation (1) [41].

$$dm/dt = (4\pi r^3/3)d[X_w]/dt = 4\pi r^2(k_f[X_w] - k_b[X_o]) \tag{1}$$

where $[X_w]$ and $[X_o]$ are the respective concentrations of X in the water phase and in the oil droplet. k_f and k_b are the mass transfer rate constants from the water phase into the droplet and from the droplet into the water phase, respectively. $[X_w]$ after the extraction of X can be assumed to be equal to the X concentration in the water during the sample preparation because of the small volume ratio of the single microdroplet ($10^{-15} - 10^{-11}$ dm^3) to the water phase (10^{-3} dm^3). As the distribution coefficient of X between the microdroplet and the solution phase (K_D) is defined by k_f/k_b, $[X_o]$ is given by [18,41]

$$[X_o] = [X_o]_{eq}[1 - \exp\{-3k_f t/(rK_D)\}] \tag{2}$$

where $[X_o]_{eq}$ is $[X_o]$ at the distribution equilibrium. When the extraction rate in a micro-droplet/water system is analysed by the first-order reaction type kinetics, the observed rate constant (k_{obs}) is given by $3k_f/(rK_D)$. Therefore, if the rate-determining step is the mass transfer process at the microdroplet/water interface, k_{obs} is directly proportional to r^{-1}.

Mass transfer from the water phase to the microdroplet under stationary conditions quickly reaches the steady-state because of spherical diffusion. When the extraction of X is diffusion-limited in the water phase, Equation (3) is obtained using the diffusion coefficient of X in the water phase (D_w) [18].

$$[X_o] = [X_o]_{eq}[1 - \exp\{-3D_w t/(r^2 K_D)\}] \tag{3}$$

In this case, k_{obs} is given by $3D_w/(r^2K_D)$ and k_f is equal to D_w/r in Equation (2).

For chemical and physical processes across microdroplet/solution interfaces, k_{obs} having dimensions of s^{-1} or $dm^3 \cdot mol^{-1} \cdot s^{-1}$ is often proportional to r^{-n} ($n = 0$, 1 or 2). A linear relationship between k_{obs} and r^{-2} has been reported for the extraction of a neutral compound such as ferrocene derivatives from water into a micro-oil-droplet without adsorption at the microdroplet/water interface [18,19] and for a photographic dye formation reaction in an oil-in-water emulsion [23]. The proportionality of a k_{obs} versus r^{-1} plot has been reported for a relatively slow process such as a photographic dye formation reaction [23,29,42], electron transfer [43–45] and adsorption at the micro-oil-droplet/water interface [19,20]. For the chemical reaction with the rate-determining step in a solution phase or a microdroplet in a microdroplet/solution system, k_{obs} is independent of r[23]. Based on the droplet size dependence of the reaction rate, the rate-determining step of the overall reaction processes across a microdroplet/solution interface is analysed and the reaction mechanism can be discussed in detail.

9.4. SIMPLE ION-PAIR EXTRACTION IN A SINGLE MICRO-OIL-DROPLET/WATER SYSTEM

For ion-pair extraction, a cation is extracted with an anion into oil. In this case, individual ions or the ion pair species transfer across a microdroplet/water interface and the extraction rate is expected to depend on the Galvani potential between the microdroplet and water, the ion transfer potentials across the liquid/liquid interface, the association constant of the ions in the solution and so forth [46–54]. Therefore, the mass transfer processes are complicated even in the absence of adsorption of an ion at the microdroplet/water interface. In this section, the kinetic analysis of a simple ion-pair extraction without adsorption is described and the extraction mechanism is discussed on the basis of the single microdroplet technique.

9.4.1. Direct Measurements of Ion-Pair Extraction Rate

As an example of a simple ion-pair extraction, the mass transfer of a (ferrocenylmethyl)trimethylammonium cation ($FcTMA^+$) with a hexafluorophosphate anion (PF_6^-) from water into a single tributyl phosphate (TBP) microdroplet is described [21,55]. Water-saturated TBP and TBP-saturated water containing FcTMABr and $NaPF_6$ were used as the oil and water phases, respectively. A single TBP microdroplet of $r = 16$–18 μm was contacted with a working microelectrode and completely covered the electrode surface using a single microdroplet manipulation technique. As shown in Figure 9.4, a cyclic voltammogram of a single microdroplet was sigmoidal in a potential sweep rate (v) of 5 $mV \cdot s^{-1}$, analogous to that observed in a homogeneous solution using a microelectrode. The anodic current corresponds to oxidation of $FcTMA^+$ to $FcTMA^{2+}$. Any anodic current was not observed without $NaPF_6$ so that $FcTMA^+$ is extracted with PF_6^- from water into the microdroplet. A cathodic current related to the reduction of $FcTMA^{2+}$ was not observed since the dication would rapidly distribute into the water. If the extraction rate of $FcTMA^+$ with PF_6^- is slow, a cyclic voltammogram shows a symmetrical peaked curve because of the complete electrolysis of $FcTMA^+$ in

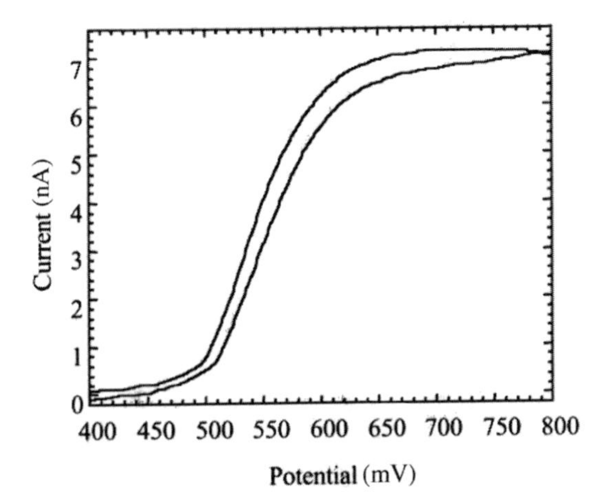

FIGURE 9.4. Cyclic voltammogram of a single TBP droplet ($r = 17$ μm) at $\nu = 5$ mV · s^{-1} in the aqueous FcTMABr (0.5 mmol · dm^{-3}) and $NaPF_6$ (1 mmol · dm^{-3}) solution.

a micrometre-sized droplet. The ion-pair extraction of $FcTMA^+$ with PF_6^- is so fast that the cyclic voltammogram has a sigmoidal curve.

A fast extraction rate was determined by the triple potential step electrolysis repeated in the single microdroplet/electrode system. Figure 9.5 shows the current (i) versus time (t) curve. The potential (E) was stepped from 0.45 V (no electrolysis of $FcTMA^+$) to 0.65 V. At 0.65 V, i rapidly decreased to a steady-state current (i_s) and $FcTMA^+$ in the microdroplet was almost oxidized except for the solute entering the droplet during the steady-state. E was then stepped to 0.45 V and $FcTMA^+$ was extracted into the microdroplet. E was changed again to 0.65 V and the i–t curve was measured for

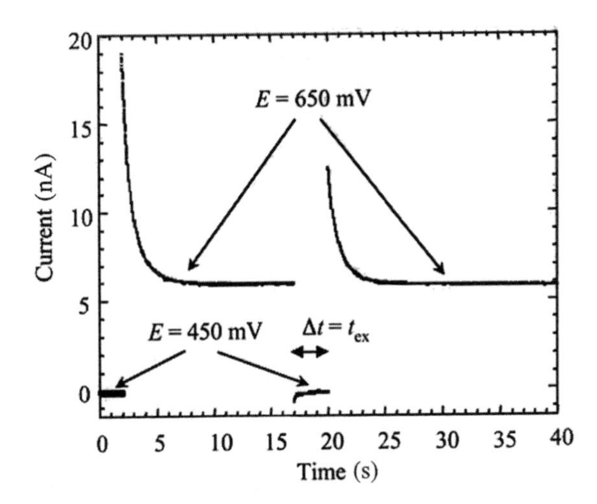

FIGURE 9.5. Response curve of triple potential step chronoamperometry for a single TBP droplet ($r = 17$ μm) in the aqueous FcTMABr (0.5 mmol · dm^{-3}) and $NaPF_6$ (1 mmol · dm^{-3}) solution.

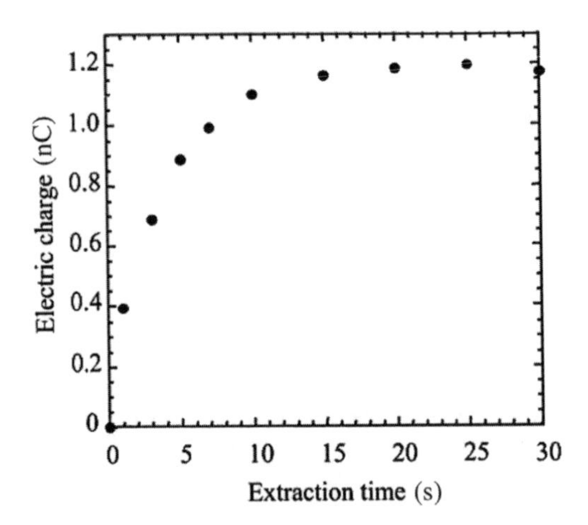

FIGURE 9.6. Extraction time dependence of Q for a single TBP droplet ($r = 17$ μm) in the aqueous FcTMABr (0.5 mmol · dm^{-3}) and $NaPF_6$ (1 mmol · dm^{-3}) solution.

calculation of the $FcTMA^+$ molecules in the microdroplet as the electric charge ($Q = \int(i - i_s)\,dt$), which depends on the extraction time (t_{ex}). The t_{ex} dependence of Q for the same microdroplet shows the relaxation curve to the distribution equilibrium (Figure 9.6). The time dependence of the $FcTMA^+$ extraction was analysed as the initial rate of the $FcTMA^+$ extraction (v_{ini}), $v_{ini} = (1/F)\,dQ/dt_{ex}$ near $t_{ex} = 0$ (F: Faraday constant and Q/F: amount of $FcTMA^+$ in the microdroplet). Figure 9.7 shows the FcTMABr and $NaPF_6$ concentration dependence of v_{ini}. The v_{ini} was directly proportional to the FcTMABr concentration, while it was independent of the $NaPF_6$ concentration. These are the first demonstrations for the direct measurements of ion-pair extraction rates.

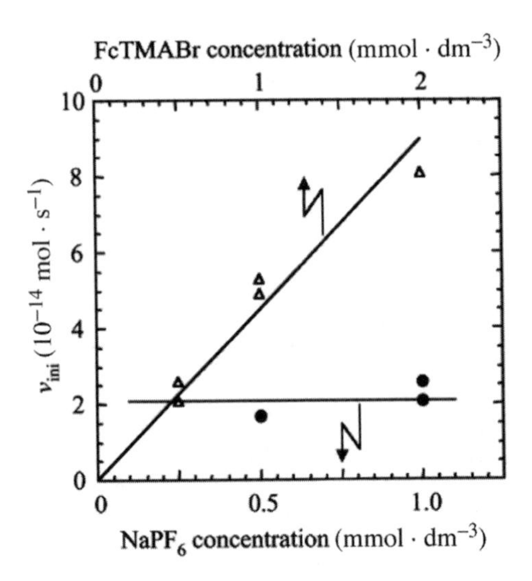

FIGURE 9.7. Initial extraction rate of $FcTMA^+$ into single microdroplets ($r = \sim 17$ μm) for various FcTMABr (1 mmol · dm^{-3} $NaPF_6$) and $NaPF_6$ (0.5 mmol · dm^{-3} FcTMABr) concentrations.

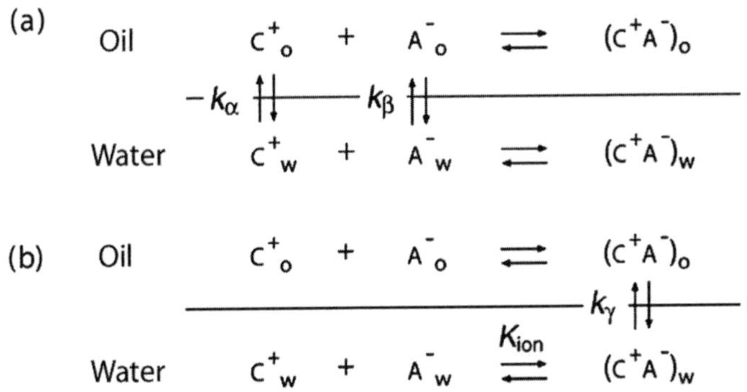

(a)

Oil $C^+_o + A^-_o \rightleftharpoons (C^+A^-)_o$

$-k_\alpha \quad k_\beta$

Water $C^+_w + A^-_w \rightleftharpoons (C^+A^-)_w$

(b)

Oil $C^+_o + A^-_o \rightleftharpoons (C^+A^-)_o$

k_γ

Water $C^+_w + A^-_w \overset{K_{ion}}{\rightleftharpoons} (C^+A^-)_w$

FIGURE 9.8. Schematic drawing of the transfer mechanism of C^+ and A^- for ion-pair extraction from water into oil.

9.4.2. Transfer Mechanism of Simple Ion-Pair Extraction

Two models of the transfer mechanism for a simple ion-pair extraction have been proposed as shown in Figure 9.8. For the model in Figure 9.8a, a cation (C^+_w) or anion (A^-_w) individually enters from the water into oil, maintaining the electroneutrality of the two phases. The driving force of the ion transfer is migration at the liquid/liquid interface, depending on the Galvani potential difference between the water and oil phases. This model is supported by the ion transfer potential data in the water/nitrobenzene and water/1,2-dichloroethane systems [50–54]. The standard ion transfer potentials of $FcTMA^+$, Na^+, PF_6^- and Br^- in a polarized water/1,2-dichloroethane interface system are −0.05, 0.59, −0.02 and −0.45 V, respectively [56]. As the half-wave potential of $FcTMA^+$ in TBP ($E_{1/2}$) is assumed to be equal to that in the water phase (0.42 V versus Ag/AgCl), the Galvani potential difference ($\Delta_0^w\phi$) in the TBP-microdroplet/water system is estimated to be −0.1 V using $E_{1/2} = E_{1/2}^{obs} + \Delta_0^w\phi$ ($E_{1/2}^{obs}$: observed half-wave potential, 0.55 V versus Ag/AgCl in water (Figure 9.4)). At $\Delta_0^w\phi = -0.1$ V, ion transfer of PF_6^- from water to the microdroplet efficiently proceeds while that of $FcTMA^+$, Na^+ or Br^- does not proceed. Besides, the ion transfer of $FcTMA^+$ from water into TBP is expected to be much faster than that of Na^+. The data thermodynamically predict that the ion-pair extraction of $FcTMA^+$ occurs with PF_6^- to maintain the electroneutrality in the microdroplet. In analytical chemistry, on the other hand, the mechanism of ion-pair extraction is frequently explained by the model in Figure 9.8b. In this case, an ion pair ($(C^+A^-)_w$) forms in water or at the oil/water interface, and $(C^+A^-)_w$ as a neutral species is extracted into the oil phase ($(C^+A^-)_o$) [46–49]. In any case, it is difficult to determine the extraction mechanism based on only the thermodynamic data.

Direct measurements of the ion-pair extraction rate are indispensable since chemical species (C^+_w/A^-_w or $(C^+A^-)_w$) transferring across a liquid/liquid interface can be detected by kinetic analysis. In the model of Figure 9.8a, the initial extraction rate (v_{ini}) is limited by the ion transfer of C^+_w or A^-_w so that v_{ini} is given as

$$v_{ini} = k_\alpha[C^+_w] \quad \text{or} \quad v_{ini} = k_\beta[A^-_w] \tag{4}$$

where k_α and k_β are the respective ion transfer rate constants of C^+_w and A^-_w from water into oil. In the model of Figure 9.8b, because the ion-pair formation is expected to be fast (association constant in water, K_{ion}), the rate-determining step is the $(C^+A^-)_w$ transfer across the liquid/liquid interface. Namely, the initial extraction rate is given as

$$v_{ini} = k_Y[(C^+A^-)_w] = k_Y K_{ion}[C^+_w][A^-_w] \tag{5}$$

where k_Y is the extraction rate constant of $(C^+A^-)_w$. In this model, v_{ini} depends on both the C^+_w and A^-_w concentrations in water.

In the $FcTMA^+/PF_6^-$ system, v_{ini} was directly proportional to the FcTMABr concentration in the water phase, independent of the $NaPF_6$ concentration (Figure 9.7). Therefore, the extraction rate is governed by the $FcTMA^+$ transfer across the microdroplet/water interface on the basis of the individual ion transfer model (Figure 9.8a). The extraction rate constant per unit surface area was determined to be $1.5 \times 10^{-3} cm \cdot s^{-1}$. The rate constant is smaller than the diffusion-limited rate constant of $FcTMA^+$ in water (k_d, 3.5×10^{-3} $cm \cdot s^{-1}$) using $k_d = D_w/r$ (D_w: diffusion coefficient of $FcTMA^+$ in water, 6×10^{-6} $cm^2 \cdot s^{-1}$). When sodium chloride was also contained in the water phase, the partitioning ratio of $FcTMA^+$ (P_{FcTMA}) from the water phase to the TBP droplet decreased with the increasing sodium chloride concentration. The P_{FcTMA} values at the sodium chloride concentrations of 0 and 0.1 $mol \cdot dm^{-3}$ were 16 and 5, respectively. Furthermore, the time required for the distribution equilibrium of $FcTMA^+$ increased with the increasing sodium chloride concentration. Because PF_6^- is extracted with Na^+ into TBP at a high sodium chloride concentration in the water phase, the distribution of $FcTMA^+$ would be suppressed by Na^+. Nonetheless, v_{ini} was independent of the sodium chloride concentration in the water phase, indicating that the rate-determining step during the initial stage of the ion-pair extraction is the $FcTMA^+$ transfer from water into the microdroplet [21].

We showed direct measurements of the mass transfer rate of a simple ion-pair extraction in the single microdroplet/water system and discussed the transfer mechanism on the basis of the kinetic data.

9.5. ION-PAIR EXTRACTION OF AN ANIONIC SURFACTANT WITH A CATIONIC DYE

Ion-pair extraction of an anionic surfactant with a cationic dye such as methylene blue from water into oil is often used for the quantitative analysis of the surfactant in water [57,58]. The surfactant concentration in water is then determined as the dye concentration in the oil or water phase by conventional absorption spectroscopy. Synthetic surfactants such as sulfates and sulfonates completely dissociate in water even at low pH. On the other hand, the association of fatty acid salts (traditional soaps) with H^+ depends on the pH. Therefore, the quantitative analysis of surfactants in water is performed by the ion-pair extraction at various pHs. Although quantitative analysis and thermodynamic studies have been already reported for the anionic surfactant/cationic dye extraction, kinetic analysis of the ion-pair extraction has been rarely reported and the extraction mechanism is not discussed in detail. In this section, we describe the kinetic analysis of the extraction of a dodecyl sulfate anion with methylene blue as a typical example using the single microdroplet manipulation and microabsorption methods [59]. In particular, the pH dependence of the ion-pair extraction is discussed.

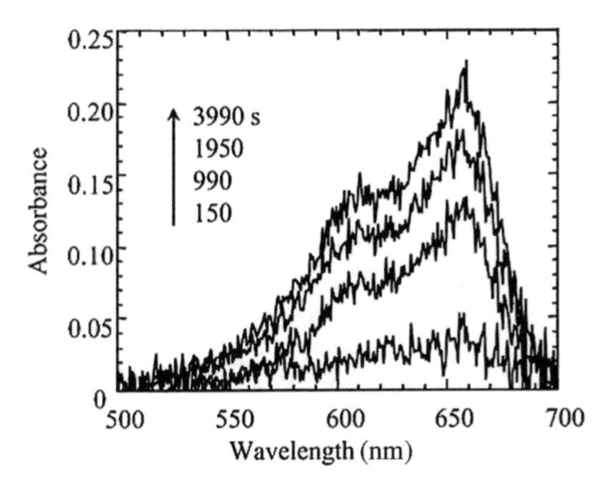

FIGURE 9.9. Absorption spectra of MB^+ in a single TBP droplet ($r = 45$ μm) after injection of the droplet into an aqueous MBCl (1×10^{-5} mol · dm^{-3}) and NaDS (5×10^{-5} mol · dm^{-3}) solution of pH 2.

9.5.1. *Extremely Slow Ion-Pair Extraction at Low pH*

Water-saturated TBP and TBP-saturated water containing hydrochloric acid (pH 2), methylene blue (MBCl, 1×10^{-5} mol · dm^{-3}) and sodium dodecyl sulfate (NaDS, 5×10^{-5} mol · dm^{-3}) were used as the oil and water phases, respectively. A single TBP microdroplet of $r = 25$–60 μm was injected into the water phase using a microcapillary manipulation method and was contacted on the tip of a glass fibre. An absorption spectrum of a single TBP microdroplet ($r = 45$ μm) was measured with time (t) after injection of the single microdroplet (Figure 9.9). The shape of the absorption spectrum was the same as that of the MB^+ monomer and MB^+ was not distributed into the microdroplet in the absence of DS^-, indicating that MB^+ is extracted with DS^- into the microdroplet. The time dependence of the absorbance at 660 nm ($A(t)$) of MB^+ extracted into single microdroplets at pH 2 is shown in Figure 9.10. Based on the $A(t)$ curve, the induction period was observed at $t = 0$–500 seconds and the extraction rate was highly dependent on r. Since the diffusion time of the solute in the microdroplet of $r = 39$ μm is shorter than several seconds, as described in Section 9.2, the solute is homogeneously distributed in the microdroplet interior. Furthermore, the mass transfer of the solute from the bulk water phase to the microdroplet surface is fast because of spherical diffusion, and the diffusion-limited rate constant (k_d) is estimated to be 1×10^{-3} cm · s^{-1} on the basis of $k_d = D_w/r$ under a steady-state condition. On the other hand, t measured in this case is $10^2 - 10^3$ s. These results indicate that the diffusion of the solute in water or the microdroplet does not become the rate-determining step of the overall extraction processes.

Sigmoidal curves having an induction period are frequently analysed on the basis of the two-step consecutive reaction-type kinetics. In this case, $A(t)$ is given by Equation (6).

$$A(t) = F_1 \exp(-k_1 t) + F_2 \exp(-k_2 t) + A_0 \qquad (6)$$

where $F_i (i = 1 \text{ or } 2)$, $k_i (i = 1 \text{ or } 2)$ and A_0 are the pre-exponential factor, the mass transfer rate constant ($k_1 > k_2$) and $A(t)$ at the extraction equilibrium, respectively. The observed $A(t)$ was satisfactorily fitted by Equation (6) (sold lines in Figure 9.10). Analogous analyses were done and k_1 and k_2 were determined for various-sized single

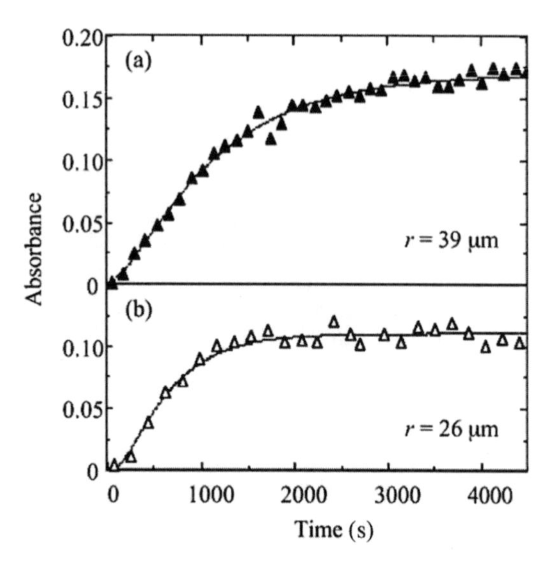

FIGURE 9.10. Time dependence on peak absorbance (660 nm) of MB^+ in single TBP droplets [r = (a) 39 or (b) 26 μm] injected into an aqueous MBCl (1×10^{-5} mol · dm^{-3}) and NaDS (5×10^{-5} mol · dm^{-3}) solution of pH 2. The solid curves represent the simulation of A(t) by Equation (6).

microdroplets (Figure 9.11). Both rate constants increased with decreasing r. When the rate constant is directly proportional to r^{-1} or r^{-2}, the elementary process is the interfacial process at the microdroplet/solution boundary or spherical diffusion from the bulk solution to the microdroplet surface, respectively. In the present system, both the k_1 and k_2 values were directly proportional to r^{-1}, indicating that the two-step processes

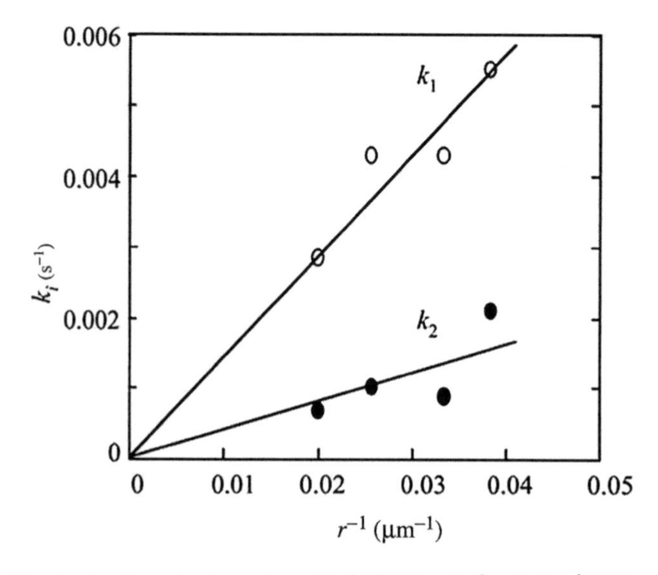

FIGURE 9.11. Droplet size dependencies on k_1 and k_2. MBCl, 1×10^{-5} mol · dm^{-3}; NaDS, 5×10^{-5} mol · dm^{-3}, pH 2.

TABLE 9.1. k_f and the partitioning ratios of both MB^+ (with DS^-) and DS^- (with H^+) in the TBP/water system.

pH	$k_f/10^{-3}$ cm · s^{-1}	P_{MB}	P_{DS}
2	0.038[a]	36	41
3	0.10	33	20
4	0.18	40	11
5	0.25	34	
6	>~1	40	4

[a] Estimated from $A(t)$ in Figure 9.10a as a single exponential rise curve.

proceed at the microdroplet/water interface. The extraction of MB^+ (1×10^{-5} mol · dm^{-3} in water) with PF_6^- (2×10^{-4} mol · dm^{-3} in water) instead of DS^- at pH 2 was also measured in the single TBP-droplet/water system. The absorbance of MB^+ in the microdroplet was saturated within 60 seconds in the MB^+/PF_6^- system. Therefore, the extraction rate in the MB^+/DS^- system is significantly slow.

9.5.2. *pH Dependence of the Ion-Pair Extraction Rate*

When the t dependence of the absorbance of MB^+ extracted into single microdroplets of $r \sim 40$ μm was measured at pH 2–6, the time required for the extraction equilibrium drastically decreased with the increasing pH. The absorbance of MB^+ in the microdroplet was saturated within 60 seconds at pH 6. Nonetheless, the MB^+ concentration in TBP at the extraction equilibrium was independent of the pH from 2 to 6 and the partitioning ratio of MB^+ (P_{MB}) was determined to be ~40 in the single microdroplet measurements (Table 9.1). Since DS^- is not known to be associated with a hydrogen ion in water at pH greater than ~2, the liquid/liquid extraction of DS^- with MB^+ for a quantitative analysis is generally performed at low pH. However, the pH in the water phase significantly influences the extraction rate of DS^- with MB^+. At a pH greater than 3, $A(t)$ monotonously increased and was saturated, different from that at pH 2. As a first approximation, the observed extraction rate was analysed by a single exponential rise curve using Equation (7).

$$A(t) = A_s\{1 - \exp(-k_s t)\} \tag{7}$$

where A_s and k_s are the saturated $A(t)$ and the apparent extraction rate constant, respectively. If a relationship between the interfacial mass transfer rate constants from water into the droplet (k_f) and from the droplet into water (k_b) is obtained by $P_{MB} = k_f/k_b$, k_s is given as in Equation (8) [59], analogous to Equation (2).

$$k_s = 3k_f/(r P_{MB}) \tag{8}$$

k_f for various pH values is summarized in Table 9.1. The rate constant increased with increasing pH. Since it took 1 minute for a single microdroplet injection and beam focusing onto the droplet, the extraction rate cannot be quantitatively discussed at pH 6. Nonetheless, k_f at pH 6 is close to k_d (1×10^{-3} cm · s^{-1}).

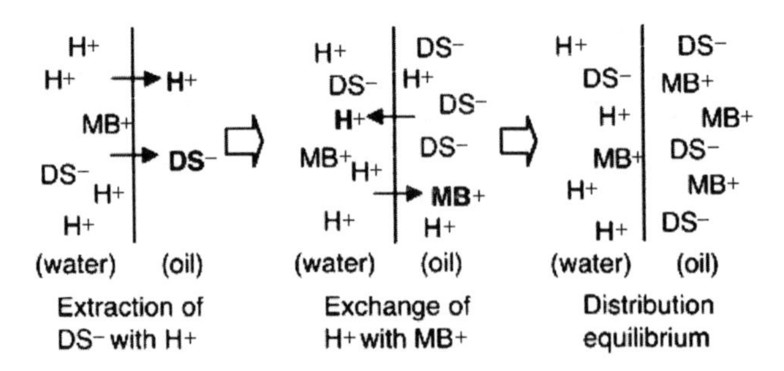

FIGURE 9.12. Ion-pair extraction scheme of DS^- in the presence of H^+ (pH 2) and MB^+.

9.5.3. *Transfer Mechanism of a Dodecyl Sulfate Anion with Methylene Blue*

Interfacial tension values of the TBP droplet in water in the absence and presence of DS^- were $\sim$8 and $\sim$5 $mN \cdot m^{-1}$ at pH 2, indicating that DS^- slightly adsorbs at the TBP/water interface. However, the interfacial tension was independent of pH so that the characteristic pH dependence of the ion-pair extraction rate in the DS^-/MB^+ system is not ascribed to the adsorption of the solute at the TBP/water interface. The standard ion transfer potentials of DS^-, Cl^-, Na^+, H^+ and MB^+ in the water/1,2-dichloroethane system were -0.08, -0.53, 0.59, 0.55 and $> \sim 0.4$ V, respectively [56]. The data predict that MB^+ is extracted with DS^-. On the other hand, DS^- will be extracted with H^+, Na^+ and MB^+. At low pH, the concentration of H^+ is much greater than that of MB^+ in the water phase (1×10^{-5} $mol \cdot dm^{-3}$) so that DS^- is likely to be extracted with H^+. By an ordinary extraction experiment in the TBP/water system without MB^+, the actual partitioning ratio of DS^- with H^+ (P_{DS}) was determined to be $\sim$40 and $\sim$4 at pH 2 and 6, respectively (Table 9.1). Therefore, DS^- is extracted with H^+ into the TBP phase in the absence of MB^+.

Based on these kinetic and thermodynamic data, the transfer mechanism of DS^- with MB^+ from water into the microdroplet is explained as follows. At pH 2, DS^- is extracted with H^+ into the microdroplet due to high H^+ concentration in water, as the first step of the distribution. However, MB^+ is a relatively hydrophobic compound compared with H^+ so that H^+ is slowly exchanged with MB^+ across the microdroplet/water interface, as the second distribution step (Figure 9.12). Therefore, the two-step consecutive reaction-type kinetics proceeding at the microdroplet/water interface could be observed (Figures 9.10 and 9.11). k_1 and k_2 correspond to the rate constants of the DS^- extraction with H^+ and the exchange of H^+ with MB^+ across the microdroplet/water interface, respectively. At pH 6, DS^- would be extracted with MB^+ on the basis of the individual ion transfer model (Figure 9.8a), so the extraction rate of MB^+ is rapid. At pH 3–5, DS^- is competitively extracted with both MB^+ and H^+ in the first step, and then H^+ partially distributed in the microdroplet is exchanged with MB^+ in water. Thus, the overall extraction processes would be analysed apparently by Equation (7). In conclusion, the characteristic pH dependence of the ion-pair extraction in the MB^+/DS^- system is ascribed to the H^+ extraction with DS^- across the microdroplet/water interface.

9.6. CONCLUDING REMARKS

Single microdroplet measurements were shown to be significant for elucidating the mass transfer processes across liquid/liquid interfaces. Using manipulation, microelectrochemistry and microspectroscopy of single microdroplets, the direct measurements of the mass transfer rates across liquid/liquid interfaces were successfully performed, so the transfer mechanism could be discussed in detail. Mass transfer in the nano-region of a liquid/liquid interface is one of the most important elementary processes in heterogeneous chemical reactions. However, the transfer mechanism of ion-pair extraction even in the absence of adsorption at a liquid/liquid interface was complicated since the ion-pair extraction rate was governed by the individual ion transfer rates, the ion transfer potentials and exchange of the ions between the two phases. For the ion-pair extraction of a surface-active ion, the adsorption/desorption rates at a liquid/liquid interface, instability of a liquid/liquid interface and so on will influence the ion-pair extraction rate. Therefore, a four-electrode technique, scanning electrochemical microscope and so forth as well as the single microdroplet technique should be used to obtain the complementary data. Furthermore, for the single microdroplet technique, simultaneous determination of different solutes using both electrochemical and spectroscopic methods is necessary for detailed mechanistic analyses. The chemical and physical processes occurring in the nano-region of liquid/liquid interfaces will be kinetically analysed by a combination of such approaches.

REFERENCES

1. Z. Samec, Chem. Rev. **88**, 617 (1988).
2. M. Senda, T. Kakiuchi and T. Osakai, Electrochim. Acta **36**, 253 (1991).
3. H.H. Girault, Charge Transfer across Liquid-Liquid Interfaces in *Modern Aspects of Electrochemistry*, Vol. 25, Eds. R.E. White, B.E. Conway and J.O. Bockris, Plenum, New York, 1993, p. 1.
4. A.G. Volkov and D.W. Deamer (Eds.), *Liquid–Liquid Interfaces, Theory and Methods*, CRC Press, Boca Raton, FL, 1996, p. 1.
5. A.G. Volkov, D.W. Deamer, D.L. Tanelian and V.S. Markin, *Liquid Interfaces in Chemistry and Biology*, John Wiley and Sons, New York, 1998, p. 1.
6. F. Reymond, D. Fermin, H.J. Lee and H.H. Girault, Electrochim. Acta **45**, 2647 (2000).
7. B. Liu and M.V. Mirkin, Anal. Chem. **73**, 670A (2001).
8. A.G. Volkov (Ed.), *Liquid Interfaces in Chemical, Biological and Pharmaceutical Applications*, Marcel Dekker, New York, 2001, p. 1.
9. H. Watarai, L. Cunningham and H. Freiser, Anal. Chem. **54**, 2390 (1982).
10. F.F. Cantwell and H. Freiser, Anal. Chem. **60**, 226 (1988).
11. H. Nagatani and H. Watarai, Anal. Chem. **70**, 2860 (1998).
12. T. Tokimoto, S. Tsukahara and H. Watarai, Chem. Lett. 204 (2001).
13. H. Yamada, S. Akiyama, T. Inoue, K. Koike, T. Matsue and I. Uchida, Chem. Lett. 147 (1998).
14. C.J. Slevin, J.A. Umbers, J.H. Atherton and P.R. Unwin, J. Chem. Soc., Faraday Trans. **92**, 5177 (1996).
15. A.L. Barker, J.V. Macpherson, C.J. Slevin and P.R. Unwin, J. Phys. Chem. B **102**, 1586 (1998).
16. T. Shioya, S. Nishizawa and N. Teramae, Langmuir **14**, 4552 (1998).
17. K. Nakatani, T. Uchida, H. Misawa, N. Kitamura and H. Masuhara, J. Phys. Chem. **97**, 5197 (1993).
18. K. Nakatani, T. Uchida, N. Kitamura and H. Masuhara, J. Electroanal. Chem. **375**, 383 (1994).
19. K. Nakatani, M. Wakabayashi, K. Chikama and N. Kitamura, J. Phys. Chem. **100**, 6749 (1996).

20. K. Nakatani, M. Sudo and N. Kitamura, J. Phys. Chem. B **102**, 2908 (1998).
21. T. Negishi and K. Nakatani, Phys. Chem. Chem. Phys. **5**, 594 (2003).
22. C.J. Slevin and P.R. Unwin, Langmuir **15**, 7361 (1999).
23. T.H. James, *The Theory of the Photographic Process*, Macmillan, New York, 1977, p. 339.
24. K. Sasaki, M. Koshioka and H. Masuhara, Appl. Spectrosc. **45**, 1041 (1991).
25. M. Koshioka, H. Misawa, K. Sasaki, N. Kitamura and H. Masuhara, J. Phys. Chem. **96**, 2909 (1992).
26. N. Tamai, T. Asahi and H. Masuhara, Chem. Phys. Lett. **198**, 413 (1992).
27. N. Tamai, T. Asahi and H. Masuhara, Rev. Sci. Instr. **64**, 2496 (1993).
28. S. Funakura, K. Nakatani, H. Misawa, N. Kitamura and H. Masuhara, J. Phys. Chem. **98**, 3073 (1994).
29. K. Nakatani, T. Sudo, M. Wakabayashi, H.-B. Kim and N. Kitamura, J. Phys. Chem. **99**, 4745 (1995).
30. T. Matsue, S. Koike, T. Abe, T. Itabashi and I. Uchida, Biochim. Biophys. Acta **1101**, 69 (1992).
31. D.J. Leszczyszyn, J.A. Jankowski, O.H. Viveros, E.J. Near and R.M. Wightman, J. Biol. Chem. **265**, 14736 (1990).
32. K. Nakatani, T. Uchida, H. Misawa, N. Kitamura and H. Masuhara, J. Electroanal. Chem. **367**, 109 (1994).
33. K. Nakatani, T. Sekine and T. Negishi, Microanalysis of Mass Transfer Processes on a Single Particle in *Encyclopedia of Surface and Colloid Science*, Vol. 3, Ed. A. Hubbard, Marcel Dekker, New York, 2002, p. 3391.
34. K. Nakatani and T. Sekine, Langmuir **16**, 9256 (2000).
35. K. Nakatani, K. Chikama and N. Kitamura, Laser Trapping-Spectroscopy-Electrochemistry of Individual Microdroplets in Solution in *Advances in Photochemistry*, Vol. 25, Ed. D.C. Neckers, D.H. Volman and G.V. Bunau, John-Wiley & Sons, New York, 1999, p. 173.
36. A. Ashkin, Phys. Rev. Lett. **24**, 156 (1970).
37. S.M. Block, D.F. Blair and H.C. Berg, Nature **338**, 514 (1989).
38. H. Misawa, M. Koshioka, K. Sasaki, N. Kitamura and H. Masahara, J. Appl. Phys. **70**, 3829 (1991).
39. K. Sasaki, M. Koshioka, H. Misawa, N. Kitamura and H. Masuhara, Jpn. J. Appl. Phys. **30**, L907 (1991).
40. N. Terui, K. Nakatani and N. Kitamura, J. Electroanal. Chem. **494**, 41 (2000).
41. T. Shimbashi, Bull. Chem. Soc. Jpn. **52**, 2832 (1979).
42. K. Nakatani, T. Suzuki, S. Shitara and N. Kitamura, Langmuir **14**, 2286 (1998).
43. K. Nakatani, K. Chikama, H.-B. Kim and N. Kitamura, Chem. Phys. Lett. **237**, 133 (1995).
44. K. Chikama, K. Nakatani and N. Kitamura, Chem. Lett. 665 (1996).
45. K. Chikama, K. Nakatani and N. Kitamura, Bull. Chem. Soc. Jpn. **71**, 1065 (1998).
46. S. Motomizu and K. Toei, Anal. Chim. Acta **120**, 267 (1980).
47. G.H. Morrison and H. Freiser, *Solvent Extraction in Analytical Chemistry*, John Wiley & Sons, New York, 1957, p. 59.
48. H.L. Friedman and G.R. Haugen, J. Am. Chem. Soc. **76**, 2060 (1954).
49. G.R. Haugen and H.L. Friedman, J. Phys. Chem. **72**, 4549 (1968).
50. Y. Yoshida, M. Matsui, O. Shirai, K. Maeda and S. Kihara, Anal. Chim. Acta **373**, 213 (1998).
51. F. Reymond, G. Steyaert, P.-A. Carrupt, B. Testa and H.H. Girault, J. Am. Chem. Soc. **118**, 11951 (1996).
52. J. Koryta and M. Skalicky, J. Colloid Interface Sci. **124**, 44 (1988).
53. F. Scholz, S. Komorsky-Lovric and M. Lovric, Electrochem. Commun. **2**, 112 (2002).
54. R. Gulaboski, V. Mirceski and F. Scholz, Electrochem. Commun. **4**, 277 (2002).
55. K. Nakatani and T. Negishi, Anal. Sci. **17**, 1109 (2001).
56. H.H. Girault, see the Web site dcwww.epfl.ch for a comprehensive list of free energy of ion transfer at various liquid/liquid interfaces, 2000.
57. A.S. Weatherburn, Can. Textile **16**, 71 (1964).
58. M.J. Rosen and H.A. Goldsmith, *Systematic Analysis of Surface-Active Agents*, Wiley-Interscience, New York, 1972, p. 17.
59. K.Chikama, T. Negishi and K. Nakatani, Bull. Chem. Soc. Jpn. **76**, 295 (2003).

10

Single Molecule Diffusion and Metal Complex Formation at Liquid/Liquid Interfaces

Hitoshi Watarai and Satoshi Tsukahara[1]
Osaka University, Toyonaka, Osaka 560-0043, Japan

10.1. INTRODUCTION

Honaker and Freiser reported the first fundamental kinetic mechanism of chelate extraction in 1962 [1]. They elucidated that the rate-determining step for the extraction of divalent metal ions with dithizone was the formation of their 1:1 complexes in the aqueous phase. They proposed that a simple batch extraction method could be used as an alternative method of the complicated stopped-flow method, which was the only method available at the time, to measure such a fast reaction rate. Since the 1970s, hydrometallurgy has been developed in many countries, and extensive kinetic studies on the metal extraction have been conducted in an effort to improve the extraction rate as well as to develop effective and reusable extractants. The extractants used in hydrometallurgy are required to be highly hydrophobic and readily coordinative with various metal ions. On the basis of the interfacial adsorptivity of the extractant, Flett *et al.* [2] expected an interfacial reaction mechanism in the chelate extraction process. There was, however, no experimental evidence to prove the interfacial mechanism directly [3].

The solvent extraction process of metal ions inherently depends on the mass transfer across the interface and the reaction that occurs at the interfacial region. Therefore, the elucidation of the kinetic role of the interface was very important in order to clarify the extraction mechanism and to control the extraction rates. In 1982, Watarai and Freiser invented the high-speed stirring (HSS) method [4,5]. Figure 10.1 shows the schematic drawing of the HSS method [6]. When a two-phase system is vigorously stirred in a

[1] Present address: Hiroshima University, Higashi-Hiroshima, Hiroshima 739-8526, Japan.

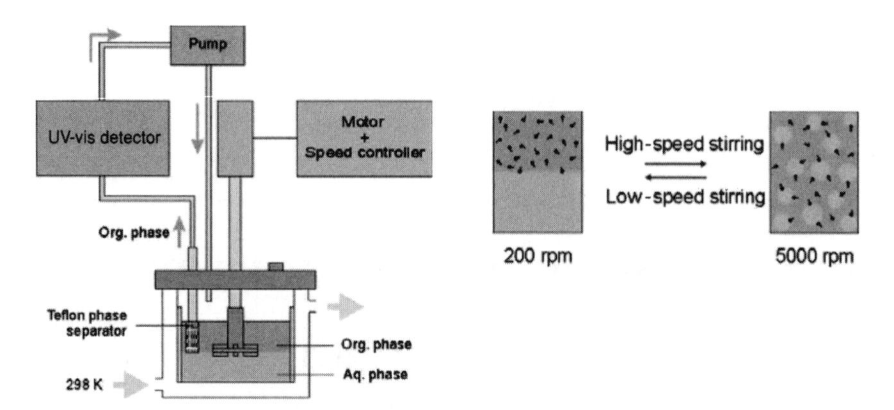

FIGURE 10.1. Schematic drawing of the high-speed stirring (HSS) apparatus. The increase in the interfacial area by the stirring can amplify the amount of the interfacial adsorbate, which results in the decrease in the bulk concentration.

vessel at a rate of 4000–5000 rpm and the interfacial area is extremely large, the amount of interfacial adsorbed compound is increased and thus the concentration in the organic phase is decreased. The change in the organic phase species and its concentration can be measured using spectrometry. The interfacial area can easily be increased to 500 times that of the standing condition. The specific interfacial area under HSS conditions can attain values as high as 400 cm^{-1}. For the first time, this method enabled the simultaneous determination of the interfacial concentration and the interfacial reaction rate. This simultaneous determination was a breakthrough not only in the studies of the extraction mechanism, but also in elucidating of the role of the interface in two-phase reaction kinetics. Thereafter, a number of new methods have been successively proposed: the two-phase stopped-flow method [7], total internal reflection (TIR) spectrometry [8,9], the centrifugal liquid membrane (CLM) method [10] and the two-phase sheath flow method [11]. Very recently, the micro-two-phase flow/MS method [12], the single-molecule probe method [13] and CLM-Raman microscope spectrometry [14] have been proposed. Magnetophoretic velocimetry of single droplets [15] is also an attractive new principle for the migration analysis of interfacial species.

What is the specific feature in the reaction at the liquid/liquid interface? The catalytic role of the interface is of primary importance in solvent extraction and other two-phase reaction kinetics. In solvent extraction kinetics, the adsorption of the extractant or an intermediate complex at the liquid/liquid interface significantly increased the extraction rate. Secondly, interfacial accumulation or concentration of adsorbed molecules, which very often results in interfacial aggregation, is an important role played by the interface. This is because the interface is available to be saturated by an extractant or metal complex, even if the concentration of the extractant or metal complex in the bulk phase is very low. Molecular recognition or separation by the interfacial aggregation is the third specific feature of the interfacial reaction and is thought to be closely related to the biological functions of cell membranes. In addition, molecular diffusion of solute and solvent molecules at the liquid/liquid interface has to be elucidated in order to understand the molecular mobility at the interface. In this chapter, some examples of specific interface reactions are reviewed and the future developments of the interfacial reactions are discussed.

10.2. INTERFACIAL DIFFUSION DYNAMICS

Chemical reaction kinetics depends on the thermal fluctuation and diffusion of all reactants and solvent molecules. In addition, the specific reaction kinetics at the liquid/liquid interface must be considered. The diffusion to the interface and the diffusion within the interface are important processes in determining the reaction rate at the interface. Several aspects of interfacial diffusion and the rotational motion of the interfacially adsorbed molecules are introduced here as results of recent studies.

10.2.1. Diffusion Limited Reaction Rate at the Liquid/Liquid Interface

Interfacial protonation of tetraphenylporphin (TPP) was investigated by means of a two-phase stopped-flow method [7] and a CLM method [10]. The protonation of TPP at the interface, which can be detected from a definite shift in its absorption maximum from 416 to 438 nm, is very fast. The rate of interfacial protonation was spectrophotometrically determined from the rate of formation of the protonated TPP at the interface of dodecane/aqueous trichloroacetic acid. The rate of protonation was so fast that the observed rate was governed by the TPP diffusion rate in the organic phase. In the two-phase stopped-flow study, the rate of protonation was observed less than 200 milliseconds after two-phase mixing. The observed first-order rate constant of 19 s^{-1}, which was independent of both acid and TPP concentrations, was analysed on the basis of the simple diffusion layer model [3]:

$$k_{obs} = \frac{D}{\delta}\frac{S_i}{V_0} \tag{1}$$

where D, δ, S_i and V_o stand for the diffusion coefficient of TPP in dodecane, the thickness of the diffusion layer, the interfacial area and the volume of bulk phase, respectively. The ratio of S_i/V_o, called a specific interfacial area, was 777 cm^{-1} for the dispersed system. On the basis of these parameters with $D = 3.5 \times 10^{-6}\ cm^2 \cdot s^{-1}$, the thickness in the dispersed system was estimated as $\delta = 1.4\ \mu m$. This is thought to be almost the minimum stagnant layer thickness at the interface.

To investigate the relationship between the interfacial rate constant and the diffusion layer thickness, the protonation reaction rate was measured by the CLM method. The schematic diagram of the CLM apparatus is shown in Figure 10.2. In the CLM method, the thickness of the dodecane phase in the aqueous hydrochloric acid phase can be controlled by the initial volume of each. When the thickness of the dodecane phase was changed from 53 to 132 μm, the observed interfacial diprotonation rate constant decreased from $9.9 \times 10^{-2}\ s^{-1}$ to $1.5 \times 10^{-2}\ s^{-1}$. The experimental results were well reproduced by Equation (1), substituting δ with the organic phase thickness [10].

Recently, the micro-two-phase sheath flow method (Figure 10.3) has been invented. Fast interfacial reactions, which proceeded in less than 1 millisecond, were measured using this method, coupled with fluorescence microspectroscopy. An inner organic micro-flow was generated in the aqueous phase flowing from the tip of a capillary (i.d.,

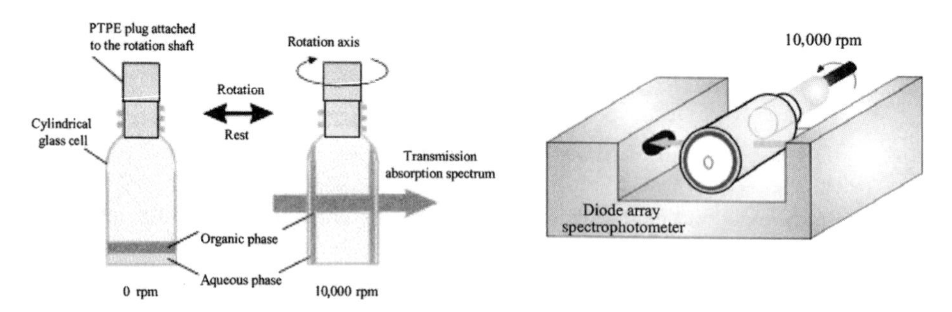

FIGURE 10.2. Schematic diagram of the centrifugal liquid membrane (CLM) method apparatus. The glass cylinder cell (i.d., 19 cm) was rotated at a speed of 7000–10,000 rpm. Introduction of 50–200 μl of the organic and the aqueous phase generated a two-liquid layer system with a thickness of 50–100 μm for each phase.

10–30 μm) with the same linear velocity. The fluorescence spectrum at the interface was observed as a function of the distance from the tip of the inner capillary, which was the starting point of the reaction. The distance from the tip was converted to the reaction time. Thus, a very fast interfacial or diffusion reaction (less than 1 mllisecond) could be measured. This technique enabled the detection of a fast interfacial reaction of the order of 10 μs. This technique is analogous to the continuous-flow method developed for a homogeneous fast reaction [16].

The fast complexation rate of Zn(II) ion with 8-quinolinol (Hqn) or 5-octyloxymethyl-8-quinolinol (Hocqn) at the 1-butanol/water interface was measured by the micro-two-phase sheath flow method [17]. The formation of a fluorescence complex at the interface was measured within a period of less than 2 milliseconds after the contact of the two phases. The depth profile of the fluorescence intensity observed across the inner organic phase flow proved that the fluorescence complex was formed only at the interface and it increased in proportion to the contact time. The diffusion length of Hocqn in the 1-butanol phase for 2 milliseconds was calculated as 0.8 μm, which is smaller than the experimental resolution depth of 2 μm in the microscopy used. Therefore, the observed rate constant was analysed by taking diffusion and reaction rates into account between Zn(II) and Hocqn at the interfacial region by a digital simulation method. The digital simulation has been used in the analysis of electrode reactions,

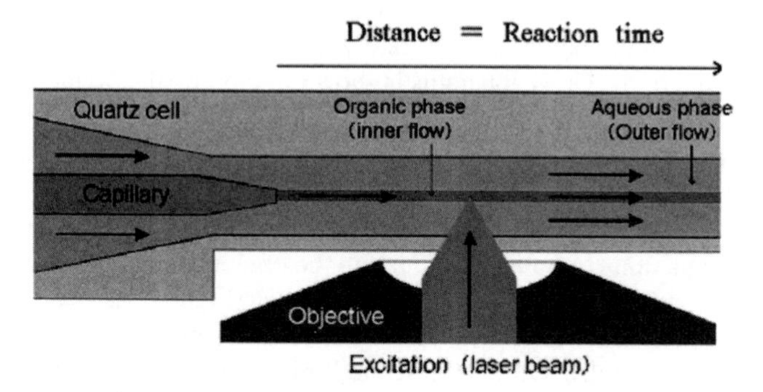

FIGURE 10.3. Schematic drawing of the laser fluorescence microscopy method for the measurement of fast interfacial reactions in micro-two-phase sheath flow systems.

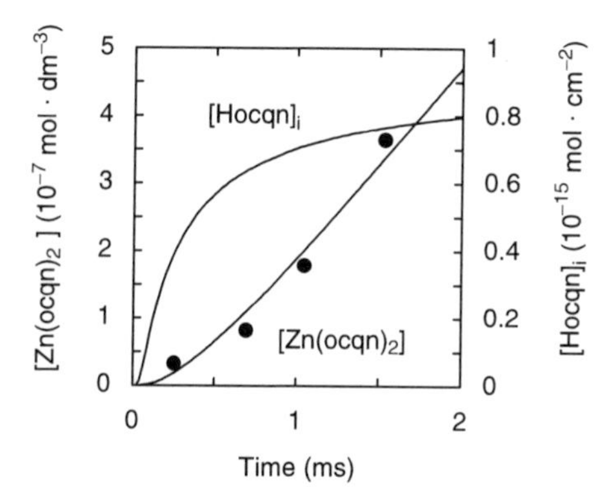

FIGURE 10.4. Result of the digital simulation of the Zn(II)–Hocqn complexation at the interfacial region. The solid lines represent the simulated concentration and the dots represent the observed concentration. The initial concentration for Hocqn was 1.1×10^{-4} M.

which was modified for the present simulation at the liquid/liquid interface [17]. The result of the digital simulation for the Zn(II)—Hocqn system in 1-butanol/water is shown in Figure 10.4. The interfacial concentration of $Zn(ocqn)_2$ gradually increased, though that of Hocqn increased rapidly. This might be due to the retardation effect of 1-butanol which is adsorbed at the interface.

10.2.2. Diffusion of Single Molecule in the Liquid/Liquid Interface

A single molecule probing is an ultimate method for measuring the properties of the nano-environment at the liquid/liquid interface. A number of studies on single molecule detection in solutions [18–20] and at the solid/liquid interface [21–23] were reported, but there were no studies on single molecule detection at the liquid/liquid interface.

The first observation of the single molecule detection at the liquid interface has been recently accomplished by means of the TIR fluorescence microscopy [13]. Figure 10.5 shows the optical arrangement of the measurement method and the microcell used in the study.

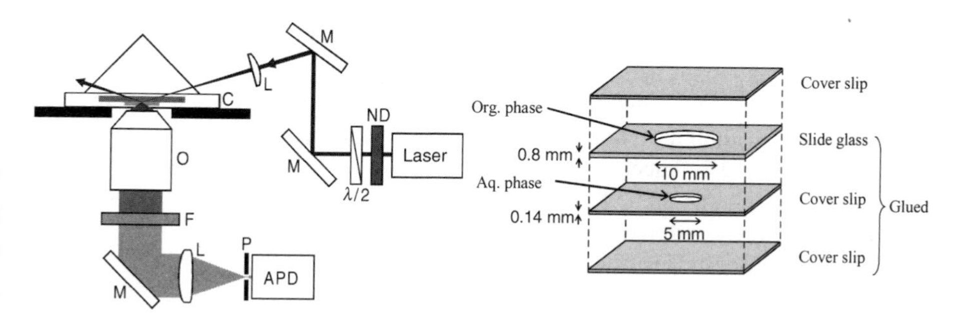

FIGURE 10.5. The left figure shows the schematic illustration of laser-induced fluorescence microscopy under the total internal reflection for the detection of single DiI molecules at the dodecane–water interface. Abbreviations: ND, ND filter; λ/2, λ/2 plate; M, mirror; L, lens; C, microcell containing dodecane and aqueous phases; O, objective (60 ×); F, bandpath filter; P, pinhole; APD, avalanche photodiode detector. The right portion of the figure shows the composition of the microcell.

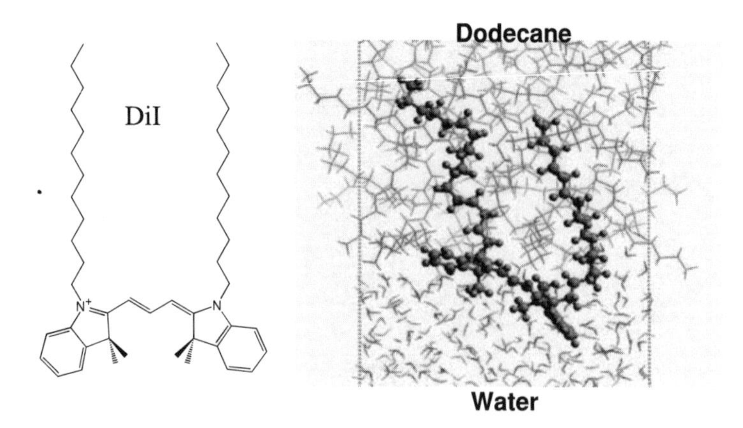

FIGURE 10.6. Molecular structure of DiI (left) and a snapshot of the MD simulation of DiI adsorbed at the dodecane/water interface (right).

The apparatus consisted of an inverted microscope, an oil immersion objective (PlanApo 60 ×, NA 1.4, working distance 0.21 mm), a cw-Nd:YAG laser (532 nm) and an avalanche photodiode detector (APD), which provided a quantum efficiency of about 65–67% at 570–600 nm. A pinhole, 50 μm in diameter, was attached just in front of the photodiode, restricting the observation area to a diameter of 830 nm (d_{obs}). The laser beam was focused to a lens at the interface through a quartz rectangular prism in a flat two-phase microcell. The shape of the laser beam at the interface was an ellipse, about 30 μm × 100 μm in size. P-polarized laser light was irradiated at an angle of incidence 73° from the interface, which was larger than the critical angle. Fluorescence emitted by interfacial molecules was collected by the objective lens and focused on the pinhole after passing through a band pass filter (path range, 587.5–612.5 nm). Time-resolved photon counting was carried out using a multichannel scaler. The overall detection efficiency of fluorescence at the interface was calculated to be 3.3%.

The influence of two kinds of surfactants, sodium dodecyl sulfate (SDS) and dimyristoyl phosphatidylcholine (DMPC), on the lateral diffusion dynamics of single molecules at the interface was investigated. 1,1′-Dioctadecyl-3,3,3′,3′-tetramethylindocarbocyanine (DiI, Figure 10.6) was used as a fluorescent probe molecule. DiI is a monovalent cation with two C_{18} alkyl-chains. Thus, it has high adsorptivity at the dodecane/water interface. A two-phase microcell, shown in Figure 10.5, was fabricated by adjoining a bored slideglass, a bored coverslip and another non-bored coverslip in this order. Aqueous phase (2.7 mm^3) was placed in the thin lower compartment of the cell and the DiI dodecane solution (63 mm^3) was added on top of the aqueous layer. A new coverslip was placed on the dodecane phase in order to close the cell. Fluorescence of the interfacial DiI was observed in the range of 587.5–612.5 nm.

DMPC was dissolved in chloroform, and the solution was mixed with pure diethyl ether at a ratio of 1:19 (chloroform/diethyl ether) by volume. Pure water was placed in the lower container, and the DMPC solution was subsequently (5 mm^3) spread carefully on the water. After evaporation of chloroform and diethyl ether, the DiI dodecane solution was added on the DMPC layer. Since DiI has a high adsorptivity at the interface, no DiI

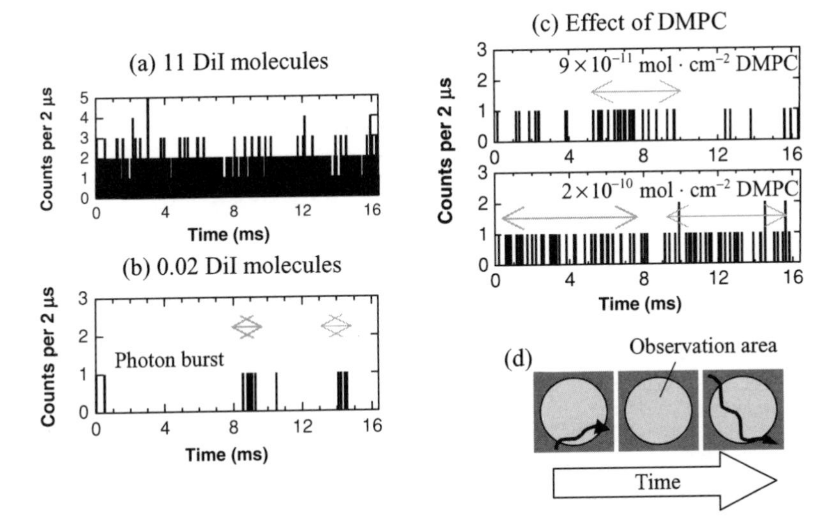

FIGURE 10.7. Total internal reflection fluorescence microscopy of the micro-two-phase system of dodecane/water. (a) Continuous photons with an average of 11 molecules in the observation area, (b) photon burst with an average of 0.02 molecules in the observation area and (c) extension of the photon burst upon the addition of DMPC. A model for the photon burst upon the observation using single molecule diffusion (d).

molecules remained in the dodecane phase in the samples at an initial DiI concentration lower than 1×10^{-7} M. Under the low concentrations of interfacial DiI, the contribution of the photobleaching was found to be negligibly small in the present study.

When there were many DiI molecules (11 molecules on an average) in the observation area, continuous photon signals were observed (Figure 10.7a). However, in the presence of a few of the DiI molecules (0.02 molecules on an average), a clear intermittent photon bundle was observed (Figure 10.7b). Given that DiI molecules are adsorbed at the interface, the diffusion of DiI is thought to be restricted in the lateral direction at the interface. Therefore, the maximum duration of the photon bundles (t_{max}) corresponds to the period in which the single DiI molecule is moving in the observation area. The t_{max} is the time for a single molecule to move along the diameter of the round observation area.

The lateral diffusion coefficient (D_l) of DiI was calculated using the following equation:

$$D_l = \frac{d_{obs}^2}{2t_{max}} \tag{2}$$

The radius (r) and the diffusion coefficient (D) of a spherical molecule are related to the viscosity (η) of the medium surrounding the molecule by the Einstein–Stokes equation:

$$D = \frac{kT}{6\pi r \bar{\eta}_i} \tag{3}$$

where k is the Boltzmann constant and T is the absolute temperature. The DiI molecule was estimated from the molecular volume of DiI to be a sphere with a radius of 0.70 nm.

TABLE 10.1. Effects of surfactants on the lateral diffusion coefficient (D_l) of single DiI molecules, apparent viscosity ($\bar{\eta}_i$) and intrinsic viscosity (η_i) at the dodecane/water interface.

Surfactant	$\Gamma_s{}^a$(mol · cm^{-2})	D_l(cm^2 · s^{-1})	$\bar{\eta}_i$(mPa · s)	η_i(mPa · s)
Free		2.3×10^{-6}	1.4	
SDS	2.0×10^{-10}	1.7×10^{-6}	1.8	–[b]
	2.5×10^{-10}	1.6×10^{-6}	1.9	–[b]
DMPC	2.0×10^{-12}	2.1×10^{-7}	15	37
	2.0×10^{-11}	4.3×10^{-8}	72	270
	9.0×10^{-11}	2.7×10^{-8}	120	470
	2.0×10^{-10}	1.8×10^{-8}	170	750

[a] Interfacial concentration of surfactant.
[b] Too low to be evaluated.

The apparent interfacial viscosities ($\bar{\eta}_i$) evaluated using Equation (3) are listed in Table 10.1 [24]. In the surfactant-free system, the calculated viscosity of 1.4 mPa · s was close to the value for dodecane (1.4 mPa · s) and was higher than that of water (0.89 mPa · s). This result suggested that the two long alkyl groups of DiI were deeply immersed in the dodecane phase, as shown in Figure 10.6.

SDS did not have a significant effect, despite the fact that the interfacial concentration of SDS was as high as the saturated interfacial concentration, 2.5×10^{-10} mol · cm^2. The D_l and $\bar{\eta}_i$ values, listed in Table 10.1, were only slightly affected by the coexistence of SDS. This indicates that the translational motion of interfacial SDS molecules is similar to that of dodecane molecules, because both are almost the same size and possess dodecyl chains.

On the other hand, DMPC had a remarkable effect, as shown in Figure 10.7c. Since the phase transition temperature of DMPC is 23°C, DMPC is in the liquid-like state under the present experimental temperature (25°C). The average number of DiI molecules in the observation area was less than 0.1. In Figure 10.7c, the width of the photon bundle increased with the increase in the interfacial DMPC concentration. A histogram of the photon number per channel was consistent with that predicted by the Poisson distribution with a high background and a small signal peak from a single DiI molecule. The D_l and $\bar{\eta}_i$ values of DiI are also listed in Table 10.1. The maximum $\bar{\eta}_i$ value in the DMPC system was 0.17 Pa · s, which was higher than that at the surfactant-free interface by two orders of magnitude. In the DMPC molecule, two-alkyl chains are bound and not exchangeable. This is thought to be the cause of the stiff monolayer generated by DMPC molecules.

One of the primary compounds that the cell membrane is composed of is DMPC, because of its surface activity caused by a hydrophobic zwitterion. A hydrodynamic model of the DMPC membrane can evaluate the intrinsic viscosity (η_i) of the surfactant monolayer, eliminating the contribution of the viscosity of dodecane and aqueous phases. The η_i values listed in Table 10.1 are about 2–4 times higher than the apparent $\bar{\eta}_i$. The maximum η_i value, 0.75 Pa · s, is comparable to that of a common viscous liquid such as glycerin (0.945 Pa · s). This study demonstrated that a single molecule probing method could successfully measure the hydrodynamic properties of the interface.

10.2.3. Rotational Dynamics at the Liquid/Liquid Interface

The adsorption equilibria of fluorescent *N*-octadecylrhodamine B ($C_{18}RB$) at a free toluene/water interface have been studied using steady-state TIR fluorometry [25]. Both the zwitterion ($C_{18}RB^{\pm}$) form and the protonated form ($C_{18}RBH^{+}$) exhibit fluorescence; they are present only at the interface, whereas the non-fluorescent lactone form ($C_{18}RB$) exists only in the toluene phase. There is no distribution of $C_{18}RB$ in the aqueous phase. There were investigated the effects of three surfactants, neutral Triton X-100 (TX-100), anionic sodium dodecyl sulfate (SDS) and anionic dihexadecyl hydrogen phosphate (DHP, $pK_a \approx 2$). The adsorption of TX-100 prevented that of $C_{18}RB$ from the organic phase. The adsorption of SDS and deprotonated DHP promoted that of $C_{18}RBH^{+}$ by electrostatic interactions. The increase in the bulk concentration of a surfactant higher than 10^{-6} $mol \cdot dm^{-3}$ reduced the interfacial concentration of $C_{18}RBH^{+}$ as a result of the interfacial occupation of the surfactant [26].

In-plane and out-of-plane rotational dynamics of $C_{18}RB$ at the toluene/water interface was evaluated using time-resolved TIR fluorescence spectroscopy [27]. The known transition dipole moment for the absorption of rhodamine B(RB) at about 530 nm ($S_0 \rightarrow S_1$) is almost parallel to that for the emission at about 570 nm ($S_1 \rightarrow S_0$) [28]. Time-resolved in-plane fluorescence anisotropy ($r_{\Phi}(t)$) was obtained using

$$r_{\Phi}(t) \equiv \frac{I_{ss}(t) - I_{sp}(t)}{I_{ss}(t) + I_{sp}(t)} = 4\langle \cos^2 \Phi_a(0) \cos^2 \Phi_e(t) \rangle - 1 \tag{4}$$

where $I(t)$ is the time-resolved fluorescence intensity. The first subscript, s, corresponds to the excitation with s-polarized light, and the second subscripts, s and p, represent the detection of s- and p-polarized emission light, respectively. Φ is the rotational angle of the transition dipole at the interface, and subscripts a and e are absorption and emission transition dipole moments, respectively. <>is the average for all molecules over the entire Φ angle. When molecules rotate with a rotational correlation time (τ or τ_{Φ}), $r(t)$ of the molecules can be expressed as

$$r(t) = \{r(0) - r(\infty)\}e^{-t/\tau} + r(\infty) \tag{5}$$

The in-plane rotation of $C_{18}RBH^{+}$ was not retarded by the interfacial TX-100 molecules. The in-plane rotational relaxation of $C_{18}RBH^{+}$ was, however, slowed down by the anionic surfactants (SDS, DHP).

Regardless of the surfactants, time-resolved out-of-plane fluorescence anisotropy was independent of time, showing a constant value of –0.23. This indicates that the $C_{18}RBH^{+}$ molecules are tilted at an angle of $65 \pm 2^{\circ}$ to the toluene/water interface. The $C_{18}RBH^{+}$ xanthene ring would thus be tilted by this angle from the normal to the interface.

The in-plane rotational correlation time (τ_{Φ}) obtained in the SDS and DHP systems may be reflected by interfacial characteristics. In a bulk solution, the rotational correlation time (τ) of a fluorophore is expressed as

$$\tau = \frac{\nu\eta}{k_B T} \tag{6}$$

where η is the viscosity of the solvent, k_B the Boltzmann constant, T the absolute temperature and v the molecular volume of the solute. The molecular volume of $C_{18}RB$ was estimated to be 0.95 nm^3 [27]. The viscosities for water and toluene were 8.90×10^{-4} and 5.53×10^{-4} Pa · s at 25°C, respectively, and τ values of $C_{18}RB$ in these solvents could be calculated as 0.21 and 0.13 ns, respectively. These estimations were consistent with the experimental results that the τ_{Φ} value was less than 0.5 ns.

Under the same concentration, 1.0×10^{-6} M, for the surfactants of SDS and DHP, τ_{Φ} values were 5.2 ± 1.4 ns and 21 ± 5 ns, which gave η values of 0.023 Pa · s and 0.091 ns for the SDS and DHP systems, respectively. These η values were clearly higher than those of toluene and water, while lower than that of a typical viscous solvent, glycerol (0.945 Pa · s).

The fluorescent molecule, $C_{18}RBH^+$, can be used as an alternative probe in the measurement of the interfacial viscosity of microparticles (about 0.5 mm in size) or rotating disks (several, 10 cm in radius) [29,30]. In comparison with the microparticles, the molecular probe has definite advantages. One of those is that the interference of the bulk viscosity is negligibly small and that information about the nano-environment around the probe can be obtained.

10.3. INTERFACIAL CATALYSIS

10.3.1. Interfacial Catalysis in Chelate Extraction

The extraction kinetics of Ni(II) and Zn(II) with n-alkyl-substituted dithizone (HL) was the first system in which the interfacial kinetics was elucidated using the HSS method [4,5]. A schematic drawing of the apparatus is shown in Figure 10.1. Usually the measurements are carried out using 50 ml for each phase at a stirring speed of 5000 rpm. The interfacial area can be increased to 500 times larger than that in the standing state. The specific interfacial area under the HSS conditions becomes as high as 400 cm^{-1}. The maximum interfacial concentration of an ordinary compound is of the order of 10^{-10} mol · cm^2. This predicts that the maximum amount of an adsorbable solute at the interface corresponds to the amount of a solute in 50 ml of solution at a concentration of 10^{-4} M. This concentration is ready to be measured using an ordinary spectroscopic method such as spectrophotometry or fluorometry. In the original HSS method, the concentration depression in the organic phase was measured continuously by spectrophotometry using a polytetrafluoroethylene (PTFE) phase separator and a photodiode array spectrometer. The interfacial amount of a ligand or a complex was observed as a function of time, and the extraction rate was determined from the spectral change caused by stirring.

The observed extraction rate constants linearly depended on both the metal ion concentration $[M^{2+}]$ and the hydrogen ion concentration in the aqueous phase. However, the observed extraction rate constant (k') did not decrease with an increase in the distribution constant (K_D) of the ligand as was expected from the mechanism in the aqueous phase. Furthermore, the HSS method revealed that the dissociated form of the n-alkyl-dithizone was adsorbed at the interface generated by vigorous stirring [5]. The following scheme was proposed based on the experimental results, considering both the aqueous

phase reaction and the interfacial reaction between M^{2+} and the dissociated form of the ligand L^-:
$M^{2+} + L^- \rightarrow ML^+$ k, formation rate constant in bulk aqueous phase
$M^{2+} + L_i^- \rightarrow ML_i^+$ k_i , interfacial formation rate constant.
The rate law for the extraction was obtained as

$$\frac{d[M^{2+}]}{dt} = k' \frac{[M^{2+}][HL]_o}{[H^+]} \tag{7}$$

and the observed extraction rate constant (k') is represented by

$$k' = \frac{K_a}{K_D}\left(k + \frac{k_i K_L' A_i}{V}\right) \tag{8}$$

where A_i/V refers to the specific interfacial area, K_a the dissociation constant and K_L' the adsorption constant of L^- from the aqueous phase to the interface. These equations were experimentally proved and the interfacial formation rate constants were determined as the $\log(k_i/M^{-1} \cdot s^{-1}) = 8.08$ for the Zn^{2+} system and $\log(k_i/M^{-1} \cdot s^{-1}) = 5.13$ for the Ni^{2+}system, respectively [6].

A criterion of the interfacial reaction in the chelate extraction was that the interfacial ligand concentration had to follow an adsorption isotherm. This was proved in the extraction systems of Ni(II) with 2-hydroxy oxime such as 5-nonylsalicylaldehyde oxime (P50) [31], 2-hydroxy-5-nonylacetophenone oxime (SME529) [32] and 2′-hydroxy-5′-nonylbenzophenone oxime (LIX65N) [33]. These extractants were adsorbed at the interface in their neutral forms, obeying the Langmuir isotherm:

$$[HL]_i = \frac{aK'[HL]_o}{a + K'[HL]_o} \tag{9}$$

where $[HL]_i$ and $[HL]_o$ refer to the concentrations at the interface and the organic phase, respectively, a the saturated interfacial concentration and K' the interfacial adsorption constant defined by $K' = [HL]_i/[HL]_o$ under the condition of $[HL]_o \rightarrow 0$. The interfacial adsorptivity of the ligands was confirmed also by the interfacial tension measurement. The initial rate for the extraction (r^0) is represented by

$$r^0 = k_i K_a' n_i [Ni^{2+}]/[H^+] \tag{10}$$

where K_a' is the dissociation constant at the interface and n_i is the initial interfacial amount of the ligand. A linear relationship between r^0 and n_i was experimentally confirmed. It was subsequently concluded that the reaction between the dissociated form (L^-) and the metal ions at the interface governed the extraction rate. The complex formed at the interface was not adsorbed at the interface, but was extracted into the organic phase. The adsorption constants of the three 2-hydroxy oximes and the complexation rate constants with Ni(II) and Cu(II) ions at the interface were determined by means of the HSS method as listed in Table 10.2.

The adsorption constants (K') of the neutral forms were all of the order of 10^{-3} cm and the complexation rate constants of the dissociated form with the Ni(II) ion at the interface were of the order of 10^5 $M^{-1} \cdot s^{-1}$ for the three extractants, whereas the

TABLE 10.2. Kinetic parameters obtained in the adsorption and extraction of Ni(II) and Cu(II) with 2-hydroxy oxime in heptane/0.1 M(H,Na)ClO_4 solution at 25°C.

2-Hydroxy oxime	pK_a	log K_D	log a mol · cm^{-2}	log A_i (cm^2)	log K' (cm)	log k_i M^{-1} · s^{-1}
Ni(II)/5-Nonylsalicyl aldehyde oxime (P50)	9.00	3.36	−9.60	4.11	−3.34	4.57
Ni(II)/2′-Hydroxy-5′-nonylacetophenone oxime (SME529)	9.79	3.99	−9.70	4.29	−3.29	5.11
Ni(II)/2-Hydroxy-5-nonylbenzophenone oxime (LIX65N)	8.70	5.69	−9.76	4.28	−2.97	5.14
Cu(II)/LIX65N	8.70	5.69	−9.76	4.28	−2.97	9.87

distribution constants (K_D) were significantly different. The complexation rate constants at the heptane/water interface were not significantly different from the values in a bulk aqueous solution.

In Table 10.2, it is important to note that the adsorption constant from the organic phase to the interface (log K') is not affected by the distribution constant (log K_D). This can be explained from the free energy relationship between the adsorption and distribution. The distribution process, which refers to the transfer free energy (G_{tr}) of a solute molecule from the aqueous phase to the organic phase, can be divided into the free energy of adsorption (G_{ad}) and the free energy of dehydration (G_{deh}) by the following equations:

$$\begin{aligned} \Delta G_{tr} &= \Delta G_{ad} + \Delta G_{deh} \\ \Delta G_{tr} &= -RT \ln K_D \\ \Delta G_{deh} &= RT \ln K' \end{aligned} \tag{11}$$

then

$$RT \ln K' = \Delta G_{tr} - \Delta G_{ad} = -RT \ln K_D + RT \ln K'' \tag{12}$$

where K'' stands for the adsorption constant of HL from the aqueous phase to the interface under infinitely diluted conditions, which also corresponds to the process of transfer of a hydrophobic moiety on the solute molecule from the aqueous phase to the organic phase at the interface. These equations reveal that since both K_D and K'' depend on the hydrophobicity of the solute molecule, K' is governed by the hydration energy of the polar moiety of the solute molecule. Equation (12) can be used to predict the adsorption constant from the log K_D for an analogous solute.

The dynamic behaviour of the 2-hydroxy oxime and its adsorptivity at the interface were well depicted by the molecular dynamics (MD) simulations [34]. It was revealed that the polar groups of −OH and =N−OH of the adsorbed 2-hydroxy oxime molecule were accommodated in the aqueous phase side so as to react with the Ni(II) ion in the aqueous phase [35]. This was thought to explain that the magnitude of the reaction rate constants of Ni(II) at the heptane/water interface and that in the aqueous phase were similar to each other. The diffusive and adsorptive behaviour of LIX65N around the interface was also simulated for 1 ns. The molecule was active around the interfacial region, moving

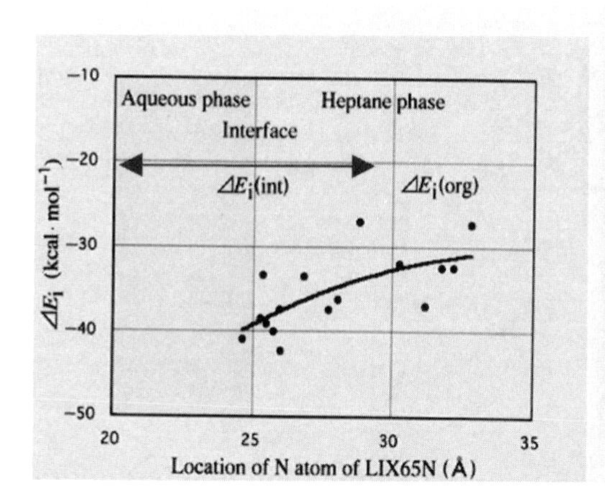

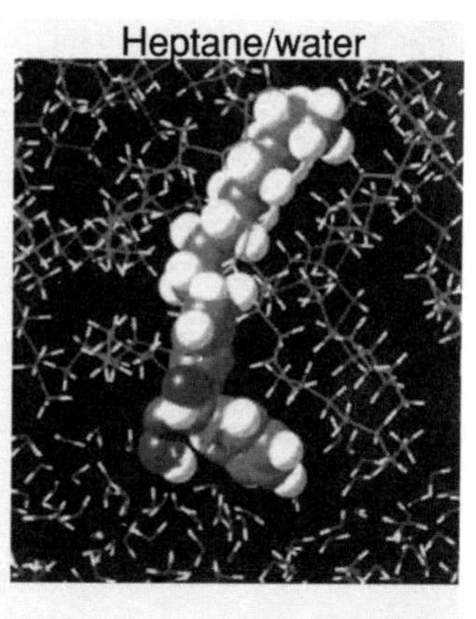

FIGURE 10.8. Photograph of the adsorbed 2-hydroxy-5-nonylbenzophenoneoxime (LIX65N) in the heptane/water system and the change in the energy of solvation with the position of the N atom of the ligand molecule calculated using the MD simulation for 1 ns. The energy of solvation is lowest in the interface.

between the interface and the heptane phase. Estimated interaction energies of LIX65N with surrounding solvent molecules are plotted in Figure 10.8 as a function of the position of the N atom of the LIX65N molecule. Figure 10.8 clearly shows that LIX65N is more stable at the interface than in the heptane phase in terms of the interaction energy.

Pyridylazophenol ligands have been widely used in the extraction photometry of various metal ions. For example, 1-(2-pyridylazo)-2-naphthol (Hpan) is one of the most well-known reagents, but it has a slow extraction rate for some metal ions such as Ni(II) and Pd(II). 2-(5-Bromo-2-pyridylazo)-5-diethylaminophenol (5-Br-PADAP) is more sensitive than Hpan for Cu(II), Ni(II), Co(II) and Zn(II), yielding metal complexes with high molar absorptivities, of the order of 10^5 $M^{-1} \cdot cm^{-1}$. 5-Br-PADAP showed a significant adsorption at the heptane/water interface under HSS conditions (5000 rpm). On the other hand, the adsorptivity at the toluene/water interface was very low. Hpan was not adsorbed at the toluene/water interface at all. The adsorption constants of 5-Br-PADAP (HL) at the heptane/water and toluene/water interfaces were obtained as log $K'A_i(\text{cm}^3) = 1.64$ and log $K'A_i$ $(\text{cm}^3) = -0.367$, respectively [6]. The solvent effect on the adsorptivity of the ligand directly affected the interfacial reaction rate. In the heptane/water system, the Ni(II) complex was not extracted into the heptane phase. On the other hand, in the toluene system, the complex was extracted very slowly. Recently, the extraction rates of Ni(II) and Zn(II) with 5-Br-PADAP have been investigated by means of a CLM [36]. Direct observation of the interface by means of the CLM method clearly showed the formation of the complex only at the interface of heptane/water systems. Based on reaction Scheme 10.1, the initial formation rate is represented by

$$r^0 = \left\{ k[\text{HL}]\frac{V_\text{a}}{V_\text{o}} + \left(\frac{k_1 k_2 [\text{HL}]_\text{i}}{k_2 + k_{-1}[\text{H}^+]} \right) \frac{A_\text{i}}{V_\text{o}} \right\} [\text{M}^{2+}] \tag{13}$$

where V_a and V_o refer to the aqueous and the organic phase volume, respectively, and the definitions of k, k_1, k_2 and k_{-1} are shown in Scheme 10.1.

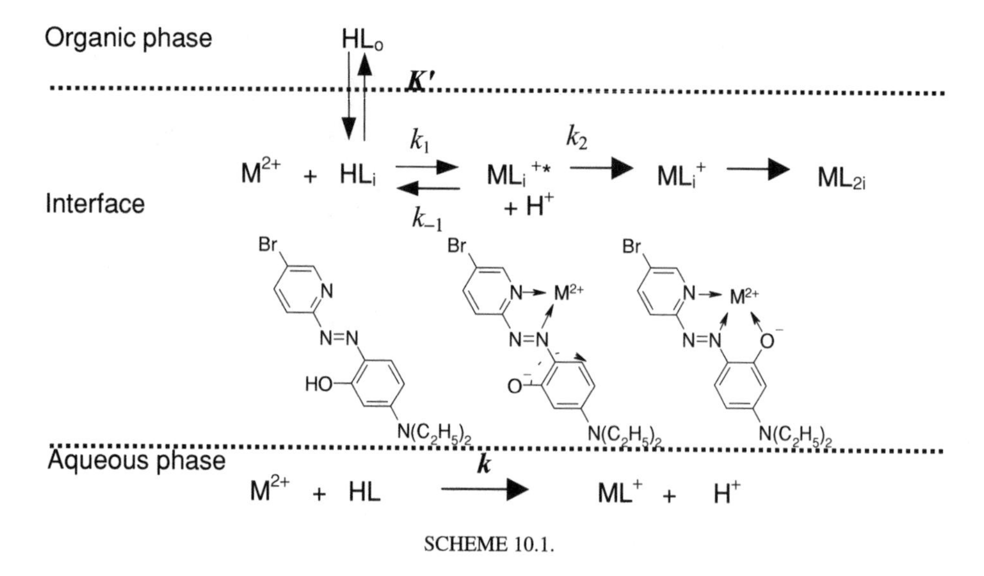

SCHEME 10.1.

The complexation proceeded almost completely at the interface. The values of the interfacial complexation rate constants are listed in Table 10.3. The rate constant, $k = 5.3 \times 10^2\ M^{-1} \cdot s^{-1}$, was determined in the aqueous solution using stopped-flow spectrometry in the region where the formation rate was independent of pH. The conditional interfacial rate constants represented by $k_i = k_1 k_2[HL]_i/(k_2 + k_{-1}[H^+])$ were larger at the heptane/water interface than at the toluene/water interface for both metal ions. The MD simulations of the adsorptivities of 5-Br-PADAP at the heptane/water and toluene/water interfaces suggested that 5-Br-PADAP could be adsorbed at the interfacial region more closely to the aqueous phase, but 5-Br-PADAP at the toluene/water interface was still surrounded by toluene molecules, which lowered the probability of reaction with aqueous Ni(II) ions [6].

CLM-Raman microscope spectrometry was applied in order to measure the rate of complex formation between Pd(II) and 5-Br-PADAP (HL) at the heptane/water interface and it was demonstrated that this method was highly useful for the kinetic measurement of the interfacial reaction [37]. A schematic diagram of the measurement system is shown in Figure 10.9.

Figure 10.10 shows typical spectra depicting the spectral change that occurs upon interfacial complex formation of PdLCl. Raman intensities at 1599, 1408 and 1303 cm^{-1}

TABLE 10.3. Kinetic parameters for the interfacial complexation of Ni(II) and Zn(II) with 5-Br-PADAP in heptane/water and toluene/water systems.

System	Metal ion	$k_1 (M^{-1} \cdot s^{-1})$	$k_{-1} (M^{-1} \cdot s^{-1})$	$k_2 (s^{-1})$
Heptane/water	Ni(II)	1.1×10^3	4.3×10^5	4.4×10^{-2}
	Zn(II)	4.4×10^4	6.9×10^5	1.1×10^{-1}
Toluene/water	Ni(II)	3.2×10	2.9×10^4	3.5×10^{-1}
	Zn(II)	5.1×10^4	2.0×10^6	4.8×10^{-2}

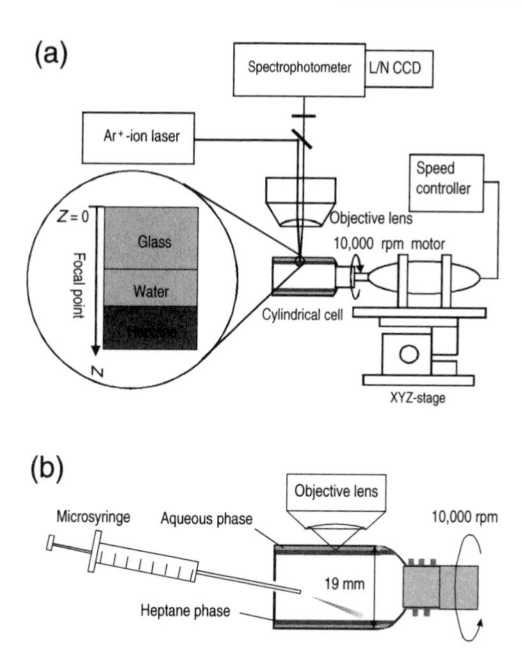

FIGURE 10.9. Schematic drawing of the apparatus for the centrifugal liquid membrane resonance Raman microprobe spectroscopy (a) and the centrifugal liquid membrane (CLM) cell with a sample injection hole at the bottom (b).

assigned to PdLCl clearly increased at the interface with increasing reaction time. The initial rate of complexation of Pd(II) with 5-Br-PADAP at the interface obtained from the Raman intensity change was in good agreement with that obtained from the absorbance change observed by CLM spectrophotometry. Furthermore, the resonance Raman spectra of the interfacial PdLCl provide information on the nano-environment of the complex.

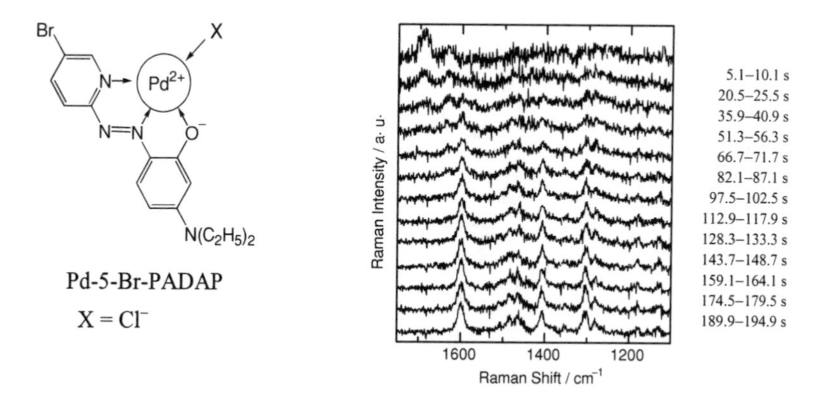

FIGURE 10.10. Change in resonance Raman spectra upon complexation of Pd(II) with 5-Br-PADAP at the heptane/water interface. Aqueous phase: $PdCl_2$, 8.0×10^{-5} M; HCl, 0.1 M; pH 1.0; heptane phase: 5-Br-PADAP, 7.8×10^{-6} M. 5-Br-PADAP was injected at 4 seconds. Final volumes of the aqueous and heptane phases were 0.250 and 0.150 cm^3, respectively. Laser power (514.5 nm) was 40 mW and the integration time for each spectrum was 5 seconds.

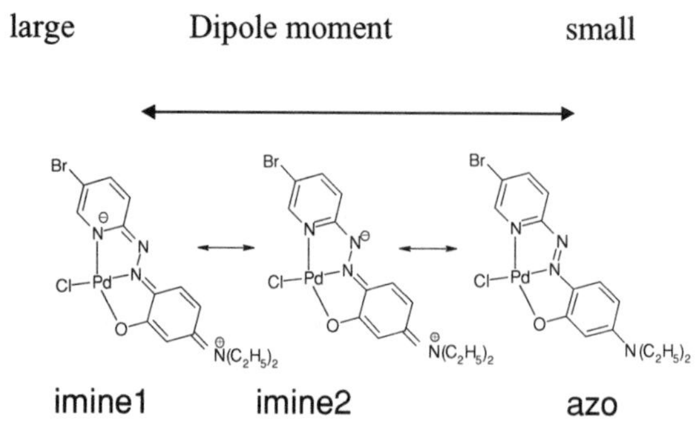

large Dipole moment small

imine1 imine2 azo

SCHEME 10.2.

The resonance Raman spectra of PdLCl adsorbed at the heptane/water and toluene/water interfaces were not in agreement with those in toluene and chloroform with the lower dielectric constants, but close to those in alcohol and aqueous alcohol mixed solvents with the higher dielectric constants. These results suggested that the PdLCl complex at the interface was partially surrounded by water molecules. The solvent effect on the resonance Raman spectra of PdLCl reflected a change in the ratio of the azo and imine resonance structures as shown in Scheme 10.2. The imine contains two forms, imine 1 and imine 2, depending on the extent of the charge separation, imine 1 being more dipolar. The bands at 1482, 1463, 1307 and 1284 cm^{-1} were assigned to ν(C=C) of pyridine ring in the azo form, imine ν(CNNC), azo ν(CNNC) and imine ν(CNNC), respectively. The azo/imine intensity ratios at 1482 and 1463 cm^{-1} (I_{1482}/I_{1463}) and at 1307 and 1284 cm^{-1} (I_{1307}/I_{1284}) increased with a decrease in the dielectric constant of the solvent. The values of I_{1482}/I_{1463} and I_{1307}/I_{1284} in the resonance Raman spectra of PdLCl adsorbed at the toluene/water interface were 0.63 and 1.47, respectively, and those of PdLCl complex formed at the heptane/water interface were 0.86 and 1.85, respectively (Figure 10.11).

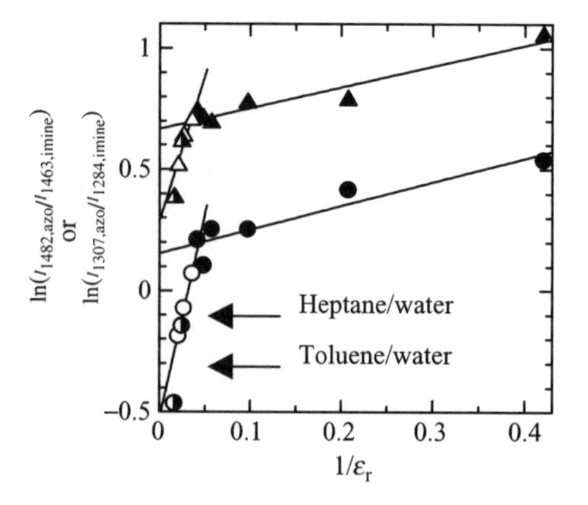

FIGURE 10.11. Variation of the logarithmic intensity ratio of the azo and imine signal intensities with the reciprocal of the dielectric constant. Circles: ln(I_{1482}/I_{1463}). Triangles: ln(I_{1307}/I_{1284}). Open keys refer to ethanol/water mixture, half closed keys the interfaces and closed keys the inert solvents.

TABLE 10.4. Rate constants ($M^{-1} \cdot s^{-1}$) for the reaction of Pd(II) with the neutral and the protonated 5-Br-PADAP in the aqueous phase and at the heptane/water and toluene/water interfaces.

Reaction site	Neutral 5-Br-PADAP (HL)	Protonated 5-Br-PADAP (H_2L^+)
Aqueous phase	5.7×10^2	5.7×10^2
Heptane/water interface	5.3×10	5.1×10^2
Toluene/water interface	6.6×10	3.3×10^2

In the ordinary chelate extraction kinetics with an acidic ligand, an acidic condition decreases the extraction rate of metal ions, because the dissociation of the ligand is suppressed under these conditions. However, in the extraction of Pd(II) chloride with 5-Br-PADAP in toluene, the decreased pH was observed to accelerate the extraction rate [38]. The protonation of 5-Br-PADAP at the N_β atom of the azo group increased its adsorptivity at the interface, but it did not decrease the reactivity with Pd(II). The rate constants obtained in heptane/water and toluene/water interfaces using the CLM method are listed in Table 10.4 [39]. The rate constant is larger for the reaction with the protonated 5-Br-PADAP, because Pd(II) is coordinated with a chloride ion and negatively charged as $PdCl_3^-$ and $PdCl_4^{2-}$. This type of acid-catalysis is observed only at the liquid/liquid interface. This finding provides a strategy for the design of a new type of catalytic extractant, which is expected to possess properties such as the interfacial activity gained by protonation and the strong coordination ability with a negatively charged complex ion in the acidic aqueous phase.

These results revealed that the liquid/liquid interface produced by agitation or stirring could catalyse the extraction rate by increasing the interfacial concentration of extractant and facilitating the interfacial complexation rate, similar to gas/solid or liquid/solid catalysis.

10.3.2. Ion-Association Extraction and Ion-Association Adsorption

Complexes of Fe(II), Cu(II) and Zn(II) with 1,10-phenanthroline (phen) and its hydrophobic derivatives exhibited remarkable interfacial adsorptivities, although the ligands themselves could be hardly adsorbed at the interface [40–42]. The extraction rate profiles of Fe(II) with phen and its dimethyl (DMP) and diphenyl (DPP) derivatives into chloroform were investigated using the HSS method [43]. Both the formation rate of the phen–derivative complex and the interfacial adsorptivity of the complex were remarkably dependent on the hydrophobicity of the ligand and the hydration tendency of anions. The initial extraction rate of the iron(II) complex is described by the following equation:

$$(d[\mathrm{FeL_3}X_2]/dt)_{t=0} = k_1[\mathrm{Fe}^{2+}][\mathrm{L}] + k_1'[\mathrm{Fe}^{2+}][\mathrm{L}]_i A_i/V_o \qquad (14)$$

where k_1 and k_1' stand for the 1:1 rate constant formation of FeL^{2+} in the aqueous phase and at the interface, respectively. The apparent extraction rate constant (k_{obs}) is written as

$$k_{obs} = \frac{1}{1 + K'_C A_i / V_o} \left(\frac{k_1}{K_D} + k_i K'_L \frac{A_i}{V_o} \right) \tag{15}$$

where K'_L and K'_C refer to the adsorption constants of L and FeL_3X_2 from the organic phase to the interface, respectively. Experimental results revealed that the rate-determining step was the 1:1 complex formation both in the aqueous phase and at the interface, depending on the hydrophobicity of the ligand. The adsorption of ligand accelerated the extraction (a positive catalytic effect), but the adsorption of the complex apparently suppressed the extraction rate (a negative catalytic effect). The effect of the anion and solvent on the extraction rate was explained by the change of the adsorption constant (K'_C) of the triscomplex ion.

Ion-association adsorption of water-soluble porphyrin was studied using time-resolved TIR fluorometry. Interfacial adsorption of anionic or cationic surfactants succeeded in attracting the oppositely charged porphyrin from the mixture of cationic and anionic porphyrins in the aqueous phase, tetrakis(sulphonatophenyl)porphyrin (TPPS) adsorbed with the hexadecyltrimethylammonium ion and tetrakis(*N*-methylpyridyl)porphyrin (TMPyP) with the hexadecanesulfonate [44]. In addition, it was found that the application of an external electric field across the interface enabled to control the interfacial ion-association adsorption of ionic porphyrin [45]. Furthermore, TIR Raman microspectroscopy revealed the interfacial concentration of *meso*-tetrakis(*N*-methylpyridyl) porphyrinato-manganese(III) ($Mn(tmpyp)^{5+}$) adsorbed with dihexadecyl hydrogenphosphate (DHP) at the toluene/water interface and also revealed that the orientation of the tilt angle was 65° from the interface normal [46].

10.3.3. Interfacial Ligand-Substitution Reaction

A typical kinetic synergism in the extraction of metal ions was reported for the Ni(II)-dithizone(Hdz)-phen chloroform system [47]. The extraction equilibrium constant ($\log K_{ex}$) was enhanced from −0.7 to 5.3 upon the addition of phen:

$$Ni^{2+} + 2Hdz_o + phen \rightleftharpoons Nidz_2phen_o + 2H^+ \tag{16}$$

At the same time, the extraction rate was accelerated by the reactions in aqueous phase [48]. The catalytic effect of phen was explained by the formation of the interfacial adsorptive complex, $Nidzphen^+$:

$$Ni^{2+} + phen \rightarrow Niphen^{2+} \qquad k = 6.8 \times 10^3 M^{-1} \cdot s^{-1} \tag{17}$$

$$Niphen^{2+} + dz^- \rightarrow Nidzphen^+ \qquad k = 9.2 \times 10^4 M^{-1} \cdot s^{-1} \tag{18}$$

The synergistic effect of DPP on the extraction of Ni(II) with dithizone was also evaluated and the formation of an interfacial complex was confirmed [49].

Results indicate that a kinetic synergism was attained upon formation of an interfacial intermediate complex, which was successively neutralized by a ligand-substitution reaction at the interface, to form a more extractable complex. The catalytic effect of *N*,*N*-dimethyl-4-(2-pyridylazo)aniline (PADA) on the extraction of $Ni(pan)_2$ was evaluated in accordance with this principle. The extraction rate of $Ni(pan)_2$ into toluene was very

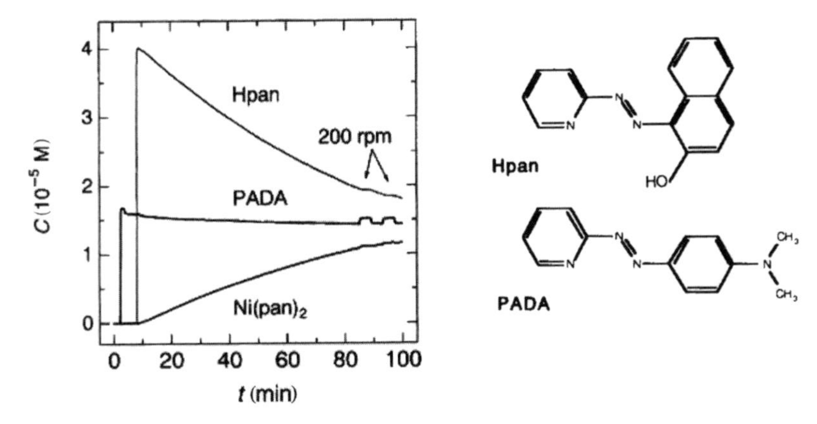

FIGURE 10.12. Concentration changes of Hpan and PADA in the catalytic extraction of $Ni(pan)_2$ into the toluene phase. Initial concentrations are $[Ni^{2+}] = 1.0 \times 10^{-5}$ M, [Hpan] $= 4.0 \times 10^{-5}$ M and [PADA] $=$ 1.7×10^{-5} M at pH 5.6.

slow even under HSS conditions. However, upon addition of PADA at a concentration as low as 10^{-5} M, the extraction rate was accelerated by about 10 times. Figure 10.12 shows the changes in concentration of the extraction system observed using the HSS method. $Ni(pan)_2$ alone was extracted into the organic solvent and PADA was not observed to be consumed after the extraction, though no significant decrease in the organic phase concentration of PADA was observed during the extraction under HSS conditions. These results were analysed using the mechanism of the interfacial ligand substitution [50]:

$$\mathrm{Ni(pada)_i^{2+} + Hpan_o \rightarrow Ni(pan)(pada)_i^{+} + H^{+}} \tag{19}$$

$$\mathrm{Ni(pan)(pada)_i^{+} + Hpan_o \rightarrow Ni(pan)_{2,o} + pada_o + H^{+}} \tag{20}$$

The value for the rate constant of formation of Equation (19) (k') was obtained to be $90\ M^{-1} \cdot s^{-1}$. The key processes for the catalytic extraction of Ni(II)-pan with PADA are the fast aqueous phase formation of $Ni(pada)^{2+}$ and the adsorption of $Ni(pan)(pada)^{+}$, followed by the ligand substitution of pada with pan [50]. This scheme is generally of importance as a guideline for the acceleration of the extraction rate using the interfacial reaction.

10.4. INTERFACIAL AGGREGATION

Another unique feature of the interfacial reaction for metal complexes is the formation of the aggregate of the complex. As observed in many cases, the liquid/liquid interface can be saturated by any active surface molecules at the interfacial concentration of the order of 10^{-10} mol $\cdot$ cm^2, which can be attained even at a diluted bulk concentration as in the case of highly hydrophobic solutes. Thus, the interface is prepared to become a two-dimensionally concentrated state for the solute. This situation very often results in the formation of the metal complex aggregates at the liquid/liquid interface. During the solvent extraction procedure for metal ions, the precipitate at the interface was observed, which was called crud or scum in the field of hydrometallurgy. This crud or scum must be suppressed. Therefore, the formation of this precipitate should be investigated from the viewpoint of the interfacial aggregation of metal complexes.

10.4.1. Interfacial Aggregation of Porphyrin Complex

A typical example of the interfacial aggregation was observed in the interfacial protonation of tetraphenylporphyrin (TPP). The aggregation rate of H_2tpp^{2+} in the dodecane/trichloroacetic acid solution was measured using a two-phase stopped-flow method [7] and a CLM spectrophotometric method [10].

Assemblies of the Pd(II)–5,10,15,20-tetra(4-pyridyl)porphyrin (tpyp) complex were formed spontaneously at the toluene/water interface when a toluene solution with tpyp was put into contact with an aqueous solution of $PdCl_2$ under acidic conditions [51]. The interfacial assemblies of tpyp were formed more effectively with Pd(II) than with other divalent metal ions such as Ni(II), Cu(II) or Zn(II). The Pd(II) ion in the complex was bound to the nitrogen of the pyridyl group of tpyp, not to the pyrrole nitrogen. In situ fluorescence microscopy elucidated the formation of two kinds of complex assemblies: assembly 1 (AS1) was generated at low tpyp concentrations and assembly 2 (AS2) at high tpyp concentrations. Fluorescence excitation and emission spectra were measured using the TIR method. The fluorescence spectra of interfacial AS2 showed a red-shift of 7–11 nm at the maximum wavelength ($\lambda_{max} = 668$ nm) relative to that of tpyp in toluene, suggesting a weakstacking interaction between tpyp molecules in AS2. Time-resolved TIR fluorometry was used to determine the fluorescence lifetimes of AS1 ($\lambda_{max} = 656$ nm) and AS2 as 0.15 ± 0.05 ns and 1.1 ± 0.1 ns, respectively. The average stoichiometric compositions for Pd/tpyp were 3:1 and 1:1 for AS1 and AS2, respectively, but the fluorescence quenching effect of Pd(II) indicated that each tpyp in AS1 was bound to four Pd(II) atoms. These results suggested that two Pd(II) atoms in AS1 were shared with two tpyp molecules and that the other two Pd(II) atoms were bound to the pyridyl nitrogens of the tpyp molecule as shown in Figure 10.13. Four pyridyl groups of one tpyp in AS1 were occupied by four Pd(II), two of which were bridged by two other tpyp molecules. This structure suggests that the random distribution of tpyp at the out-of-plane orientation is more favourable than that at a fixed orientation. Contrastingly, the tpyp molecule in AS2 was bound to one Pd(II) atom. The tpyp molecules in AS2 were weakly associated with each other and the average aggregation number was about 3. It was, however, found that the energy migration between tpyp molecules in AS2 rarely

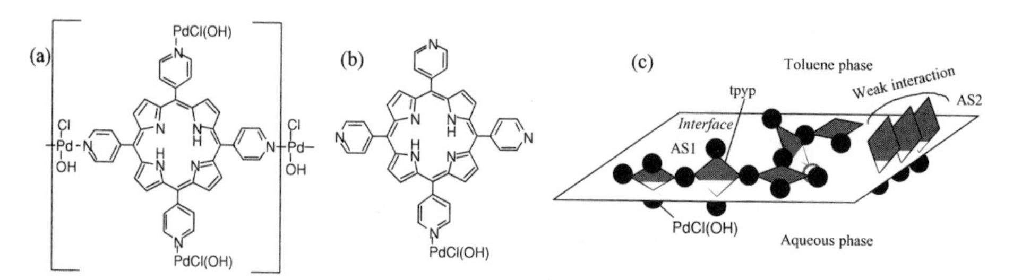

FIGURE 10.13. Unit structures of AS1 (a) and AS2 (b) formed under the low and high total interfacial tpyp concentrations, respectively. Aggregation at the interface is illustrated in (c).

affected the lifetime of its fluorescence. The lifetimes of tpyp in AS1 and AS2 were primarily determined by the quenching of Pd(II) directly bound to pyridyl groups. Tpyp in AS2 showed a halfway orientation angle (Θ_{eq}) of about 43° at the lower total interfacial tpyp concentration (2.6×10^{-11} mol · cm^{-2}), where the AS1 and AS2 assemblies coexisted at the interface. Tpyp in AS2 tended to lie at the interface ($\Theta_{eq} \leq 61°$) as the total interfacial tpyp concentration increased to 1.3×10^{-10} mol · cm^{-2}, where AS2 was primarily present at the interface.

10.4.2. Molecular Recognition of Interfacial Aggregation

The palladium(II) ion forms a 1:1 complex with 5-Br-PADAP, leaving one site for the coordination of another ligand. The Pd(II)-5-Br-PADAP (PdL) complex cation has specific properties such as an extremely high molar absorptivity ($\varepsilon_{564} = 4.33 \times 10^4$ $M^{-1} \cdot cm^{-1}$ in toluene), a high adsorptivity to the liquid/liquid interface and a soft Lewis acid easy to be bound to a soft Lewis base. Therefore, PdL^+ is expected to function as an interfacial molecular recognition reagent of Lewis bases.

The molecular recognition ability of Pd(II)-5-Br-PADAP for the isomers of diazine derivatives has been evaluated. Figure 10.14 summarizes the molecular structures of the diazines (Dzs or N) studied, the acid-dissociation constant (K_a) and the distribution constant values between toluene and water (K_D). PdLCl reacted with a neutral N with one or two nitrogen atoms, forming a $PdLN^+$ complex at the toluene/water interface. The interfacial formation constant (β_i) of the $PdLN^+$ complex was determined as follows [52]:

$$PdLCl_i + N \rightleftharpoons PdLN_i^+ + Cl^- \qquad (21)$$

$$\beta_i = \frac{[PdLN^+]_i[Cl^-]}{[PdLCl]_i[N]} \qquad (22)$$

A linear relationship was obtained between the logarithmic interfacial formation constant (log β_i) for PdL^+-diazine derivative complexes and the logarithmic ratio of

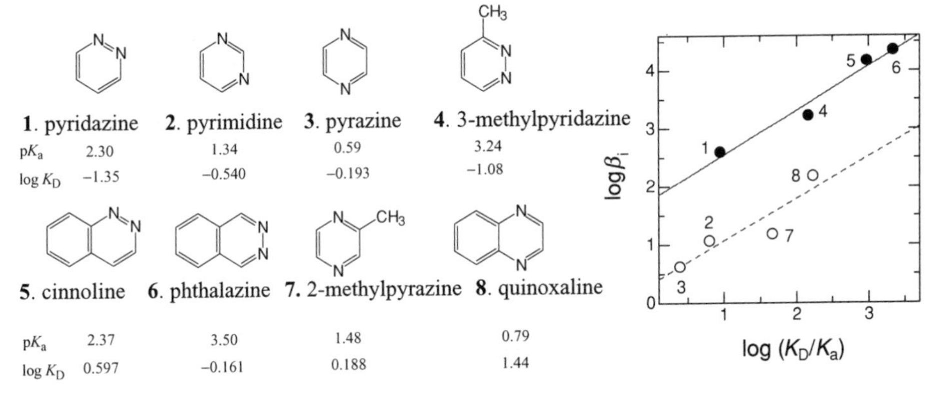

FIGURE 10.14. Diazines structures, pK_a's and log K_D values in toluene/water (25°C). The plots of log β_i vs. log(K_D/K_a) are higher for the pyridazine derivatives 1,4,5 and 6.

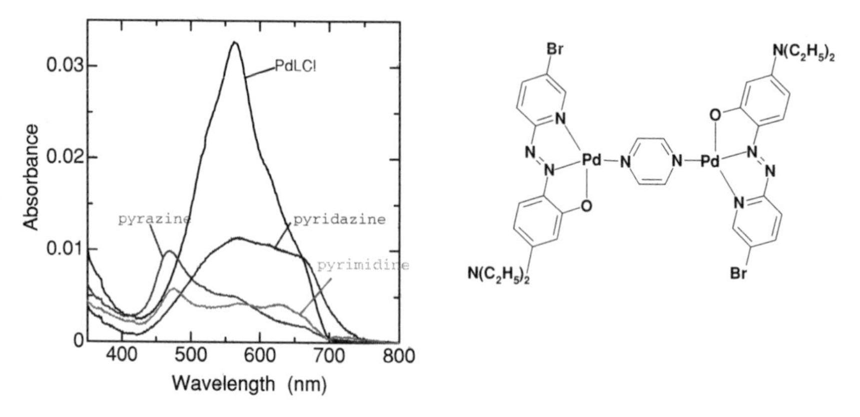

FIGURE 10.15. Absorption spectra of PdLCl and the interfacial aggregates of ternary PdL-Dz complexes measured using the CLM method. $[PdLCl]_{init} = 5.6 \times 10^{-5}$ M, [pyridazine] $= 2.0 \times 10^{-4}$ M, [pyrimidine] $= 8.0 \times 10^{-3}$ M, [pyrazine] $= 8.0 \times 10^{-5}$ M, $[ClO_4^-] = 0.1$ M, $[Cl^-] = 0$ M, pH 2.0. The right diagram shows a probable unit structure of the membrane-like aggregate formed in pyrazine complexes.

the distribution constant (K_D) to the acid-dissociation constant (K_a) for the two groups, as shown in Figure 10.14. PdL^+-pyridazine derivative complexes showed much higher stability at the interface than pyrimidine and pyrazine derivative complexes. This result suggests that pyridazine derivative complexes become more liable to be adsorbed at the interface than pyrimidine and pyrazine derivative complexes, since the bonding of the pyridazine derivative to PdL^+ increases its hydrophobicity by decreasing its dipole moment, which was evaluated by molecular orbital calculations.

At low concentrations of chloride ions, PdL-Dz complexes formed aggregates at the interface [53]. Figure 10.15 shows the absorption spectra of PdLCl and the PdL-Dz complex aggregates. The formation of each interfacial aggregate of PdL-Dz was accompanied with a drastic absorbance decrease at 564 nm of PdLCl and the change in the spectral shape. Difference in the compositions of the interfacial aggregates of PdL-Dz was observed among the Dz isomers: the PdL-pyridazine aggregate formed blue domains at the interface and the composition ratio was pyridazine/PdL = 1:1, whereas the PdL-pyrazine aggregate was a thin membrane with the composition of pyrazine/PdL = 1:2 and the PdL-pyrimidine aggregate showed blue domains with the unknown composition.

In spite of the fact that pyrazine has the lowest stability among Dz isomers in the formation of PdL complex at the toluene/water interface, the stability for the formation of the interfacial aggregate was the highest for the pyrazine complex. This result indicates that the formation of the interfacial aggregate of PdL-Dz isomers was governed by the geometric structure of Dz, not by its basicity.

The resonance Raman spectra for the interfacial aggregates were measured by means of CLM-Raman microspectroscopy, and the Raman intensity ratio of the imine and azo forms (I_{azo}/I_{imine}) was determined as shown in Table 10.5. The ratio was smaller in more polar solutions such as at the interface. This observation indicated that the nano-environment of the azo-group in the membrane-like aggregate of cationic PdL-1,4-Dz was more polar than that of the crystal-like PdL-1,2-Dz cation, although both were in more polar environment than the toluene solutions.

TABLE 10.5. Raman shift and the imine/azo intensity ratio of the Pd(II)-5-Br-PADAP complex and its aggregates.

Complex	State	$\nu N=N$	$\nu CNNC_{azo}$	$\nu CNNC_{imine}$	I_{azo}/I_{imine}
PdLCl	Crystal	1400	1301	1279	2.19
PdLCl	Aggregate (crystal-like)	1402	1300	1285	3.52
PdL-1,2-Dz	Aggregate (crystal-like)	1408	1309	1286	2.11
PdL-1,4-Dz	Aggregate (membrane-like)	1409	1301	1286	1.16
PdLCl	Toluene/water interface	1408	1303	1280	1.47
PdLCl	Heptane/water interface	1408	1303	1284	1.85
PdLCl	Bulk toluene	1401	1307	1283	2.87

The results from VIS and Raman spectroscopy suggest that the drastic decrease in the molar absorptivity of PdL in the PdL-1,2-Dz and PdL-1,4-Dz aggregates may be ascribable to the electrostatic interaction between the ligand and an incorporated anion and to the stacking interaction between the ligands in the aggregates. These interactions prevent resonance in the charged quinone structure of the ligand in the Pd(II)-5-Br-PADP complex. Similar interfacial aggregation and a decrease in the absorbance were observed in the interfacial aggregation between Pd(II)-5-Br-PADAP and the purine bases of the nucleic acids, adenine and guanine. Figure 10.16 also shows the absorbance decrease of PdL^+ upon addition of adenine and guanine. The compositions of the aggregates were determined as $[PdL^+]/[Ade] = 6:1$ and $[PdL^+]/[Gua] = 7:1$.

The interfacial aggregation of PdL with pyridazine, pyrazine, adenine and guanine occurred at very low concentrations of Dzs and the purine bases, not observed in the bulk phases. This indicates that the interfacial aggregation could be amplified by the presence of a very minute amount of the base.

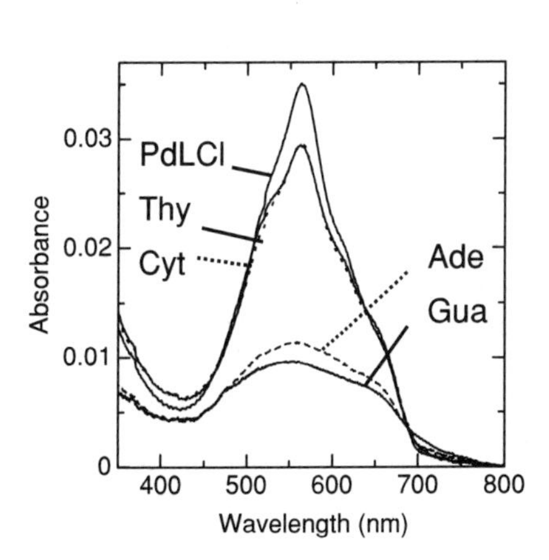

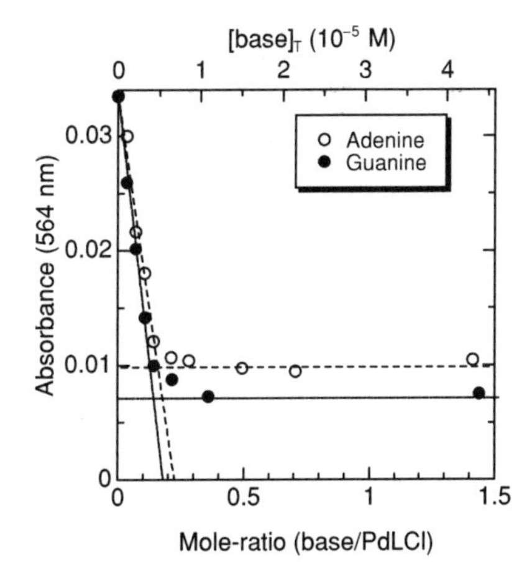

FIGURE 10.16. (Left) Absorption spectra measured using the CLM method in four base systems. $[PdLCl]_{init} = 5.6 \times 10^{-5}$ M, [adenine] $= 7.7 \times 10^{-5}$ M, [guanine] $= 4.4 \times 10^{-5}$ M, [cytosine] $= 8.8 \times 10^{-4}$ M, [thymine] $= 8.8 \times 10^{-4}$ M, $[ClO_4^-] = 0.1$ M, $[Cl^-] = 0$ M, pH 2.0. (Right) Mole-ratio plots for the interfacial aggregation of PdL^+ with adenine and guanine, which yielded $[PdL^+]/[Ade] = 6:1$ and $[PdL^+]/[Gua] = 7:1$.

Similar interfacial aggregation phenomena were observed in the complexation of Cu(II) with octadecylthiazolylazophenol ($C_{18}TAR$, HR) at the heptane/water interface [54]. The CuR^+ complex was formed at the interface with the absorption maximum at 560 nm. The increase in pH promoted the formation of an aggregate of $Cu_2R_3^+$ shifting the absorption maximum from 560 to 510 nm.

The composition of $Cu_2R_3^+$ was suggested by the stoichiometric analysis of the interfacial absorption spectra measured using the CLM method and it was confirmed by the micro-two-phase flow/direct electro-spray ionization mass spectrometry [54] as well. The addition of adenine and guanine into the aqueous phase disaggregated the interfacial complex by forming new interfacial complex, CuR-base, accompanied with the recovery of the interfacial spectra of the monomer complex. Cytosine and thymine were not so susceptible to being bound to the Cu(II) aggregate as adenine and guanine. This strongly suggested that interfacial aggregation was necessary for molecular recognition. The interfacial molecular recognition aggregation found in these systems is a promising new means of the future design for a sensitive and selective detection method on various compounds such as nucleic acid bases, DNA, drugs and pesticides.

Very recently, the second harmonic generation—circular dichroism (SHG-CD) method has been applied, for the first time, to the measurement of the CD spectra of the interfacial complex [55]. Interfacial aggregation of tetra-*p*-sulfonatophenylporphyrin (TPPS) induced by the coexistence of a cationic surfactant was studied using the SHG-CD method. Cetyltrimethylammonium bromide (CTAB) was used as a cationic surfactant. The SHG signal increased with the increase in the concentration of CTAB. Under acidic conditions (pH 3.0), the absorption spectrum of the aqueous solution of TPPS showed a maximum at 435 nm, which was assigned to the diprotonated TPPS (H_4TPPS^{2-}). On the other hand, the p-polarized interfacial SHG spectrum obtained from the irradiation of the p-polarized fundamental beam at pH 3.0 gave two maxima at 400 and 415 nm. The maxima in the SHG spectrum were assigned to the mixture of the H-aggregate of TPPS (405 nm) and the aqueous unprotonated TPPS (413 nm). It was, therefore, suggested that these two unprotonated species coexisted at the heptane/water interface. It has been reported that J- and H-aggregates are successively formed with an increase of CTAB in an acidic aqueous solution [56]. However, the present results suggested that the formation of the H-aggregate was favoured at the interface in contrast with the results in aqueous solutions. The chirality of the interfacial aggregate of TPPS has been investigated using SHG-CD spectroscopy for the first time. This study enabled the chirality measurements of the interfacial complex and the interfacial aggregates, which were very difficult to be measured using ordinary CD spectroscopy, because of the very low interfacial concentrations. In the near future, the chiral selectivity related to the molecular recognition ability of the interfacial aggregates will be discussed in detail.

10.5. CONCLUDING REMARKS

The specific features of the interfacial reactions, notably the complexation of metal ions, were reviewed. The dynamic nano-properties of the interface were also discussed from the experimental results of single molecule probing and other dynamic microscopic experiments. The following points can be summarized as the specific features of the interfacial properties and the interfacial reactions:

1. The study on the rotational and translational dynamics of molecules at the interface revealed the rheological properties of the interface. For example, the difference in the effects of a single-chained surfactant (SDS) and a double-chained surfactant (DHP and DMPC) on the diffusion or reactivity of the coadsorbed molecules was indicated. The double-chained surfactant more significantly reduced the reactivity or diffusivity of the solute molecule coadsorbed at the interface. Single molecule probing at the interface is a promising technique to measure the properties and reactivity at the liquid/liquid interface. This approach will be extended to the study of the reaction mechanism on the biological cell-membrane systems including proteins and enzymes.
2. The catalytic role of the interface was recognized in various liquid/liquid extraction systems. Interfacial adsorption of reactants was the key step in the interfacial catalysis in the extraction of metal ions. The interfacial ligand-substitution mechanism has great importance in the kinetic synergism. Some essential guidelines proposed here are highly useful, not only in solvent extraction but also in interfacial synthesis.
3. Interfacial adsorption of molecules or ions can produce a highly concentrated environment for the adsorbed species. It is, therefore, prepared for species aggregation at the interface. Aggregates of metal complexes exhibited a unique function as molecular recognition reagents. The aggregated state is similar to a clustered state, in which the monomer molecules are closely packed at the two-dimensional interface. In such a state, the selectivity in the structural recognition for isomers can be significantly enhanced.
4. Microscopic spectrometry and evanescent spectroscopy are very useful for the investigation of the liquid/liquid interfacial reaction. The fluorescence method is now the most versatile and sensitive detection method for the interfacial species, but the resonance Raman spectroscopy is highly promising as a more deterministic method. However, a more innovative method is needed for the measurement of the chemical reaction at the nano-region of the interface. In the future measurement methods, the development of a nano-probing technique and the generation of a micro-field gradient will be the key points as well as the creation of an extremely small, thin two-phase system.

As for the liquid/liquid interface of an inert solvent/water system like that of heptane/water, the interface is thought to be flat on the molecular scale. In the systems including substantial mutual dissolution, however, solvent clusters can be generated in the nano-region of the liquid/liquid interfaces [57]. The solvent cluster or nano-droplets will show some different behaviour from those of the ordinary solvent molecules, because of the long-range van der Waals and drag forces. This kind of situation may occur in the formation of vesicles and lipids in the interior of biological cells. Thus, the liquid/liquid interface will continue to be a fundamental subject of investigation, especially in relation to the biological reactions. Extensive studies from the physical, chemical and biological viewpoints are required in this field of research.

ACKNOWLEDGEMENTS

This study was supported in part by a Grant-in-Aid for Scientific Research on Priority Area "Nano-Chemistry at the Liquid-Liquid Interfaces", a Grant-in-Aid for Scientific Research (A) (No. 12304045) and the 21 COE project of Osaka University "Creation of Integrated Eco-Chemistry" from the Ministry of Education, Culture, Sports, Science and Technology of Japan.

REFERENCES

1. C.B. Honaker and H. Freiser, J. Phys. Chem. **66**, 127 (1962).
2. D.S. Flett, D.N. Okuhara and D.R. Spink, J. Inorg. Nucl. Chem. **35**, 2471 (1973).
3. P.R. Danesi, Solvent Extraction Kinetics in *Principles and Practices of Solvent Extraction*, Eds. J. Rydberg, C. Musikas and G.R. Choppin, Marcel Dekker, New York, 1992, p. 157.
4. H. Watarai and H. Freiser, J. Am. Chem. Soc. **105**, 189 (1983).
5. H. Watarai and H. Freiser, J. Am. Chem. Soc. **105**, 191 (1983).
6. H. Watarai, M. Gotoh and N. Gotoh, Bull. Chem. Soc. Jpn. **70**, 957 (1997).
7. H. Nagatani and H. Watarai, Anal. Chem. **68**, 1250 (1996).
8. H. Watarai and Y. Saitoh, Chem. Lett. 283 (1995).
9. L. E. Morrison and G. Waier, Biophys. J. **52**, 367 (1957).
10. H. Nagatani and H. Watarai, Anal. Chem. **70**, 2860 (1998).
11. T. Tokimoto, S. Tsukahara and H. Watarai, Chem. Lett. 204 (2001).
12. H. Watarai, A. Matsumoto and T. Fukumoto, Anal. Sci. **18**, 367 (2002).
13. F. Hashimoto, S. Tsukahara and H. Watarai, Anal. Sci. **17**, i81 (2001).
14. A. Ohashi and H. Watarai, Chem. Lett. **2001**, 1238 (2001).
15. M. Suwa and H. Watarai, J. Chromatogr. A **1013**, 3 (2003).
16. G.D. Owens, R.W. Taylor, T.Y. Ridley and D.W. Margerum, Anal. Chem. **52**, 130 (1980).
17. T. Tokimoto, S. Tsukahara and H. Watarai, Bull. Chem. Soc. Jpn. **76**, 1569 (2003).
18. S. Nie, D.T. Chiu and R.N. Zare, Anal. Chem. **67**, 2849 (1995).
19. T. Funatsu, Y. Harada, M. Tokunaga, K. Saito and T. Yanagida, Nature **374**, 555 (1995).
20. D.C. Nguyen, R.A. Keller, J.H. Jett and J.C. Martin, Anal. Chem. **59**, 2158 (1987).
21. M. Ishikawa, K. Hirano, T. Hayakawa, S. Hosoi and S. Brenner, Jpn. J. Appl. Phys. **33**, 1571 (1994).
22. M.J. Wirth and D.J. Swinton, Anal. Chem. **70**, 5264 (1998).
23. 23. S. Ho Kang, M.R. Shortreed and E.S. Yeung, Anal. Chem. **73**, 1091 (2001).
24. F. Hashimoto, S. Tsukahara and H. Watarai, Langmuir **19**, 4197 (2003).
25. H. Watarai and F. Funaki, Langmuir **12**, 6717 (1996).
26. S. Tsukahara, Y. Yamada, and H. Watarai, Langmuir **16**, 6787 (2000).
27. S. Tsukahara, Y. Yamada, T. Hinoue and H. Watarai, Bunseki Kagaku **47**, 945 (1998).
28. K. Kemnitz, N. Tamai, I. Yamazaki, N. Nakashina, and K. Yoshihara, J. Phys. Chem. **90**, 5094 (1986).
29. J.T. Petkov, K.D. Danov and N.D. Dankov, Langmuir **12**, 2650 (1996), and references therein.
30. P.R. Briley, A.R. Deemer and J.C. Slattery, J. Colloid Interface Sci. **56**, 1 (1976).
31. H. Watarai and K. Satoh, Langmuir **10**, 3913 (1994).
32. H. Watarai, M. Takahashi and K. Shibata, Bull. Chem. Soc. Jpn. **59**, 3469 (1986).
33. H. Watarai and M. Endo, Anal. Sci. **7**, 137 (1991).
34. Y. Onoe and H. Watarai, Anal. Sci. **14**, 237 (1998).
35. H. Watarai and Y. Onoe, Solvent Extr. Ion Exch. **19**, 155 (2001).
36. Y. Yulizar, A. Ohashi and H. Watarai, Anal. Chim. Acta **447**, 247 (2001).
37. A. Ohashi and H. Watarai, Chem. Lett. **32**, 218 (2003).
38. A. Ohashi, S. Tsukahara and H. Watarai, Anal. Chim. Acta **364**, 53 (1998).
39. A. Ohashi and H. Watarai, Anal. Sci. **17**, 1313 (2001).
40. H. Watarai, J. Phys. Chem. **89**, 384 (1985).

41. H. Watarai, Talanta **32**, 817 (1985).
42. H. Watarai and Y. Shibuya, Bull. Chem. Soc. Jpn. **62**, 3446 (1989).
43. H. Watarai, K. Sasaki and N. Sasaki, Bull. Chem. Soc. Jpn. **63**, 2797 (1990).
44. Y. Saitoh and H. Watarai, Bull. Chem. Soc. Jpn. **70**, 351 (1997).
45. R. Okumura, T. Hinoue and H. Watarai, Anal. Sci. **12**, 393 (1996).
46. K. Fujiwara and H. Watarai, Langmuir **19**, 2658 (2003).
47. B. Freiser and H. Freiser, Talanta **17**, 540 (1970).
48. H. Watarai, K. Sasaki, K. Takahashi and J. Murakami, Talanta **42**, 1691 (1995).
49. H. Watarai, K. Takahashi and J. Murakami, Solvent Extr. Res. Dev. Jpn. **3**, 109 (1996).
50. Y. Onoe, S. Tsukahara and H. Watarai, Bull. Chem. Soc. Jpn. **71**, 603 (1998).
51. N. Fujiwara, S. Tsukahara and H. Watarai, Langmuir **17**, 5337 (2001).
52. A. Ohashi and H. Watarai, Langmuir **18**, 10292 (2002).
53. A. Ohashi, S. Tsukahara and H. Watarai, Langmuir **19**, 4645 (2003).
54. H. Oyama and H. Watarai, Anal. Sci, in press.
55. K. Fujiwara, H. Monjushiro, and H. Watarai, Chem. Phys. Lett. **349** (2004).
56. N.C. Maiti, S. Mazumudar and N. Periasamy, J. Phys. Chem. B **102**, 1528 (1998).
57. A. Pfennig, Chem. Eng. Sci. **55**, 5333 (2000).

11

Molecular Recognition of Ions at Liquid/Liquid Interfaces

Norio Teramae[1], Seiichi Nishizawa[1], Akira Yamaguchi[1]
and Tatsuya Uchida[2]
[1]*Tohoku University, Sendai 980-8578, Japan,*
[2]*Tokyo University of Pharmacy and Life Science, Hachioji 192-0392, Japan*

11.1. INTRODUCTION

The chemical processes occurring at the interface between two immiscible liquids are well known to play an important role in many fields such as analytical chemistry, organic chemistry and biochemistry [1,2]. Specific reactions of molecules (adsorbed) at the liquid/liquid (L/L) interface have been reported in various systems that include photo-induced electron transfer reactions [3], organic synthesis [4], charge transfer [5], phase-transfer catalysis [6], ion-pair extraction [7] and molecular association [8]. These characteristic features of molecules at the L/L interface should be responsible for the large differences in physical properties compared to bulk solutions. Various analytical techniques have been applied to reveal characteristics of L/L interfaces. In particular, linear and non-linear laser spectroscopies represented by time-resolved total internal reflection (TIR) fluorescence [9–11] and second harmonic generation (SHG) [12,13] spectroscopies have been used to obtain information on L/L interfaces such as polarity and interfacial roughness. As for the polarity of L/L interfaces, our group [9] and others [10,13] have experimentally revealed that the interface is a layer having intermediate polarity between those of the organic and aqueous phases, which well agree with the prediction by molecular dynamics simulations [14]. Our group has recently reported solvation dynamics of fluorophores with an anthroyloxy group at the heptane/water interface, revealing that preferential solvation occurs for fluorophores at the interface based on the observation of time-dependent fluorescence spectral shift [15].

In addition to these studies to elucidate physical properties of L/L interfaces, our group has successfully found several kinds of specific complexation reactions that originate from the unique properties of L/L interfaces [16–25]. For example, from the analysis of complexation kinetics of 5-alkyloxymethyl-8-quinoinols at the L/L interface by dynamic interfacial tensiometry and ion transfer polarography, it is found that the kinetics for the interfacial complexation with Ni^{2+} is significantly affected by the alkyl chain length of the chelating reagents, and that the complexation rate at the interface is enhanced as compared to that in bulk aqueous solutions [18]. Our group has also studied electrochemical anion transfer reactions across the L/L interface as facilitated by hydrogen-bonding ionophores [20]. The ionophore-assisted anion transfer process was successfully observed for the first time, indicating that, in spite of significant interference from anion hydration, complementary hydrogen bonding at phase boundaries is indeed effective for analyte recognition.

This chapter is focused on our recent research topics regarding the analysis of complexation reactions at L/L interfaces. We first describe the hydrogen-bond-mediated anion recognition as studied by ion transfer polarography and interfacial tensiometry [22,23], and then alkali metal ion recognition as studied by SHG spectroscopy [24,25].

11.2. HYDROGEN-BOND-MEDIATED ANION RECOGNITION AT L/L INTERFACES

The potential applications in biomedical and environmental research have led to great interest in the design and synthesis of abiotic ionophores for phosphates and their derivatives [26]. Selective binding and sensing of phosphate species have been successfully demonstrated in some cases by sophisticated ionophores in bulk organic and/or aqueous media [27]. On the other hand, in the case of ionophore-based two-phase-transfer systems such as ion-selective electrodes (ISEs) and membrane transports, it still remains a challenge to achieve a useful selectivity for very hydrophilic phosphate anions over hydrophobic anions [28,29]. This is due to the difficulty in compensating large differences in Gibbs free energies of transfer between phosphates and interfering anions by selective complexation [30]. Phosphate is indeed the most hydrophilic anion, when considering its end position in the Hofmeister series, which reflects the free energy of hydration of anions: $ClO_4^- > Br^- > Cl^- > CH_3COO^- > H_2PO_4^-/HPO_4^{2-}$.

Among the variety of possible approaches towards the appearance of phosphate selectivity in two-phase-transfer systems, of our particular interest is the use of hydrogen bonds to provide a binding affinity for this anion as has been utilized in biological systems [31]. X-ray analysis [31] clearly revealed that a phosphate binding protein could recognize HPO_4^{2-} through the formation of 12 hydrogen bonds, by which high stability and selectivity were achieved for the transport of this anion across the biological membrane. Significantly, hydrogen-bond forming guanidinium derivatives [32] and uranyl salophenes [33] are the only examples of abiotic ionophores, which have been identified as exhibiting good phosphate selectivities in ISEs and membrane transport experiments. In these abiotic ionophores [32,33], multiple hydrogen bonds are indeed involved in the binding event with phosphate anions, demonstrating the potential use of hydrogen bonds for the design of phosphate-selective ionophores.

In this section, our recent approaches based on the use of hydrogen-bonding ionophores [22,23,34,35] are described to achieve the phosphate selectivity in two-phase-transfer systems.

11.2.1. Highly Efficient Transport of $H_2PO_4^-$ Anions by a Phosphate/Ionophore 2:1 Complex Formation

We first describe the use of complexation-induced phosphate–phosphate interactions as a novel extension of the design strategy based on a hydrogen-bonding motif. While most previous studies have focused on ionophores based on a 1:1 complex formation, here we discuss a hydrogen-bonding ionophore that binds two $H_2PO_4^-$ ions so as to form hydrogen bonds between the complexed phosphates in a 2:1 complex. Since the complexation-induced anion–anion interaction is known for phosphate anions [36], such a unique property could be utilized to distinguish phosphate anions from other interfering anions. Also, a significant positive cooperativity is expected to appear in the phosphate/ionophore 2:1 complex formation, i.e., a high stability of the 2:1 complex would be achieved by forming additional hydrogen bonds between the complexed phosphates. To investigate binding behaviours of hydrogen-bonding ionophores across the L/L interface, we have employed ion transfer polarography [37], which can quantitatively provide kinetic and thermodynamic parameters of the ion transfer process. From the analysis of anion transfers across the nitrobenzene (NB)/water interface as facilitated by thiourea-based ionophores as shown in Figure 11.1, bis-thiourea **1c** is found to most

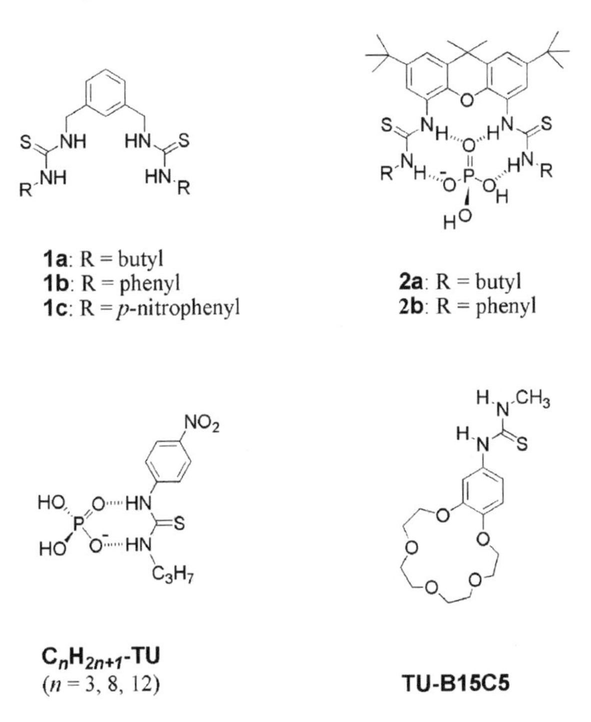

FIGURE 11.1. Structures of ionophores examined in our studies.

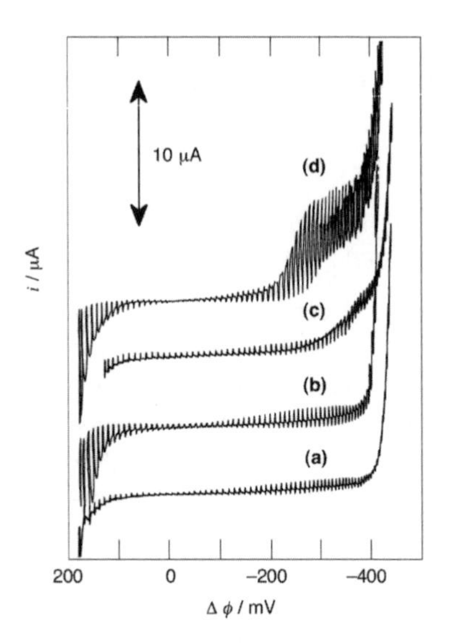

FIGURE 11.2. Ion transfer polarograms of $H_2PO_4^-$ across the NB/water interface. Aqueous phase: 0.5 M NaH_2PO_4. Organic phase: 0.05 M $N(C_7H_{15})_4TPB$ and (a) 0 mM ionophore, (b) mono-thiourea C_3H_7-TU, (c) bis-thiourea **2a** or (d) bis-thiourea **1c**. [ionophore]: 0.5 mM. Flow rate: 25 ml · h^{-1}. Reprinted from Ref. [22] with permission.

strongly facilitate the transfers of various hydrophilic anions from the water to the NB phase [22]. The $H_2PO_4^-$ transfer by **1c** is indeed based on the formation of the phosphate/ionophore 2:1 complex, by which a highly efficient transport of $H_2PO_4^-$ can be achieved across the NB/water interface.

Figure 11.2 shows ion transfer polarograms for $H_2PO_4^-$ across the NB/water interface as facilitated by ionophores **1c**, **2a** and C_3H_7-TU. As compared to results in the absence of ionophores (curve a), only a slight change is observed even in the presence of mono-thiourea C_3H_7-TU (curve b), indicating that C_3H_7-TU has little ability to facilitate $H_2PO_4^-$ transfer from the aqueous to the NB phase. By contrast, new waves appear in the presence of bis-thiourea **1c** or **2a**. While the effect of bis-thiourea **2a** is not so dramatic (curve c), a well-defined wave is clearly observed when bis-thiourea **1c** was present in the NB phase (curve d). Obviously, the $H_2PO_4^-$ transfer across the interface is facilitated by the complexation with bis-thiourea **1c** or **2a**.

The analysis of these transfer polarograms reveals that the $H_2PO_4^-$ transfer assisted by **1c** is based on the formation of a 2:1 complex between $H_2PO_4^-$ and ionophore, and the transfer reaction is more stable by over -12 kJ · mol^{-1} than the case of C_3H_7-TU. The stabilization of the $H_2PO_4^-$ transfer for **1c** is even stronger by -11 kJ · mol^{-1} than that for bis-thiourea **2a**, which forms a 1:1 complex through the formation of four hydrogen bonds. Since mono-thiourea C_3H_7-TU, the counterpart of bis-thiourea **1c**, shows little ability to facilitate the $H_2PO_4^-$ transfer across the interface, a significant positive cooperativity should appear in the 2:1 complexation between $H_2PO_4^-$ and bis-thiourea **1c**, i.e., the binding of the second $H_2PO_4^-$ should be accelerated by the formation of the 1:1 complex ($K_{21} > K_{11}$). If the two binding sites are independent of the formation of a 2:1 complex ($K_{21} = 1/4K_{11}$) [38], the $H_2PO_4^-$ transfer based on the 2:1 complex

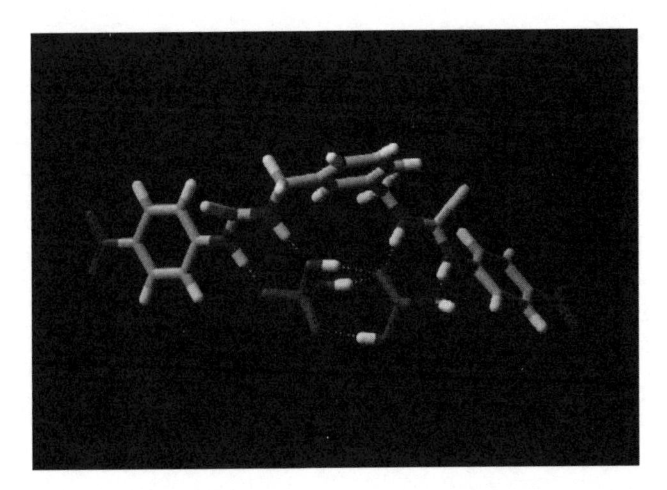

FIGURE 11.3. A possible conformation of the 2:1 complex between $H_2PO_4^-$ and bis-thiourea **1c**, based on Pure LowMode conformation search performed using MacroModel ver. 8.0. The hydrogen bonds are shown as broken lines. Reprinted from Ref. [22] with permission.

should be destabilized by 1.7 kJ · mol^{-1} as compared to the transfer based on the 1:1 complex with a binding constant of K_{11}. Thus, it should be noted again that the strong facilitating ability of bis-thiourea **1c** is indicative of the significant positive cooperativity in the 2:1 complex formation.

^{1}H NMR binding studies support the formation of the 2:1 complex between $H_2PO_4^-$ and **1c** in NB-d_5. Furthermore, the origin of the positive cooperativity can be understood from molecular modelling simulations. Figure 11.3 shows a possible structure of the 2:1 complex between $H_2PO_4^-$ and bis-thiourea **1c** as obtained by MacroModel with OPLS-AA* force field [39]. From this calculated structure, it is evident that two $H_2PO_4^-$ anions are complexed with the thiourea moeity of **1c** via hydrogen bonding. More importantly, two complexed $H_2PO_4^-$ anions are connected by three hydrogen bonds in the ternary complex. Similar dimerization of $H_2PO_4^-$ can be seen in crystal structures of uranyl salophene $H_2PO_4^-$ complexes [36], where two short hydrogen bonds are formed between phosphates. It is therefore likely that such additional hydrogen bonds between two $H_2PO_4^-$ anions are responsible for the significant positive cooperativity in the 2:1 complex formation.

It is interesting to compare the stability of the 2:1 complex with stability values obtained by electrically neutral hydrogen-bonding ionophores in the literature. The strongest $H_2PO_4^-$ binding in DMSO is achieved by bis-thiourea **2b** with a 1:1 binding constant of 2.0×10^5 M^{-1}[40]. Its complex stability, however, seems rather moderate as compared to that of bis-thiourea **1c**. With respect to the stability of the $H_2PO_4^-$ complex with bis-thiourea **2a**(K_{11} : 5.55×10^4 M^{-1} in DMSO) [40], the ratio of the 1:1 binding constants (or $(\beta_{21})^{1/2}/K_{11}$) is only 3.6 for **2b**, while the ratio is indeed 75 for bis-thiourea **1c**. Clearly, the stability of the 2:1 complex between $H_2PO_4^-$ and **1c** is extremely high even though direct comparison of the complex stabilities is not possible due to the difference in the organic solvent.

TABLE 11.1. Effect of anions on the adsorption behaviours of TU-B15C5 at the 1,2-dichloroethane/water interface.[a]

Anion[b]	$K_{ad}(M^{-1})$[c]	$\Gamma_{sat}(10^{-6}\ mol \cdot m^{-2})$[d]	A_{occ} ($Å^2$)[e]
No guest	1,200 ± 100	0.80 ± 0.1	210 ± 30
CH_3COO^-	2,200 ± 700	0.60 ± 0.1	280 ± 50
Br^-	2,300 ± 400	0.66 ± 0.06	250 ± 20
ClO_4^-	3,700 ± 1,300	0.68 ± 0.1	240 ± 40
Cl^-	6,100 ± 2,300	0.58 ± 0.09	290 ± 50
$H_2PO_4^-$	15,000 ± 3,000	0.42 ± 0.03	400 ± 30

[a] Reproduced from Ref. [23] with permission.
[b] 0.1 M solution of Na^+ salts.
[c] The adsorption constant.
[d] The saturated interfacial concentration.
[e] The apparent molecular occupied area.

11.2.2. Selective $H_2PO_4^-$ Recognition by Ionophores Adsorbed at the 1,2-Dichloroethane/Water Interface

As described in Section 11.2.1, we have succeeded in a highly efficient transport of $H_2PO_4^-$ anions across the NB/water interface based on the formation of the phosphate/ionophore 2:1 complex. However, the observed selectivity for phosphates is still moderate: The selectivity, as evaluated from the reversible half-wave potentials of anion transfers, follows in the order of $Cl^-(-0.077\ V) > HPO_4^{2-}(-0.17\ V) > H_2PO_4^-(-0.23V)$ [22]. Therefore, the further design and synthesis should be required for sufficiently selective ionophores in two-phase-transfer systems.

We here describe an alternative approach for the appearance of useful phosphate selectivity, utilizing the complexation-induced interfacial adsorption of ionophore/anion complexes. From the examination of the adsorption behaviours of three types of hydrogen-bonding ionophores (bis-thiourea **2a** [40], C_nH_{2n+1}-TU [21,41,42] and TU-B15C5 [43]) by interfacial tensiometry [44], we found that thiourea-functionalized benzo-15-crown-5 (TU-B15C5), a bifunctional receptor for simultaneous binding of anions and sodium ions, can work as a selective receptor for $H_2PO_4^-$($H_2PO_4^- > Cl^-$, ClO_4^-, Br^-, CH_3COO^-) at the 1,2-dichloroethane (DCE)/water interface [23].

Table 11.1 summarizes the adsorption parameters of TU-B15C5 at the DCE/water interface. While the adsorption constants K_{ad} for CH_3COO^- (2200 ± 700), Br^- (2300 ± 400), ClO_4^- (3700 ± 1300) and Cl^- (6100 ± 2300) are almost equal within the experimental errors, the adsorption constant for $H_2PO_4^-$ (15, 000 ± 3000) is indeed larger by a factor of 13 than that in the absence of these anions (1200 ± 100). Similarly, clear changes in the saturated interfacial concentration Γ_{sat} and the apparent molecular occupied area A_{occ} are observed only in the presence of $H_2PO_4^-$ in the aqueous phase. TU-B15C5 at the interface therefore shows high selectivity for $H_2PO_4^-$ over other anions ($H_2PO_4^- > Cl^-$, ClO_4^-, Br^-, CH_3COO^-): $H_2PO_4^-$ is the only anion that induces significant changes in the adsorption parameters of TU-B15C5 at the DCE/water interface. If the strength of interaction between anions and TU-B15C5 is controlled only by the free energies of anion adsorption to the interface without specific binding, the selectivity should obey the Hofmeister series as has been found in the case of anion adsorption to hexadecyltrimethyl-ammonium monolayers at the NB/water interface ($Br^- > Cl^- > F^-$) [45]. Apparently, the observed dependence of the adsorption parameters on anions is indicative of the interfacial binding between anions and the thiourea moiety of TU-B15C5 via the formation of hydrogen bonds.

It is interesting to note that, for the appearance of such anion binding functions of TU-B15C5 at the interface, the simultaneous binding of cations at the crown moiety is essential. The complex formation between TU-B15C5 and $H_2PO_4^-$ is observable only when Na^+ is present in the aqueous phase; no evidence for the binding with $H_2PO_4^-$ is obtained in the case of Li^+, K^+ and Rb^+ (K_{ad} (M^{-1}): LiH_2PO_4: 1000 ± 100; KH_2PO_4: 1200 ± 400; RbH_2PO_4: 840 ± 100). The role of Na^+ binding at the crown moiety is ascribed to (i) the enhancement of hydrogen bond donation ability of the thiourea NH protons [43], and/or (ii) neutralization of the negative charge of the TU-B15C5/$H_2PO_4^-$ complex to stabilize its adsorption [45,46].

The results presented here suggest that selective binding of very hydrophilic $H_2PO_4^-$ can be indeed obtained by hydrogen-bonding receptors just at the L/L interface. Similarly, the selective sensing of $H_2PO_4^-$/HPO_4^{2-} has recently been demonstrated at the interface of electrodes chemically modified with monolayers of hydrogen-bonding receptors [47,48]. Both studies indicate that, unlike bulk two-phase-transfer systems, complete dehydration of phosphate anions is not required for the interfacial binding event, which probably makes it possible to selectively recognize phosphates over hydrophobic anions. This finding would offer a novel approach to the sensing of very hydrophilic anions such as phosphates that are difficult to detect by conventional ionophore-based chemical sensors.

11.3. MOLECULAR RECOGNITION AT L/L INTERFACES AS STUDIED BY SECOND HARMONIC GENERATION SPECTROSCOPY

Second harmonic generation (SHG) spectroscopy has been used for the investigation of interfacial phenomena in the past decades, because it has inherent sensitivity towards molecules at interfaces [12,49]. SHG is based on the second-order non-linear process through which a fundamental light with a frequency of ω undergoes conversion into a light with a frequency of 2ω. The non-linear processes are forbidden for a medium with inversion symmetry under the electric dipole approximation. Even though molecules are oriented randomly in a bulk liquid medium, they have possibility of being oriented at interfaces. Thus, SHG spectroscopy allows the detection of molecules at interfaces without interference from molecules in a bulk liquid medium, and its interface selectivity is higher than that of linear spectroscopies using an evanescent wave under the condition of TIR.

The SHG signal is significantly enhanced when the incident laser wavelength, and/or its second harmonic (SH) wavelength, is in resonance with an electronic transition between molecular states [50]. The SHG spectra, which are obtained by measuring the SHG signal as a function of the incident laser wavelength, correspond to electronic spectra of interfacial species [51]. The peak intensity of SHG spectrum is related to the amount of molecules adsorbed at interfaces, and its input and/or output polarization dependence allows us to determine the absolute molecular orientation of adsorbates

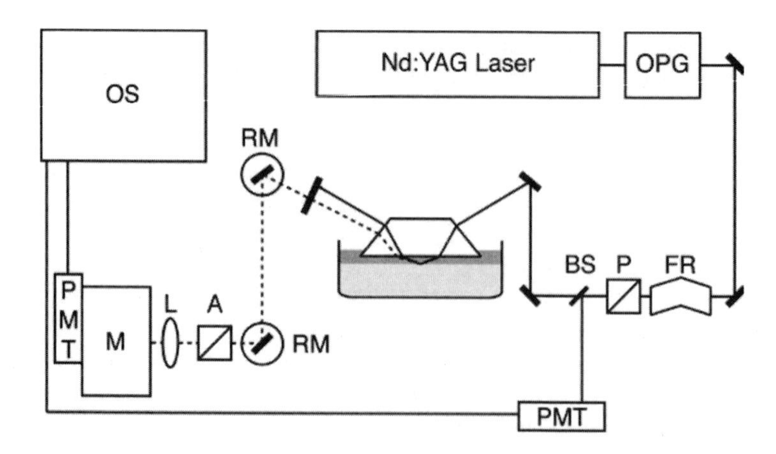

FIGURE 11.4. Schematic illustration of the measurement system for SHG spectroscopy.

[24]. SHG spectroscopy, therefore, can provide information on a molecular level about molecular orientation [52], chemical equilibria [53], molecular ordering [54] and molecular association [24,55] at L/L interfaces. We have developed the measurement system for SHG spectroscopy [56] and have applied it to the analysis of molecular recognition processes occurring at L/L [24,25] and S/L interfaces [57].

11.3.1. Measurement System for SHG Spectroscopy

The experimental set-up for measuring SHG spectra is shown in Figure 11.4. Tunable visible laser light, used as fundamental light, was generated by an optical parametric generator (OPG) pumped by a Q-switched Nd:YAG laser, covering the spectral region of 420–680 nm with a 30-ps pulse duration at a repetition rate of 10 Hz. The polarization of fundamental light was controlled by a Fresnel Rhomb.

It is well known that the intensity of SH light generated at an interface is remarkably enhanced when the incident angle of fundamental light is near the critical angle [58]. Thus, SHG spectra were measured under the TIR condition. The fundamental light was incident on the interface at an angle of 74.3° through a quartz prism. The beam diameter at the entrance face of the prism was ca. 2 mm, and the power of fundamental light was typically 0.6 mJ · cm^2.

The direction of SH light generated at the interface is expressed as [59]

$$n_{2\omega} \sin \theta_{2\omega} = n_\omega \sin \theta_\omega \tag{1}$$

where n_ω and $n_{2\omega}$ are refractive indices of the heptane phase at wavelengths of fundamental and SH light, and θ_ω and $\theta_{2\omega}$ are incident and output angles of fundamental and SH light. This equation means that the direction of SH light depends on its wavelength according to dispersion of the refractive index. In order to measure SHG spectra continuously compensating the effect of dispersion of the refractive index, SH light was guided to an entrance slit of a monochromator through an analyser by using a set of two rotatable mirros that were synchronously controlled to keep right direction of SH light by a computer.

The residual fundamental light and unnecessary emission from the interface were eliminated by a combination of short pass filters and a monochromator. Both of SH light and 1% of fundamental light were simultaneously detected using two photomultiplier tubes and stored in a digital storage oscilloscope, and the SH intensity was divided by the square of the fundamental intensity to correct wavelength dependence of the OPG output. Data were collected at every 2 or 3 nm by averaging 30 pulses during scanning the OPG. The processed data were plotted against the wavelength of fundamental light, and thus, SHG spectra were obtained.

11.3.2. Molecular Association at the Heptane/Water Interface

Molecular assembly and association are considered as essential for molecular recognition processes at L/L interfaces, because a cooperative binding of host and guest molecules is expected when host and/or guest molecules is highly ordered at the interface with high surface coverage. We demonstrated that SHG spectroscopy is suitable to observe molecular association at the L/L interface [24].

Figure 11.5a shows SHG spectra of rhodamine B (RB) adsorbed at the heptane/water interface. A single major peak assigned to a resonance with the S_0–S_1 electronic transition of RB is recognized in each spectrum. The peak positions in these spectra are determined by fitting data points to a combination of Lorentzian functions based on the two-state model [60].

$$I_{2\omega} \sim \left| \frac{a}{\frac{1}{\lambda_{\max}} - \left(\frac{1}{\lambda + i\Gamma}\right)} + b \right|^2 \tag{2}$$

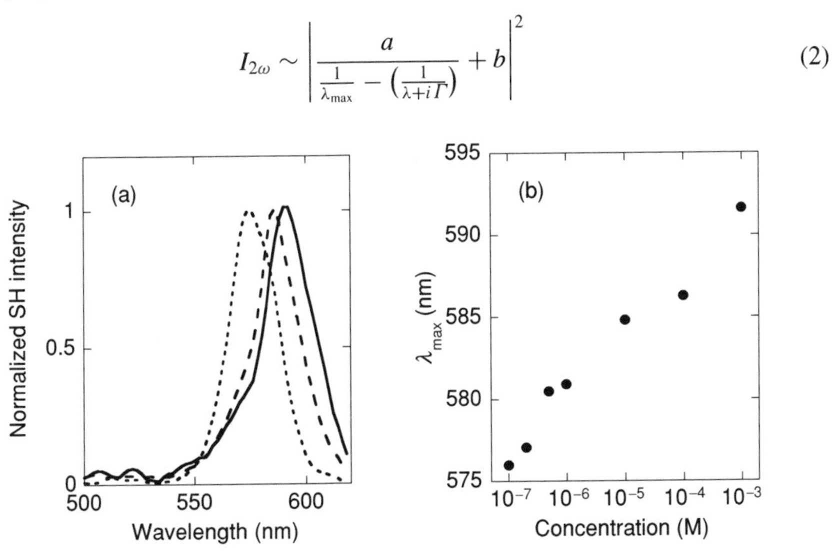

FIGURE 11.5. (a) SHG spectra of RB at the heptane/water interface for 1.0×10^{-7} M (---), 1.0×10^{-4} M (– –) and 1.0×10^{-3} M (—) of RB in the bulk water phase; (b) peak maxima determined from fitting analysis of the SHG spectra by Equation (2). The spectra were taken using the s-polarized fundamental input and p-polarized SHG output.

where a and b are the amplitudes of resonant and non-resonant parts of the non-linear susceptibility component, respectively. λ_{max} and Γ are the peak wavelength and the line width, respectively. The estimated peak wavelength of each SHG spectrum at the interface is shown in Figure 11.5b. The wavelength of the peak maximum is red-shifted from 575 to 592 nm as the concentration of RB in the water phase increases from 1.0×10^{-7} to 1.0×10^{-3} M. These peak wavelengths are long compared with the peak maximum (557 nm) in the absorption spectrum of RB monomer in the aqueous phase. Considering the dipole–dipole interaction between monomers, the red-shifted peaks found in SHG spectra are ascribed to the formation of in-plane associates in which the electronic transition dipoles of each monomer are arranged parallel to one another based on a simple molecular exciton theory [61,62].

The conformation of the in-plane associates formed at the heptane/water interface was further examined by polarization dependence of SH light. The SH intensity curves as a function of a polarization angle do not show any significant difference between p- and s-input polarizations, when the concentration of RB is below 1.0×10^{-5} M, suggesting that a molecular orientational angle of RB does not change in this concentration range. From the theoretical fitting of the polarization curves, information on molecular orientation at the interface can be obtained [63–65], and the tilt angle between the transition dipole parallel to the long axis of the xanthane ring and the interface normal is estimated as ca. 70°.

Figure 11.6 shows adsorption isotherms for RB and R110, which are obtained by plotting the square root of the SH intensities at peak maximum against the bulk concentrations of the dyes. These isotherms can be fitted well by a Langmuir adsorption isotherm, and the adsorption constants for RB and R110 are estimated as 4.9×10^{6} M^{-1} and 4.4×10^{5} M^{-1}, respectively. The difference in the adsorption constant between RB

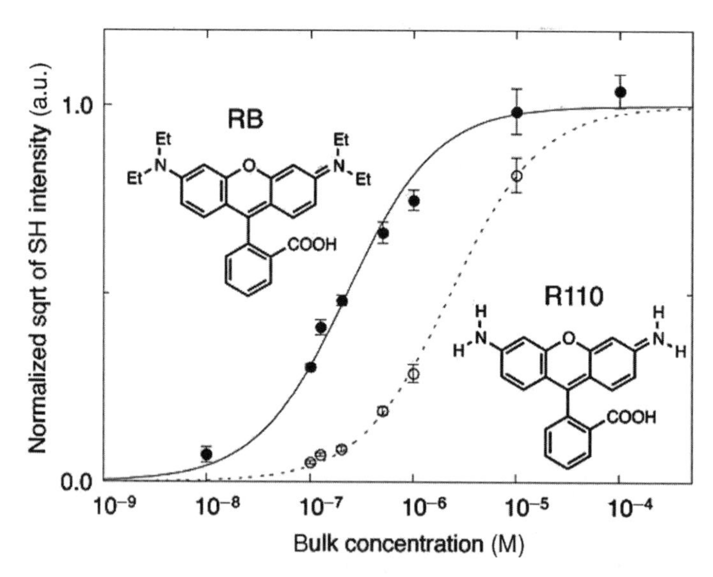

FIGURE 11.6. Adsorption isotherms of RB (filled circles) and R110 (open circles) for the heptane/water interface fitted by a Langmuir model.

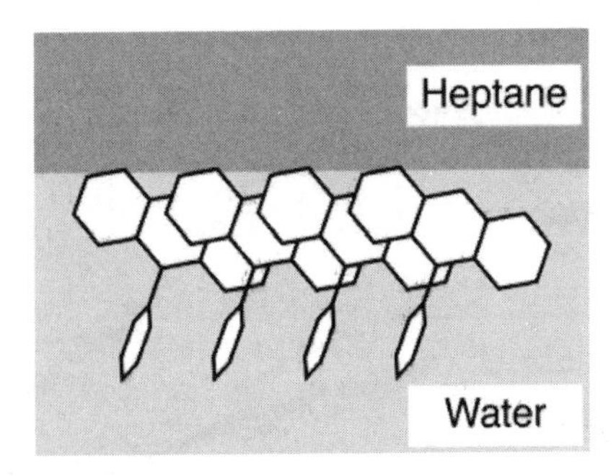

FIGURE 11.7. Schematic illustration of the in-plane associate of RB at the heptane/water interface

and R110 can be ascribed to the hydrophobicity of these dyes, because R110 has no alkyl chains in the amino groups of the xanthane moiety and RB has two diethylamino groups. The hydrophobicity of xanthane moiety of RB should be higher than that of R110. The hydrophobic interaction of the xanthane moiety of a rhodamine dye with heptane molecules would be responsible for their adsorption at the interface.

From the results of SHG spectroscopy, the rhodamine dyes adsorb and form the in-plane associates at the heptane/water interface, pointing the xanthane moiety towards the heptane phase with a tilt angle as shown schematically in Figure 11.7. The structure of the associates at the interface differs from that in a bulk aqueous phase. In an aqueous solution, RB makes a sandwich dimmer when the concentration is above 1.0×10^{-6} M [66], and the molar fraction of the dimmer is 0.56 at 1.0×10^{-3} M. The absorption spectra of the sandwich dimmer show blue-shifted peak around 525 nm. In contrast, red-shifted peaks found in SHG spectra indicate that the in-plane associates are predominantly formed at the interface despite the presence of the sandwich dimmer in the water phase. The hydrophobic property arising from heptane molecules at the heptane/water interface would be crucial for the predominant formation of in-plane associates.

11.3.3. Alkali Metal Recognition at the Heptane/Water Interface

To understand molecular recognition at liquid interfaces on a molecular level, the adsorption and orientation changes of host molecules induced by the recognition process must be clarified. Such information would provide a novel design concept for molecular recognition at liquid interfaces in relation to mimicing biological membrane processes. It is described in this section that a molecular orientation of a crown ether derivative changes by complexation with alkali metal cation at the heptane/water interface.

Figure 11.8 shows the SHG spectra of [2-hydroxy-5-(4-nitrophenylazo)phenyl]-methyl-15-crown-5 (azoprobe **1**) upon addition of alkali metal salts in the aqueous phase. The wavelength of the peak maximum of each spectrum is ca. 540 nm, which indicates the red-shift of ca. 45 nm from the UV-vis absorption peaks in bulk water. The red-shift is ascribed to a negative solvatochromism induced by different polarities between the heptane/water interface and bulk water, since the polarity of the interface is lower than that of bulk water [9,13]. On the other hand, the peak intensity of azoprobe **1** is found to significantly increase upon addition of alkali metal salts. The salt-dependent increase

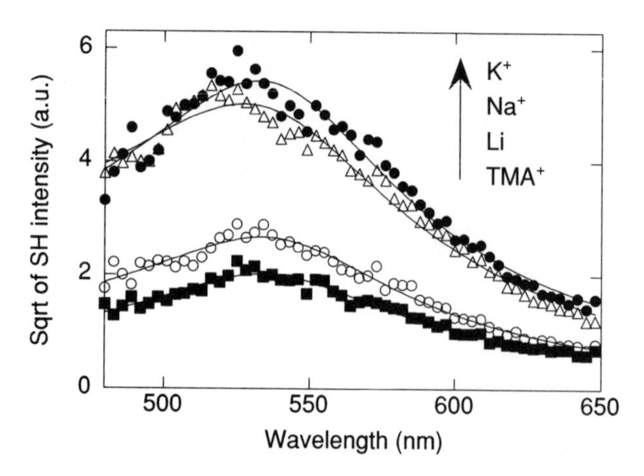

FIGURE 11.8. SHG spectra of azoprobe **1** at the heptane/water interface containing alkali metal and tetramethylammonium (TMA) ions (as chloride salts). The concentration of azoprobe **1** in bulk aqueous phase is 1.0×10^{-6} M.

of the peak intensity is ascribed to the adsorption of free anionic species (L^-) and alkali metal complex form (ML) of azoprobe **1**. The square root of SH intensity is expressed by Equation (3) [53,59]:

$$\sqrt{I_{2\omega}} \propto \chi^{(2)}_{\mathrm{s},IJK} = \left| N_{\mathrm{s,L^-}} \langle \beta_{\mathrm{L^-}} \rangle_{ijk} + N_{\mathrm{s,ML}} e^{i\Delta\phi} \langle \beta_{\mathrm{ML}} \rangle_{ijk} \right| \tag{3}$$

where $\chi^{(2)}_{\mathrm{s},IJK}$ is the element of the second-order susceptibility, and $N_{\mathrm{s,L^-}}$ and $N_{\mathrm{s,ML}}$ denote the interfacial concentrations of L^- and ML. $\beta_{\mathrm{L^-}}$ and β_{ML} are the molecular hyperpolarizabilities of L^- and ML, $\Delta\phi$ is the difference in phase angles between the hyperpolarizability tensor elements of L^- and ML. The $\langle\beta\rangle_{ijk}$ is a simple expression of $\langle T(\theta, \phi, \varphi)\rangle_{ijk}$, where $\langle T(\theta, \phi, \varphi)\rangle$ is the transformation tensor between the molecular and the laboratory frames. The molecular frame is referenced with Euler's angles θ, ϕ. and φ (Figure 11.9a).

Equation (3) clearly indicates that the factors affecting the SH intensity are (i) the difference in the molecular hyperpolarizability of ML species compared with L^-; (ii) the influence of $\Delta\phi$; (iii) the orientational change by ML complex formation; and (iv) the concentration change of L^- and ML species at the heptane/water interface. Since absorption spectral features of azoprobe **1** do not depend on alkali metal ions and TMA^+, the molecular hyperpolarizability of L^- is almost the same as that of ML complex, and hence the effect of $\Delta\phi$; upon the SH intensity is negligible. Since the influence of the orientational change upon the SH intensity is estimated to be small, the increase in the SH intensity observed in Figure 11.8 is mainly attributable to the increase in ML concentration at the heptane/water interface. The increase of intensity depends on ionic species, and the order of the intensity is $K^+ > Na^+ \gg Li^+$. This behaviour is very similar to that of the solvent extraction system. The extraction behaviour of azoprobe **1** for alkali metal ions from the aqueous phase into 1,2-dichloroethane has been examined and the order of extractabilities is $K^+ > Na^+ \gg Li^+$ [67]. The comparison of SH intensity change upon addition of alkali metal ions with the extractability of azoprobe **1** reveals that the increase of SH intensity is strongly dependent on the extraction selectivity of azoprobe

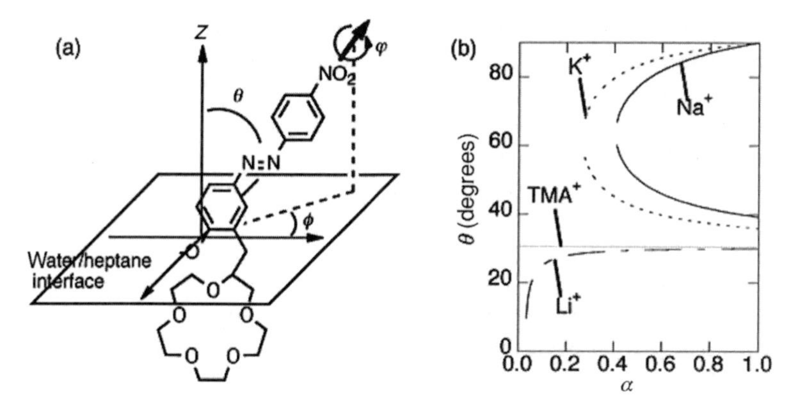

FIGURE 11.9. (a) Definition of the angles θ, ϕ and φ (b) Tilt angles of θ_{ML} as a function of the mole fraction of complex α at the interface, simulated by using the orientation parameter D. For TMA^+ system, L^- species are dominant and the tilt angle θ_{ML} is estimated as 31°.

1 for alkali metal ions. This is clear evidence that the ML complexes are selectively adsorbed at the heptane/water interface.

The input polarization dependence of p- and s-polarized SH light was measured to elucidate the molecular orientation of L^- and ML species. The results can be fitted well to the theory and the relative values of three non-zero elements of the surface susceptibility [63–65], which determine the molecular orientation, are obtained from the fitting. Since hyperpolarizability of azobenzene dyes is known to be dominated by a single element β_{zzz}, which is an element of the hyperpolarizability along the π–π^* moment direction [52], the tilt angle θ can be estimated from the relative values of the surface susceptibility,

$$D = \frac{\langle\cos^3\theta\rangle}{\langle\cos\theta\rangle} = \frac{\chi_{zzz}}{\chi_{zzz}+2\chi_{xzz}} = \frac{N_{s,L} - \beta_{zzz,L} - \langle\cos^3\theta_{L^-}\rangle + N_{s,ML}\beta_{zzz,ML}\langle\cos^3\theta_{ML}\rangle}{N_{s,L} - \beta_{zzz,L} - \langle\cos\theta_{L^-}\rangle + N_{s,ML}\beta_{zzz,ML}\langle\cos\theta_{ML}\rangle}$$
$$= \frac{(1-\alpha)\langle\cos^3\theta_{L^-}\rangle + \alpha\langle\cos^3\theta_{ML}\rangle}{(1-\alpha)\langle\cos\theta_{L^-}\rangle + \alpha\langle\cos^3\theta_{ML}\rangle} \quad (4)$$

where D is the orientation parameter, θ_{L^-} and θ_{ML} are the orientation angles of L^- and ML and α represents the mole fraction of ML at the heptane/water interface. For the TMA^+ system, only L^- species exists at the interface and the orientation angle of L^- is calculated as 31°. In the presence of alkali metal ions, the θ_{ML} values cannot be calculated because three uncertain constants (α, θ_{L^-}, θ_{ML}) remain.

Since no SHG spectral shifts in Figure 11.8 is noted upon addition of alkali metal salts for azoprobe **1**, the interaction between azoprobes at the interface must be small. Thus the orientation angle of L^- is estimated to be constant even in the presence of ML species, and the relationship between θ_{ML} and α under constant θ_{L^-} can be simulated using D values derived from the experimental results on input polarization dependence of SH light. The simulation results are shown in Figure 11.9b. The orientation angle of ML (θ_{ML}) is larger for Na^+ and K^+ as compared with θ_{L^-}, whereas the θ_{ML} for Li^+ is smaller in every α fraction. This result clearly demonstrates that the Na^+ and K^+ complexes are flatter and the Li^+ complex tends to lift up at the heptane/water interface as compared with the orientation angle of L^- species. The observed orientation changes of ML species may be ascribed to the difference in the charge distribution of alkali metal ion complexes formed at the interface. Since Na^+ and K^+ can form tight ion pairs with azoprobe **1** based on the best size-fitting with the 15-crown-5 binding site [68], charge delocalized complexes, which induce the flatter orientation at the interface, are expected. On the other hand, Li^+ ion is too small to fit inside the 15-crown-5 cavity. Thus the highly hydrated Li^+ complex makes the binding site of azoprobe **1** more hydrophilic, resulting in the lift-up conformation at the heptane/water interface. These orientation differences among the alkali metal complexes offer a unique characteristic for molecular recognition at the L/L interface, which is clarified by SHG spectroscopy. It is expected that SHG spectroscopy will provide useful information for developing novel molecular recognition processes at the L/L interfaces.

11.4. CONCLUDING REMARKS

In molecular recognition of ions and molecules at L/L interfaces using hydrogen-bonding artificial receptors, we have found that hydrogen bonding, which is not effective at all in bulk aqueous media, does work effectively at L/L interfaces, and that unusual selectivity for very hydrophilic $H_2PO_4^-$ by artificial receptors, which cannot be attained by two-phase distribution systems, is achieved via hydrogen-bonding interaction at the L/L interfaces based on measurements by ion transfer polarograms and surface tensions. In the measurements of ion recognition processes at L/L interfaces by resonant SHG spectroscopy, we have found orientation changes of receptor molecules depending on the presence of guest ions. These characteristics in molecular recognition observed at L/L interfaces should be ascribed to the partial dehydration of ions and molecular orientations of receptor and complexed molecules, and the results obtained in this study would give deeper insight to develop novel recognition systems for ions and molecules. To pursue such studies, origin of unusual selectivity should be widely examined and clarified using surface-sensitive measurements such as SHG in combination with computer simulations at L/L interfaces, and the accomplishments should also be reflected to develop practical analytical systems.

ACKNOWLEDGEMENTS

This work was partially supported by Grants-in-Aid for Scientific Research on Priority Research Areas (B), No. 13129201, and for Scientific Research (A), No. 14204074, from the Ministry of Education, Culture, Sports, Science and Technology, Japan.

REFERENCES

1. A.G. Volkov, D.W. Deamer, D.L. Tanelian and V.S. Martin, *Liquid Interface in Chemistry and Biology*, John Wiley & Sons, Inc., New York, 1998.
2. A.G. Volkov and D.W. Deamer, *Liquid–Liquid Interfaces—Theory and Methods*, CRC Press, Inc., New York, 1996.

3. D.J. Fermin, D. Duong, Z. Ding, P.-F. Brevet and H.H. Girault, J. Am. Chem. Soc. **121**, 10203 (1999).
4. Y.-T. Kong, S. Imabayashi and T. Kakiuchi, J. Am. Chem. Soc. **122**, 8215 (2000).
5. S. Amemiya, Z. Ding, J. Zhou and A.J. Bard, J. Electroanal. Chem. **483**, 7 (2000), and references therein.
6. Y. Uchiyama, T. Kitamori, T. Sawada and I. Tsuyumoto, Langmuir **16**, 6597 (2000).
7. K. Chikama, T. Negishi, K. Nakatani, Bull. Chem. Soc. Jpn. **76**, 295 (2003).
8. N. Fujiwara, S. Tsukahara and H. Watarai, Langmuir **17**, 5337 (2001).
9. K. Bessho, T. Uchida, A. Yamauchi, T. Shioya and N. Teramae, Chem. Phys. Lett. **264**, 381 (1997).
10. S. Ishizaka, H.-B. Kim and N. Kitamura, Anal. Chem. **73**, 2421 (2001).
11. S. Ishizaka and N. Kitamura, Bull. Chem. Soc. Jpn. **74**, 1983 (2001).
12. R.M. Corn and D.A. Higgins, Chem. Rev. **94**, 107 (1994).
13. H. Wang, E. Borguet and K.B. Eisenthal, J. Phys. Chem. B **102**, 4927 (1998).
14. D. Michael and I.J. Benjamin, J. Chem. Phys. **107**, 5684 (1997).
15. T. Yamashita, T. Uchida, T. Fukushima and N. Teramae, J. Phys. Chem. B **107**, 4786 (2003).
16. T. Shioya, S. Tsukahara and N. Teramae, Chem. Lett. 469 (1996).
17. T. Shioya, S. Tsukahara and N. Teramae, Chem. Lett. 695 (1997).
18. T. Shioya, S. Nishizawa and N. Teramae, Langmuir **14**, 4552 (1998).
19. T. Shioya, S. Nishizawa and N. Teramae, Langmuir **15**, 2575 (1999).
20. T. Shioya, S. Nishizawa and N. Teramae, J. Am. Chem. Soc. **120**, 11534 (1998).
21. T. Hayashita, T. Onodera, R. Kato, S. Nishizawa and N. Teramae, Chem. Commun. 755 (2000).
22. S. Nishizawa, T. Yokobori, R. Kato, K. Yoshimoto, T. Kamaishi and N. Teramae, Analyst **128**, 663 (2003).
23. K. Shigemori, S. Nishizawa, T. Yokobori, T. Shioya and N. Teramae, New J. Chem. **26**, 1102 (2002).
24. T. Uchida, A. Yamaguchi, T. Ina and N. Teramae, J. Phys. Chem. B **104**, 12091 (2000).
25. K. Nochi, A. Yamaguchi, T. Hayashita, T. Uchida and N. Teramae, J. Phys. Chem. B **106**, 9906 (2002).
26. P.D. Beer and P.A. Gale, Angew. Chem. Int. Ed. **40**, 486 (2001).
27. S.R. Adams, A.T. Harootunian, Y.J. Buechler, S.S. Taylor and R.Y. Tsien, Nature **349**, 694 (1991); A.W. Czarnik, Acc. Chem. Res. **27**, 302 (1994); M. Sato, N. Hida, T. Ozawa and Y. Umezawa, Anal. Chem. **72**, 5918 (2000); S.E. Schneider, S.N. ONeil and E.V. Anslyn, J. Am. Chem. Soc. **122**, 542 (2000); K. Choi and A.D. Hamilton, Angew. Chem. Int. Ed. **40**, 3912 (2001); S. Mizukami, T. Nagano, Y. Urano, A. Odani and K. Kikuchi, J. Am. Chem. Soc. **124**, 3920 (2002); S. Aoki, K. Iwaida, N. Hanamoto, M. Shiro and E. Kimura, J. Am. Chem. Soc. **124**, 5256 (2002); A. Ojida, Y. Mito-oka, M. Inoue and I. Hamachi, J. Am. Chem. Soc. **124**, 6256 (2002).
28. P. Bühlmann, E. Pretsch and E. Bakker, Chem. Rev. **98**, 1593 (1998).
29. Y. Umezawa, K. Umezawa, P. Bühlmann, N. Hamada, H. Aoki, J. Nakanishi, M. Sato, K.P. Xiao and Y. Nishimura, Pure Appl. Chem. **74**, 923 (2002).
30. E. Bakker, P. Bühlmann and E. Pretsch, Chem. Rev. **97**, 3083 (1997).
31. H. Luecke and F.A. Quiocho, Nature **347**, 402 (1990).
32. M. Berger and F.P. Schmidtchen, J. Am. Chem. Soc. **118**, 8947 (1996); M. Fibbioli, M. Berger, F.P. Schmidtchen and E. Pretsch, Anal. Chem. **72**, 156 (2000).
33. H.C. Visser, D.M. Rudkevich, W. Verboom, F. de Jong and D.N. Reinhoudt, J. Am. Chem. Soc. **116**, 11554 (1994); M.M.G. Antonisse, B.H.M. Snellink-Ruel, I. Yigit, J.F.J. Engbersen and D.N. Reinhoudt, J. Org. Chem. **62**, 9034 (1997); L.A.J. Chrisstoffels, F. de Jong and D.N. Reinhoudt, Chem. Eur. J. **6**, 1376 (2000).
34. S. Nishizawa, T. Yokobori, T. Shioya and N. Teramae, Chem. Lett. 1058 (2001).
35. S. Nishizawa, T. Yokobori, R. Kato, T. Shioya and N. Teramae, Bull. Chem. Soc. Jpn. **74**, 2343 (2001).
36. D.M. Rudkevich, W. Verboom, Z. Brozozka, M.J. Palys, W.P.R.V. Stauthamer, G.J. van Hummel, S.M. Franken, S. Harkema, J.F.J. Engbersen and D.N. Reinhoudt, J. Am. Chem. Soc. **116**, 4341 (1994).
37. J. Koryta, Electrochim. Acta **24**, 293 (1979); S. Kihara, M. Suzuki, K. Maeda, K. Ogura, S. Umetani, M. Matsui and Z. Yoshida, Anal. Chem. **58**, 2954 (1986); T. Kakutani, Y. Nishiwaki, T. Osakai and M. Senda, Bull. Chem. Soc. Jpn. **59**, 781 (1986); Z. Samec, Chem. Rev. **88**, 617 (1988); Z. Samec and P. Papoff, Anal. Chem. **62**, 1010 (1990); H. Matsuda, Y. Yamada, K. Kanamori, Y. Kudo and Y. Takeda, Bull. Chem. Soc. Jpn. **64**, 1497 (1991); H.H. Girault, Chapter 1 in *Modern Aspects of Electrochemistry*, Eds. J.O. Bockris, B.E. Conway and R.E. White, Plenum Press, New York, 1993, Vol. 25, p. 30.
38. K.A. Connors, Chapter 2 in *Binding Constants*, John Wiley & Sons, Inc., New York, p. 78, 1987.
39. N.A. McDonald, E.M. Duffy and W.L. Jorgensen, J. Am. Chem. Soc. **120**, 5104 (1998).

40. P. Bühlmann, S. Nishizawa, K.P. Xiao and Y. Umezawa, Tetrahedron **53**, 1647 (1997).
41. S. Nishizawa, R. Kato, T. Hayashita and N. Teramae, Anal. Sci. **14**, 595 (1998).
42. R. Kato, S. Nishizawa, T. Hayashita and N. Teramae, Tetrahedron Lett. **42**, 5053 (2001).
43. S. Nishizawa, K. Shigemori and N. Teramae, Chem. Lett. 1185 (1999).
44. J. Van Hunsel, G. Bleys and P Joos, J. Colloid Interface Sci. **114**, 432 (1986).
45. T. Kakiuchi, M. Kobayashi and M. Senda, Bull. Chem. Soc. Jpn. **61**, 1545 (1988).
46. A. Varnek, L. Troxler and G. Wipff, Chem. Eur. J. **3**, 552 (1997).
47. K.P. Xiao, P. Bühlmann and Y. Umezawa, Anal. Chem. **71**, 1183 (1999).
48. H. Aoki, K. Hasegawa, K. Tohda and Y. Umezawa, Biosensors and Bioelectronics **18**, 261 (2003).
49. Y.R. Shen, *The Principles of Nonlinear Optics*, John Wiley & Sons, New York, 1984; V. Vogel and Y.R. Shen, Annu. Rev. Mater. Sci. **21**, 515 (1991); K.B. Eisenthal, Chem. Rev. **96**, 1343 (1996).
50. T.F. Heinz, C.K. Chen, D. Ricard and Y.R. Shen, Phys. Rev. Lett. **48**, 478 (1982); N.E. Van Wyck, E.W. Koenig, J.D. Byers and W.M. Hetherington, Chem. Phys. Lett. **122**, 153 (1985); R. Yam and G. Berkovic, Langmuir **9**, 2109 (1993); J.D. Byers and J.M. Hicks, Chem. Phys. Lett. **231**, 216 (1994).
51. S.H. Lin, R.G. Alden, A.A. Villaeys and V. Pfiumio, Phys. Rev. A **48**, 3137 (1993).
52. R.R. Naujok, D.A. Higgins, D.G. Hanken and R.M. Corn, J. Chem. Soc., Faraday Trans. **91**, 1411 (1995).
53. J. Rinuy, A. Piron, P.-F. Brevet, M.B.-Desce and H.H. Girault, Chem. Eur. J. **6**, 3434 (2000).
54. J.C. Conboy, J.L. Daschbach and G.L. Richmond, J. Phys. Chem. **98**, 9688 (1994).
55. H. Nagatani, A. Piron, P.-F. Brevet, D.J. Fermin and H.H. Girault, Langmuir **18**, 6647 (2002).
56. A. Yamaguchi, T. Uchida, N. Teramae and H. Kaneta, Anal. Sci. **13**(3, Suppl.) 85 (1997).
57. A. Yamaguchi, R. Kato, S. Nishizawa and N. Teramae, Chem. Lett. **32**, 798 (2003).
58. B.U. Felderhof, A. Bratz and G. Marowsky, J. Opt. Soc. Am. B **10**, 1824 (1993).
59. T.F. Heinz, Chapter 5 in *Modern Problems in Condensed Matter Sciences*, Eds. H.-E. Ponath and G.I. Stegeman, Elsevier, Amsterdam, 1991, Vol. 29, p. 353.
60. J. Wang, E. Borguet and K.B. Eisenthal, J. Phys. Chem. B **101**, 713 (1997).
61. M. Kasha, H.R. Rawls and M.A. El-Bayoumi, Pure Appl. Chem. **11**, 371 (1965).
62. K. Kemniz, N. Tamai, I. Yamazaki, N. Nakahima and K. Yoshihara, J. Phys. Chem. **90**, 5094 (1986).
63. B. Dick, Chem. Phys. **96**, 199 (1985).
64. D.A. Higgins, S.K. Byerly, M.B. Abrams and R.M. Corn, J. Phys. Chem. **95**, 6984 (1991).
65. R.M. Corn and D.A. Higgins, Chapter 12 in *Characterization of Organic Thin Films*, Ed. A. Ulman, Butterworth-Heinemann, Stoneham, 1995, p. 227.
66. I.L. Arbeloa and P.R. Ojeda, Chem. Phys. Lett. **87**, 556 (1982).
67. Y. Katayama, K. Nita, M. Ueda, H. Nakamura and M. Takagi, Anal. Chim. Acta **173**, 193 (1984).
68. T. Hayashita and M. Takagi, Chapter 17 in *Comprehensive Supermolecular Chemistry*, Ed. G.W. Gokel, Pergamon, New York, 1996, Vol. 1, p. 635.

12

Photochemistry at Liquid/Liquid Interfaces

Shoji Ishizaka and Noboru Kitamura
Hokkaido University, Sapporo 060-0810, Japan

12.1. INTRODUCTION

The nature of an interface (liquid/liquid, solid/liquid, air/liquid and so on) should reflect on the relevant interfacial phenomena, so that detailed understandings of chemical and structural characteristics of the interface at a microscopic level are of primary importance for further advances in various sciences. In practice, solid/liquid and air/liquid interfacial systems have been studied widely by various experimental techniques, and the knowledges about the characteristics of the interfaces have been accumulated. However, very little is known about the chemical and structural characteristics of a liquid/liquid interface at a microscopic level. So far, thermodynamic and electrochemical techniques have been applied to study liquid/liquid interfacial chemistry. Nonetheless, its dynamic aspects have rarely been explored.

It is well known that both nanometre and nanosecond–picosecond resolutions at an interface can be achieved by total internal reflection (TIR) fluorescence spectroscopy. Unlike steady-state fluorescence spectroscopy, fluorescence dynamics is highly sensitive to microscopic environments, so that time-resolved TIR fluorometry at water/oil interfaces is worth exploring to obtain a clearer picture of the interfacial phenomena [1]. One of the interesting targets to be studied is the characteristics of dynamic motions of a molecule adsorbed on a water/oil interface. Dynamic molecular motions at a liquid/liquid interface are considered to be influenced by subtle changes in the chemical/physical properties of the interface, particularly in a nanosecond–picosecond time regime. Therefore, time-resolved spectroscopy is expected to be useful to study the nature of a water/oil interface.

We have studied water/oil interfaces on the basis of time-resolved TIR fluorescence spectroscopy and discussed the structures and characteristics of the interface at a molecular level [1–5]. In these studies, we employed the following experimental approaches. One is a magic angle dependence of the TIR fluorescence decay profile of dye molecules adsorbed on a water/oil interface [3–5]. The other approach is a structural dimension analysis of excitation energy transfer dynamics between dye molecules adsorbed on a water/oil interface [4,5]. We applied these methods to studying water/oil interfaces, and succeeded in obtaining invaluable information on characteristic features at water/oil interfaces.

In this chapter, we review recent results on the studies of chemical and structural characteristics at water/oil interfaces with special references to those at water/CCl_4 and water/1,2-dichloroethane (DCE) interfaces.

12.2. THEORETICAL AND EXPERIMENTAL BACKGROUNDS

In order to study interfacial phenomena at a liquid/liquid boundary at a microscopic level, surface-selective or depth-resolved measurements at an interface are absolutely necessary. Among several methods, TIR spectroscopy is a powerful means to obtain an inside look at an interfacial layer in several tens to several hundreds of nanometres. In this study, TIR fluorescence spectroscopy was employed to follow chemical and physical characteristics at liquid/liquid interfaces. Before discussing characteristic features of the structures at liquid/liquid interfaces, the basic theory of TIR of light is reviewed briefly in the following.

12.2.1. Total Internal Reflection

Whenever electromagnetic radiation is incident at an interface separating two media, a part of the beam is *reflected* back into a medium 1, while the other part continues into a medium 2, but with an altered direction of propagation. This latter phenomenon is termed *refraction*. Figure 12.1 depicts this situation, where the subscripts i, t and r refer to the

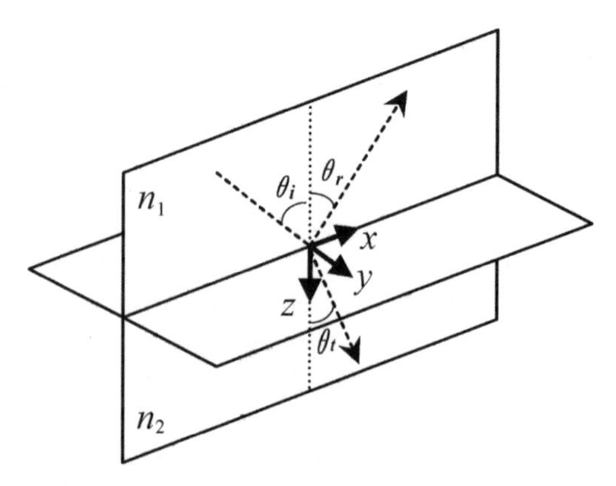

FIGURE 12.1. Schematic representation of refraction and reflection of a plane electromagnetic wave at a boundary. The surface normal is taken along the z-axis, and the incident beam is assumed to be in the x–z plane (plane of incidence).

incident, transmitted and reflected beams, respectively. The radiation is incident from a medium of a lesser refractive index (n_1) upon a medium of a greater refractive index (n_2); $n_2 > n_1$. In such a case, a phenomenon of reflection is termed *external reflection*, and a real angle of refraction, θ_t, exists for all possible choices of the angle of incidence, θ_i, with the condition of $\theta_t < \theta_i$. On the other hand, under the refractive index condition of $n_1 > n_2$, a phenomenon of reflection is termed *internal reflection*, and it is obvious from Snell's law that the angle of refraction becomes imaginary or values of the angle of incidence such that $\theta_i > \sin^{-1}(n_2/n_1)$. The angle above which this refracted wave ceases to be real is termed the *critical angle*, θ_c.

$$\theta_c = \sin^{-1}(n_2/n_1) \tag{1}$$

In the case of $\theta_i > \theta_c$, the reflectivity becomes unity for both the parallel and perpendicular components. This phenomenon is appropriately termed *total internal reflection* [6,7].

12.2.2. Evanescent Wave

In terms of the relative index of refraction, $n = (n_2/n_1) < 1$, the following expressions follow from Snell's law for angles of incidence greater than the critical angle:

$$\sin\theta_t = (1/n)\sin\theta_i \tag{2}$$

$$\cos\theta_t = [1 - (1/n^2)\sin^2\theta_i]^{1/2} = i[(1/n^2)\sin^2\theta_i - 1]^{1/2} \tag{3}$$

Using these values in Fresnel's formula, the amplitude of the reflected beam perpendicular to the plane of incidence ($A_{r\perp}$) is given by Equation (4),

$$A_{r\perp} = A_{i\perp}\frac{(1/n)\cos\theta_i - i[(1/n^2)\sin^2\theta_i - 1]^{1/2}}{(1/n)\cos\theta_i + i[(1/n^2)\sin^2\theta_i - 1]^{1/2}} \tag{4}$$

which is equivalent to

$$A_{r\perp} = A_{i\perp}\exp(-i\delta_\perp) \tag{5}$$

where

$$\tan(\delta_\perp/2) = [(1/n^2)\sin^2\theta_i - 1]^{1/2}/[(1/n)\cos\theta_i] \tag{6}$$

Thus, the perpendicular component of a reflected wave whose incidence in the x–z plane is

$$\boldsymbol{E}_{r\perp} = A_{i\perp}\exp[i\{\omega t - k(x\sin\theta_i - z\cos\theta_i) - \delta_\perp\}] \tag{7}$$

In a similar way, those of the parallel components are expressed as

$$A_{r\parallel} = A_{i\parallel}\frac{\cos\theta_i - (i/n)[(1/n^2)\sin^2\theta_i - 1]^{1/2}}{\cos\theta_i + (i/n)[(1/n^2)\sin^2\theta_i - 1]^{1/2}} \tag{8}$$

or

$$\boldsymbol{E}_{r\parallel} = A_{i\parallel}\exp[i\{\omega t - k(x\sin\theta_i - z\cos\theta_i) - \delta_\parallel\}] \tag{9}$$

where

$$\tan(\delta_{\parallel}/2) = (1/n)[(1/n^2)\sin^2\theta_i - 1]^{1/2}/\cos\theta_i \tag{10}$$

Hence, the components of $\boldsymbol{E}$ parallel and perpendicular to the plane of incidence undergo phase retardations $\delta_{\parallel}$ and $\delta_{\perp}$, respectively, although the amplitudes are unaltered by reflection. If the incident wave is plane-polarized, the incident components are in phase, while those of the reflected wave are not, so that elliptical polarization is produced. The magnitude of the phase shift for $\theta_i > \theta_c$ is

$$\tan(\delta_{\parallel}/2 - \delta_{\perp}/2) = \cos\theta_i[(1/n^2)\sin^2\theta_i - 1]^{1/2}/[(1/n)\sin^2\theta_i] \tag{11}$$

The above-mentioned phase shift for the angle of incidence greater than the critical angle implies the existence of a resultant field in the first medium at the interface. In order that the boundary conditions remain satisfied, there must be a resultant disturbance in the second phase. It can readily be shown that this disturbance is of the nature of an exponentially damped wave that penetrates into the second medium. This wave is termed an *evanescent wave*.

The depth of penetration (d_p) of an evanescent wave into the second medium can be quantitatively defined in terms of the distance required for the electric field intensity to decrease to $1/e$ of its initial value,

$$d_p = (\lambda/2\pi)\left({n_1}^2\sin^2\theta_i - {n_2}^2\right)^{-1/2} \tag{12}$$

where λ is the wavelength of the radiation in the first medium. However, we deal with the intensity of the light in TIR fluorometry rather than that of the electric field, which is proportional to the square of the electric field. More practical formula of the penetration depth of an evanescent wave is given as follows [8]:

$$d_p = (\lambda/4\pi)\left({n_1}^2\sin^2\theta_i - {n_2}^2\right)^{-1/2} \tag{13}$$

Equation (13) shows that the penetration depth depends on both the wavelength and the angle of an incident beam. Thus, an appropriate choice of the angle of an incident light beam enables depth-resolved measurements at an interface.

12.2.3. Probe Molecules for TIR Fluorescence Spectroscopy

For TIR fluorescence spectroscopy on water/oil interfaces, the choice of a probe molecule is of primary importance. For example, the penetration depth (d_p) of an incident evanescent wave at a 1,2-dichloroethane (DCE, refractive index (n); $n_1 = 1.44$)/water ($n_2 = 1.33$) interface is calculated to be ~94 nm on the basis of Equation (13), where $\lambda = 580$ nm and $\theta_i = 80°$. It has been reported that the thickness of a sharp water/oil interface represented by water/DCE is ~1 nm [9], so that d_p of the incident evanescent wave is thicker than the thickness of the interfacial layer, and the fluorescence characteristics of a probe molecule in the bulk phase are superimposed, more or less, on those at the interface [2]. Therefore, a probe molecule should be highly surface-active and adsorb on the interface, so as to exclude fluorescence of the probe molecule from the bulk phase. In the present experiments, we employed xanthene dyes as fluorescence probes throughout

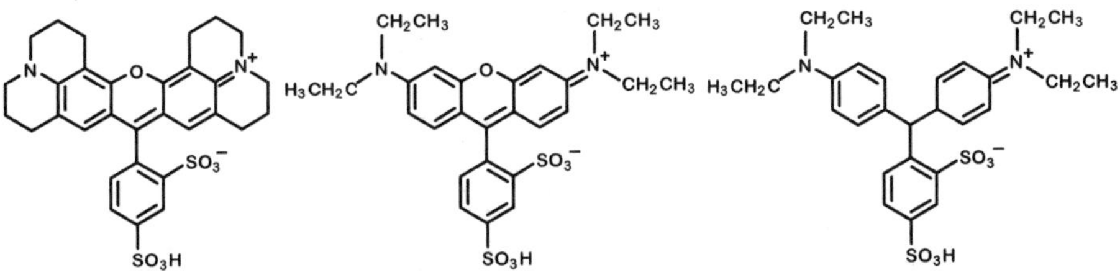

Sulforhodamine 101 (SR101) Sulforhodamine B (SRB) Acid Blue 1 (AB1)

SCHEME 12.1. Structures and abbreviations of the dye molecules used in this study. (a) Sulforhodamine 101; SR101, (b) Sulforhodamine B; SRB, (c) Acid Blue 1; AB1.

the study, since these dyes are highly surface-active and adsorb strongly on a water/oil interface. This has been confirmed by interfacial tension (γ) measurements as reported elsewhere [3–5]. The structures and abbreviations of the dye molecules used in this study are shown in Scheme 12.1.

For example, the amount of adsorbed SR101 on a water/DCE interface (Γ) is given by the Gibbs equation: $\Gamma = -(1/2.3RT)\,\mathrm{d}\gamma/\mathrm{d}\log[\mathrm{SR101}]$. The Γ value was then calculated to be 5.0×10^{-14} mol · cm^{-2} ([SR101] $= 1.0 \times 10^{-9}$ M). When the interfacial area is assumed to be 1 cm^2 for simplicity, the number of SR101 molecules adsorbed on the interface (5.0×10^{-14} mol) is 5000 times higher than that expected to be involved in the excited volume by the evanescent wave (1.0×10^{-17} mol) without adsorption. Almost the same results with those for SR101 were obtained for other dye/water/oil systems. At [SR101] $= 1 \times 10^{-9}$ M, therefore, the fluorescence response observed under the TIR conditions is ascribed essentially to that from the interface.

12.3. TIR FLUORESCENCE DYNAMIC ANISOTROPY AT A LIQUID/LIQUID INTERFACE

12.3.1. Principle

Dynamic fluorescence anisotropy is based on rotational reorientation of the excited dipole of a probe molecule, and its correlation time(s) should depend on local environments around the molecule. For a dye molecule in an isotropic medium, three-dimensional rotational reorientation of the excited dipole takes place freely [10]. At a water/oil interface, on the other hand, the out-of-plane motion of a probe molecule should be frozen when the dye is adsorbed on a sharp water/oil interface (i.e., two-dimensional in respect to the molecular size of a probe), while such a motion will be allowed for a relatively thick water/oil interface (i.e., three-dimensional) [11,12]. Thus, by observing rotational freedom of a dye molecule (i.e., excited dipole), one can discuss the thickness of a water/oil interface; the correlation time(s) provides information about the chemical/physical characteristics of the interface, including the dynamical behaviour of the interfacial structure. Dynamic fluorescence anisotropy measurements are thus expected

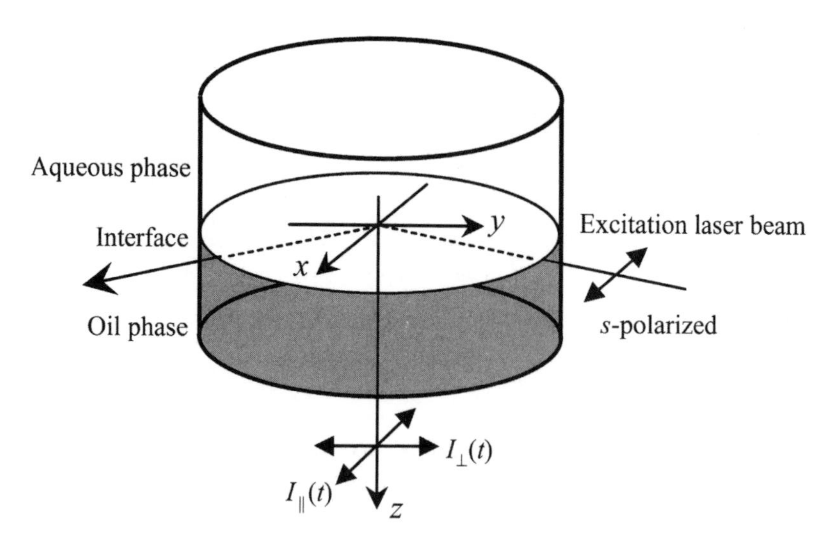

FIGURE 12.2. Coordinate system defined in the experiment. The x–y plane is the plane of the interface.

to provide new insights into the structure at a water/oil interface, not obtained by conventional spectroscopies.

The laboratory coordinate system chosen for TIR fluorescence anisotropy measurements is illustrated in Figure 12.2. SR101 molecules located at a water/oil interface (in the x–y plane) are excited by an s-polarized laser beam along the x-axis. The TIR fluorescence is then detected along the z-axis and its polarization is selected by a polarizer. The fluorescence decay profile observed under such a configuration is analysed for two limiting cases, depending on the structure of a water/oil interface: two-dimensional or three-dimensional.

12.3.1.1. Two-Dimensional Model. If the thickness of a water/oil interface is comparable to the molecular size of SR101 and the dye molecules located at the interface are strongly oriented, the rotational motions of SR101 will be strongly restricted in the interfacial layer (x–y plane of the interface, two-dimension), and the emission dipole moment of SR101 (direction of the long axis of the xanthene ring) directs within the x–y plane. In such a case, the time profile of the total fluorescence intensity of SR101 observed from the interface should be proportional to $I_\parallel(t) + I_\perp(t)$, where $I_\parallel(t)$ and $I_\perp(t)$ represent the fluorescence decay profiles observed with emission polarization parallel and perpendicular to the direction of excitation polarization, respectively. When the angle of the emission polarizer is set at 45° with respect to the x-axis (magic angle), fluorescence anisotropy is cancelled, so that the TIR fluorescence decay curve can be analysed by a single-exponential function. If a water/oil interface is very sharp, therefore, fluorescence dynamic anisotropy $r(t)$ obeys Equation (14):

$$r(t) = \frac{I_\parallel(t) - I_\perp(t)}{I_\parallel(t) + I_\perp(t)} = r(0)\exp(-t/\tau^{\mathrm{rot}}) \tag{14}$$

where $r(0)$ and τ^{rot} are initial anisotropy ($t = 0$) and the reorientation correlation time, respectively. In Case I, $r(0)$ should be equal to 0.5 [12].

12.3.1.2. Three-Dimensional Model. On the other hand, if the interfacial layer is thick enough compared to the molecular size of SR101 and if SR101 molecules adsorbed on the interface are weakly oriented, the rotational motions of SR101 take place in three dimensions, similar to those in a bulk phase. If this is the case, the contribution of the fluorescence with the excited dipole moment of SR101 directed along the z-axis cannot be neglected, so that the time profile of the total fluorescence intensity must be proportional to $I_{\parallel}(t) + 2I_{\perp}(t)$. Thus, fluorescence dynamic anisotropy is given by Equation (15), as is well known for that in a macroscopically isotropic system [10,13]:

$$r(t) = \frac{I_{\parallel}(t) - I_{\perp}(t)}{I_{\parallel}(t) + 2I_{\perp}(t)} = r(0)\exp(-t/\tau^{\mathrm{rot}}) \tag{15}$$

In this case, $r(0)$ and the magic angle are calculated to be 0.4 and 54.7°, respectively. The thickness of a water/oil interfacial layer would be evaluated through TIR fluorescence anisotropy measurements and the τ^{rot} value(s) provides information about characteristic features at a water/oil interface.

12.3.2. *A Magic-Angle Dependence of the TIR Fluorescence Decay Profile of SR101 at a Water/Oil Interface*

The experimental idea mentioned above was then applied to study water/carbon tetrachloride (CCl_4) and water/DCE interfacial systems [4]. Figure 12.3 shows fluorescence decay profiles of SR101 observed from water/CCl_4 and water/DCE interfaces at an emission polarization angle of 45° [(a) and (c)] or 54.7° [(b) and (d)], together with the relevant weighted residuals (Re) and autocorrelation trace (Cr) for each single-exponential fit. In the case of the water/CCl_4 interface [(a) and (b)], Re and Cr of the data observed at 54.7° exhibited nonrandom distributions compared to those predicted by the best fit (b), particularly those in the initial stage of excitation (<1 ns). On the other hand, the profile observed at the magic angle 45° was reasonably fitted by a single-exponential function as judged by the relevant Re and Cr (a) [10]. The χ^2 and Durbin–Watson (DW) parameters for the fitting also supported that the best fit of the observed data by a single-exponential function was attained by setting an emission polarizer at the magic angle 45° (Table 12.1). Therefore, it is concluded that the water/CCl_4 interface is sharp with respect to the molecular size of SR101 and that three-dimensional rotational motions of SR101 are inhibited at the water/CCl_4 interface.

In the case of a water/DCE interface [(c) and (d)], on the other hand, a fitting of the data by a single-exponential function cannot be attained by setting an emission polarizer at 45°, as confirmed by deviations of Re and Cr from the optimum values (c). When the fluorescence decay profile is measured by setting an emission polarizer at 54.7° (d), fluorescence anisotropy can be reasonably fitted by a single-exponential function including the time response in the initial stage of excitation (see also χ^2 and

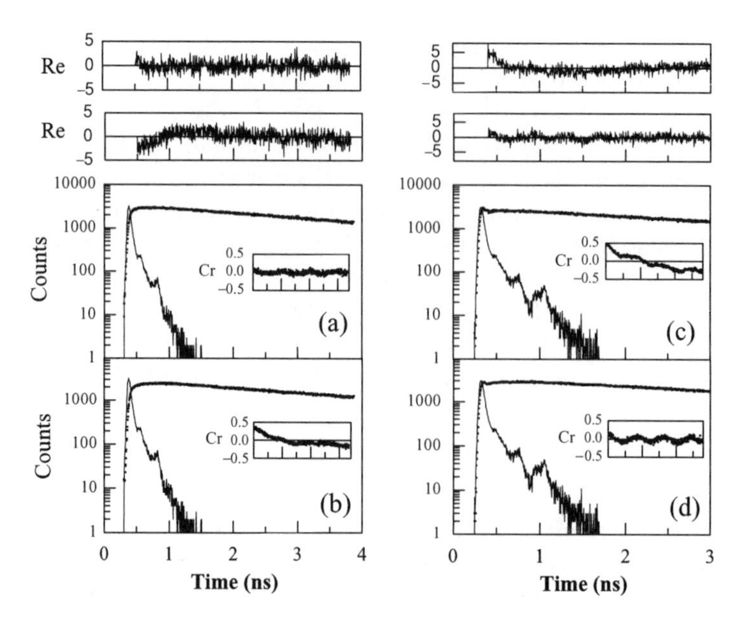

FIGURE 12.3. TIR fluorescence decay curves of SR101 at water/CCl_4 [(a) and (b)] and water/DCE [(c) and (d)] interfaces. The angle of the emission polarizer was set 45° [(a) and (c)] or 54.7° [(b) and (d)] in respect to the direction of excitation polarization. The upper and inner panels of each figure represent the plots of the weighted residuals (Re) and the autocorrelation trace (Cr) for a single-exponential fitting, respectively.

DW in Table 12.1). Therefore, the interfacial layer of the water/DCE interface is thick compared to the molecular size of SR101, and the dye molecules adsorbed on the interface are weakly oriented. Otherwise, the interface is spatially rough at the molecular size of SR101. SR101 molecules at the water/DCE interface behave similar to those in an isotropic medium, in contrast to the results at the water/CCl_4 interface.

TABLE 12.1. Fluorescence decay parameters of SR101 adsorbed on water/oil interfaces observed under the TIR conditions and in an aqueous solution.

		τ (ns)	χ^{2a}	DW[a]
Water/CCl_4	45°[b]	4.06 ± 0.02	1.11	1.88
	54.7°[b]	4.21 ± 0.03	1.61	1.37
Water/DCE	45°[b]	3.62 ± 0.02	2.28	0.90
	54.7°[b]	4.12 ± 0.03	1.09	1.73
Aq[c]		4.12 ± 0.01	1.16	1.94

[a] χ^2 and DW represent the χ-squared and Durbin–Watson parameters for the fitting, respectively.
[b] The angle of the emission polarizer in respect to that of the excitation laser beam.
[c] Determined in an aqueous SR101 solution ([SR101] = 1.7×10^{-7} M).

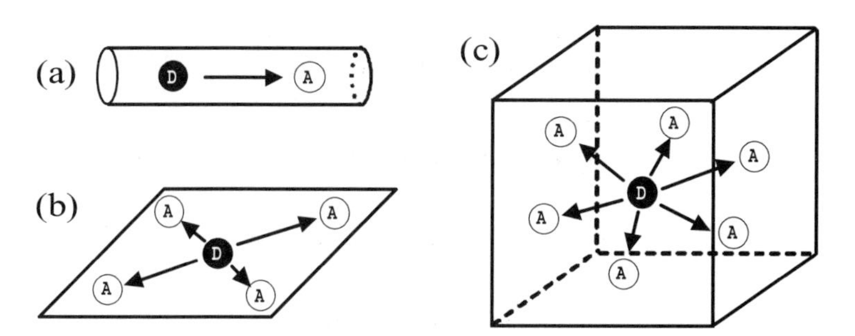

FIGURE 12.4. Schematic illustration of excitation energy transfer in Euclidean dimension. (a) 1-dimension, (b) 2-dimension (c) 3-dimension.

12.4. EXCITATION ENERGY TRANSFER AND ITS DYNAMICS AT A WATER/OIL INTERFACE

It is worth noting that water/oil interfacial structures would be governed by various factors, so that a complementary study other than fluorescence dynamic anisotropy is required to obtain further detailed information about the characteristics at a water/oil interface. As a new and novel approach, therefore, excitation energy transfer dynamics and the relevant structural (fractal) dimension analysis were introduced to elucidate the structure of a water/oil interface [4].

We consider here dipole–dipole (Förster-type) excitation energy transfer [14,15] between an energy donor (D) and an acceptor (A), both adsorbed on a water/oil interface. When diffusional motions of D and A are inhibited, as in the case for strong binding of the molecules to the surface by adsorption, excitation energy transfer quenching dynamics of D by A reflects structural dimension around D and A through spatial distributions of the components, as shown in Figure 12.4.

In such a case, fluorescence dynamics of D ($I_D(t)$) should obey the following equation, as reported by Klafter and Blumen [16,17],

$$I_{\mathrm{D}}(t) = A\exp(-(t/\tau_{\mathrm{D}}) - P(t/\tau_{\mathrm{D}})^{\bar{d}/6}) \tag{16}$$

where A is a pre-exponential factor and τ_{D} is the excited state lifetime of D without A. P is a parameter proportional to the probability that A resides within the critical energy transfer distance (R_0) of the excited donor. $\bar{d}$ is called the fractal dimension and reflects a spatial distribution of A around D. If D and A molecules adsorb uniformly on a sharp water/oil interface (two-dimensional), $\bar{d}$ should be 2.0, since excitation energy transfer takes place exclusively along the lateral direction at the interface. On the other hand, if a water/oil interface is thick and rough in respect to the molecular size of a probe molecule, $\bar{d}$ should be 2.0–3.0, since the possibility of energy transfer along the direction other than the lateral direction cannot be neglected. Therefore, we expect that a study on excitation energy transfer dynamics will provide invaluable information about the characteristics at a water/oil interface, along with a complementary study on the same system by fluorescence dynamic anisotropy.

Since the fluorescence spectrum of SR101 overlaps with the absorption spectrum of AB1 (see Scheme 12.1) [4], effective excitation energy transfer from SR101 to AB1 is

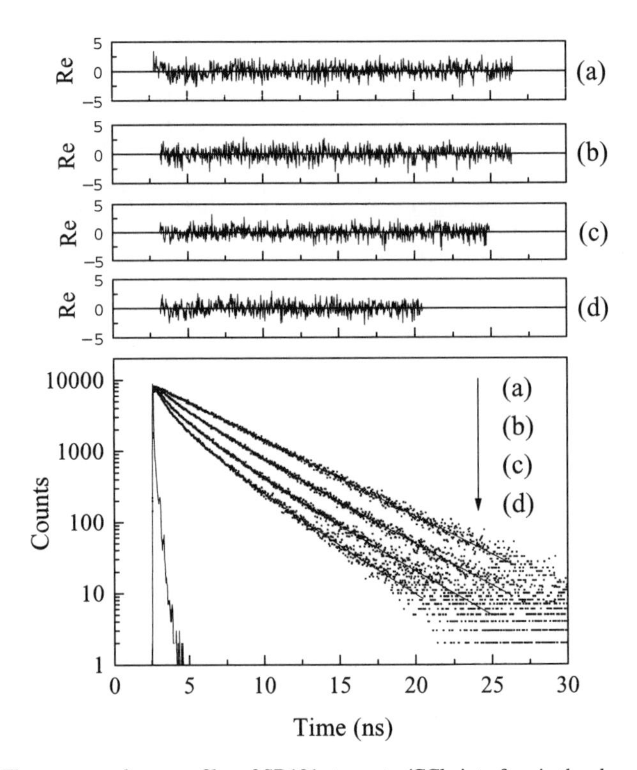

FIGURE 12.5. Fluorescence decay profiles of SR101 at a water/CCl_4 interface in the absence and presence of AB1: (a) [AB1] = 0, (b) 1.91×10^{-9}, (c) 3.82×10^{-9}, (d) 5.73×10^{-9} M. The solid curve shows the best fit by Equation (16), and the fitting parameters are listed in Table 12.2.

expected. In the present case, energy transfer between the excited singlet state of SR101 and AB1 proceeds via a Förster-type mechanism [14,15], and the critical energy transfer distance (R_0) in water is calculated to be 71 Å [4]. Therefore, on the basis of analysis of fluorescence decay curves of SR101 at a water/oil interface in the presence of AB1, the structure of the interface can be estimated with the spatial resolution being in the order of R_0 (~70 Å).

Figure 12.5 shows the fluorescence decay curves of SR101 at a water/CCl_4 interface in the absence (a) and presence of AB1 (1.91, 3.82 and 5.73×10^{-9} M for (b), (c) and (d), respectively). Analogous results were obtained for a water/DCE system (data are shown elsewhere) [1]. In the absence of AB1, the fluorescence decay of SR101 was fitted satisfactorily by a single-exponential function with the decay time of 4.21 ns. In the presence of the acceptor, the fluorescence decay profile of the donor was analysed by a Klafter–Blumen equation (Equation (16)). Simulations of the observed decay profiles in Figure 12.5 were then performed on the basis of Equation (16) with τ_D being fixed at 4.21 ns. The parameters obtained by the simulation (P and $\bar{d}$ are summarized in Table 12.2, together with the χ^2 parameters for the fittings. The χ^2 parameters and the weighted residuals (Re) of the fittings, shown in the upper traces of Figure 12.5, indicate that the fluorescence decay profiles of SR101 are well fitted by Equation (16).

TABLE 12.2. Structural dimension analysis of excitation energy transfer quenching of SR101 by AB1 at the water/oil interfaces.

	[AB1] (10^{-9} M)	P^a	$\bar{d}^b$	χ^2
Water/CCl_4	1.91	0.82	1.86	1.06
	3.82	1.46	1.94	1.00
	5.73	1.97	2.00	1.01
Water/DCE	1.91	0.50	2.47	1.05
	3.82	1.06	2.45	1.05
	5.73	1.40	2.53	1.15

[a] The P parameter (see Ref. [4]).
[b] The structural dimension determined by excitation energy transfer quenching of SR101 fluorescence by AB1 (see main text).

The structural dimension for energy transfer at the water/CCl_4 or water/DCE interface (Table 12.2) was 1.86–2.00 (average 1.93) or 2.45–2.53 (2.48), respectively. The present Klafter–Blumen analysis of the data is very meaningful and structural characteristics of the interface are reflected on the $\bar{d}$ value. The $\bar{d}$ value for the water/CCl_4 interface (1.93) indicates that the interface is sharp and can be regarded as two-dimensional, as long as the spatial resolution by the energy transfer quenching method (~70 Å or ~7 nm). Phenomenologically, the result agrees well with that derived from fluorescence dynamic anisotropy measurements; the interface is thin enough to inhibit three-dimensional rotational motions of SR101. If the interface is thin enough and modelled strictly by two-dimensions, however, the structural dimension should be equal to 2.0, since energy transfer quenching is restricted in the x–y plane of the interface. A $\bar{d}$ value smaller than 2.0 has frequently been reported for the interfacial systems such as vesicles and Langmuir–Blodgett films, and $\bar{d} < 2.0$ in these systems has been discussed in terms of a non-uniform distribution of an energy acceptor at the interface: fractal structure [18,19]. Therefore, a fractal-like distribution of AB1 at the water/CCl_4 interface might play a role in deciding the $\bar{d}$ value.

On the other hand, the structural dimension determined for the water/DCE interface was larger than 2.0: 2.48. This suggests that the water/DCE interface is thicker than the water/CCl_4 interface. The fluorescence dynamic anisotropy measurements also gave analogous results; the interface is thick enough to allow three-dimensional rotational motions of SR101. It should be noted that these results do not necessarily imply that the water/DCE interface is characterized by a three-dimensional space. Taking the results obtained by molecular dynamics simulations into account, it is supposed that the water/DCE interface is thin, but is rough in respect to the spatial resolution by the energy transfer quenching method, as discussed in the following section.

12.5. STRUCTURES AT A LIQUID/LIQUID INTERFACE

12.5.1. Water/CCl_4 and Water/DCE Interfaces

In the present study, both fluorescence dynamic anisotropy and excitation energy transfer methods were successful in providing information about the structures and thickness/roughness of the water/CCl_4 and water/DCE interfaces. However, it should be noted that, although the results obtained by the present two methods are complementary to discuss the structures of the interfaces, these cannot be compared directly with each other, since fluorescence dynamic anisotropy provides information about molecular level structures of the interface (i.e., dimension of the molecular size of SR101 ~1 nm) while the excitation energy transfer method affords relatively long-range interfacial structures (i.e., dimension of the critical energy transfer distance ~7 nm). Taking such experimental backgrounds into account, the following discussions will be possible.

A water/CCl_4 or water/DCE interface is a representative system studied by various techniques; so far, invaluable information about the structure of the interface has been accumulated [20]. As an example, Benjamin and his co-workers have demonstrated that the thickness of the water/DCE or water/CCl_4 interface is 1 nm or <1 nm, respectively [21]. The spatial resolution of fluorescence dynamic anisotropy measurements is ~1 nm, so that the present results are worth comparing with the predictions from the simulations.

The thickness of the water/DCE interface (1 nm) is comparable to the molecular size of SR101 (1.0–1.4 nm) [3]. Furthermore, SR101 molecules adsorbed on the water/DCE interface are supposed to be restricted within the two-dimensional plane, with the hydrophilic SO_3^- group being towards to the water phase while the hydrophobic xanthene ring is directed to the DCE phase. Therefore, when the water/DCE interface is very sharp, three-dimensional-like rotational reorientation of SR101 would not be observed, in contrast to our experimental observation. One possible reason for this is thermal fluctuations of the interface, by which SR101 molecules behave similar to those in an isotropic medium: SR101 on the interface is tumbled by thermal fluctuations. Similar situations are also expected for SR101 at the water/CCl_4 interface, while the dynamic anisotropy experiments suggest that the interface is two-dimensional-like. It has been reported that thermal capillary waves at an interface are related to the interfacial tension of the system [22–25]. In the present systems, the lower interfacial tension of the water/DCE system compared to that at the water/CCl_4 interface indicates that the amplitude(s) of the surface wave(s) is larger for the water/DCE system relative to that for the water/CCl_4 interface. Thus, the contribution of the thermal fluctuation to the rotational motions of SR101 would be smaller at the water/CCl_4 interface, leading to a two-dimensional-like anisotropy decay of the dye at the interface.

Short-range (~1 nm) structural information about the interface obtained by fluorescence dynamic anisotropy experiments is not contradicted by the results by the excitation energy transfer method: $\bar{d}$ (water/CCl_4) = 1.93 and $\bar{d}$ (water/DCE) = 2.48. Energy transfer at a water/oil interface is schematically shown in Figure 12.6. For simplicity, it is assumed here that an SR101 molecule (closed circle) sits on the interface (x–y plane). The molecular size of SR101 is 1.4 nm and the critical energy transfer distance is ~7 nm, so that energy transfer proceeds at a long distance. When both SR101 and AB1 are located at a flat interface, the structural dimension of energy transfer should be exactly 2.0. If the interface is rough and AB1 (closed triangle in Figure 12.6) is displaced from the x–y plane, on the other hand, the angle between the positions of the SR101 and AB1 molecules along the z-axis (φ in Figure 12.6) determines the $\bar{d}$ value. Clearly, when displacement along the z-axis (δ) is very small, $\bar{d}$ is 2.0. An increase in δ implies that the interface becomes rougher and, thus, $\bar{d}$ increases to 3.0. Since SR101 at the water/DCE interface exhibits three-dimensional-like motions, the interface has been

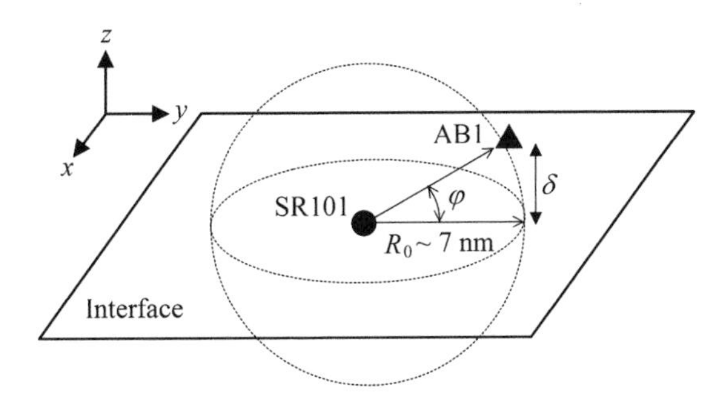

FIGURE 12.6. Schematic illustration of excitation energy transfer at a water/oil interface.

suggested to be rougher compared to the water/CCl_4 interface. Therefore, the structural dimension of the water/DCE interface ($\bar{d} = 2.48$) will be a reasonable consequence, reflecting the characteristic structure of the interface.

The structural dimension at a water/DCE interface is $\bar{d} = 2.48$, while short-range structural information about the interface obtained by the fluorescence dynamic anisotropy experiments suggests that the interface is three-dimensional-like. Taking the results obtained by molecular dynamics simulations into account, these results can be understood only by the fact that the water/DCE interface is thin (~1 nm), but is rough with respect to the spatial resolution of the excitation energy transfer quenching method (~7 nm), as shown in Figure 12.7.

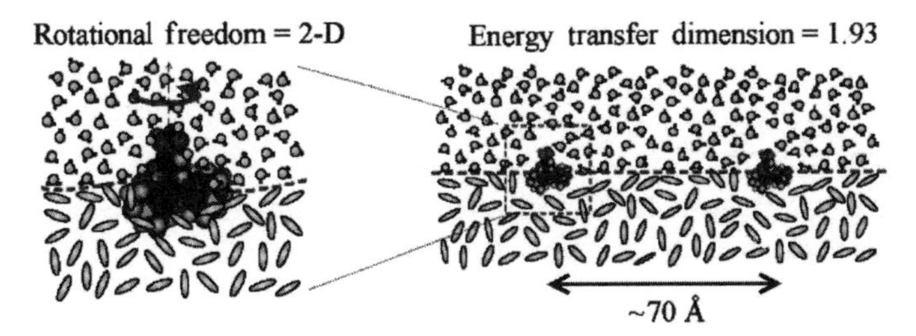

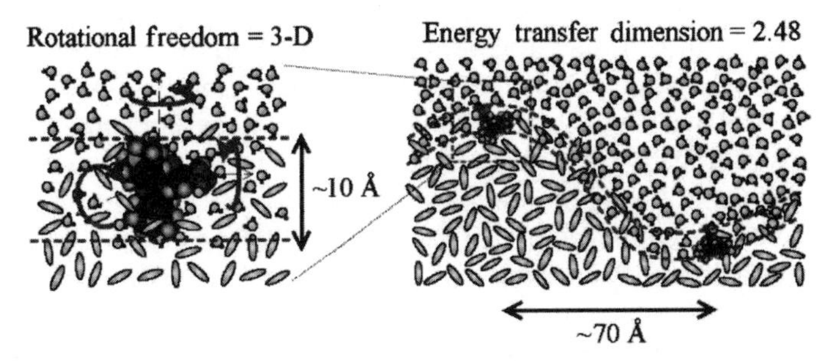

FIGURE 12.7. Schematic illustration of water/oil interfaces.

TABLE 12.3. Magic angles for the TIR fluorescence anisotropy measurements and structural dimensions at the interfaces.

Organic phase	γ^a (mN $\cdot$ m^{-1})	Aq in[b] (wt%)	In aq[b] (wt%)	Magic angle[c]	$\bar{d}^d$
Cyclohexane	51[e]	0.01	0.006	45°	1.90
CCl_4	45	0.01	0.08	45°	1.93
Toluene	33	0.03	0.05	~45°	2.13
CB[f]	37	0.03	0.05	~45°	2.20
DCB[f]	39	0.31	0.03	45°–54.7°	2.30
DCE	28	0.15	0.81	~54.7°	2.48

[a] Interfacial tension of the water/oil system, determined by a pendant drop method.
[b] Taken from Ref. [26].
[c] The angle of an emission polarizer, by which the fluorescence decay is best fitted by a single-exponential function.
[d] The structural dimension determined by excitation energy transfer quenching of SR101 fluorescence by AB1 (see main text).
[e] Ref. [26].
[f] CB and DCB represent chlorobenzene and *o*-dichlorobenzene, respectively.

Although the structural differences between the water/CCl_4 and water/DCE interfaces are not so large, the chemical and/or physical nature of the organic phase itself reflects on the photophysical properties of a probe molecule, indicating the novelty of the present experimental approaches. Systematic investigations are important to reveal factors governing structural and physical characteristics of water/oil interfaces. Therefore, we introduced fluorescence dynamic anisotropy and excitation energy transfer measurements to other water/oil interfacial systems; the data are summarized in Table 12.3. The results are discussed in terms of the relationship between the interfacial structure and the polarity at the water/oil interfaces (Section 12.6).

12.5.2. Theoretical Considerations

The simplest way to understand the difference in the interfacial structures mentioned above will be to consider the relevant interfacial tension, since an interfacial tension (mN $\cdot$ m^{-1}) is equivalent to the interfacial free energy (mJ $\cdot$ m^{-2}) [27]. As shown in Table 12.3, the interfacial free energy in the water/CCl_4 system (44.5 mJ $\cdot$ m^{-2}) is higher than that in the water/DCE system (27.9 mJ $\cdot$ m^{-2}). This implies that, since the energy necessary to construct a water/CCl_4 interface is larger than that for a water/DCE interface, the former is likely to construct a molecularly sharp interface. Therefore, interfacial roughness becomes smaller with increasing the interfacial tension (Table 12.3).

In the physical meaning, interfacial roughness is explained in terms of the thermal capillary waves propagating at a water/oil interface, which is related to the interfacial tension of the system [22–25]. It is well established that mean squares roughness at a liquid interface can be calculated from a model originally proposed by Buff *et al.* [22]. In essence, any single-valued interface can be described by a Fourier series of interfacial waves. At a liquid surface, the amplitude of each wave is dependent on both its wavelength and a temperature via a Boltzmann distribution. Macroscopic roughness of the interface can be evaluated from the wave amplitude integrated over all relevant wavelengths. The mean squared height ($\langle z^2 \rangle$) is then described by the following expression:

$$\langle z^2 \rangle = \frac{k_B T}{4\pi^2} \int_0^{2\pi} \int_0^{k_{max}} \frac{k_r}{\Delta\rho g + \gamma k_r^2} \, dk_r \, d\phi \tag{17}$$

Integration of Equation (17) yields

$$\langle z^2 \rangle = \frac{k_B T}{4\pi\gamma} \ln\left(\frac{\Delta\rho g + \gamma k_{max}^2}{\Delta\rho g}\right) \tag{18}$$

where z is the vertical displacement from the mean height, k_r is the magnitude of the radial wavevector for a particular capillary wave within the interfacial plane, ϕ is the azimuth angle of rotation within the interfacial plane, k_B is the Boltzmann constant, T is the temperature, γ is the surface tension. $\Delta\rho$ is the difference in the density between the two media and g is the gravitational constant. The upper integration limit, k_{max}, describes the largest magnitude of the interfacial wavevector, which is equivalent to the smallest capillary wavelength. Since capillary waves with wavelengths shorter than the intermolecular spacing are not expected to be stable, k_{max} is usually assumed to be in the order of π/r_{min}. In the case of a vapour–liquid interfacial system, r_{min} is the molecular radius of the liquid. In the case of a liquid/liquid interfacial system, r_{min} is assumed to be a weighted mean of the two liquid molecular radii.

According to Equation (18), the interfacial width of a water/CCl_4 or water/DCE interface is calculated to be 5 or 6.4 Å, respectively. These values are smaller than the molecular size of SR101 (14 Å). Therefore, the experimental results of three-dimensional molecular rotational reorientation of SR101 and the structural dimension larger than 2 at the water/DCE interface cannot be explained by the capillary wave theory (Equation (17)). Such a discrepancy between microscopic and macroscopic roughness at a water/oil interface has been discussed by Wirth and Burbage [28]. The thermal capillary wave theory is derived from macroscopic properties at a liquid/liquid interface, so that the theory is not necessarily sufficient to explain interfacial roughness at a molecular level. The applicability of the thermal capillary wave theory has been confirmed for interfacial roughness with tenths of micrometres or larger, and the theory is predicted to fail for that in submicrometre length scales [29].

In order to explain molecular-scale interfacial roughness, several theoretical models have also been proposed [25,30]. Simpson and Rowlen [25] introduced scale-dependent effective roughness, which described the root mean square probability of finding a particular difference in the height between any two points at the interface as a function of lateral separation. Mecke and Dietrich [30] proposed a density functional theory, in which they reconciled the two approaches, by distinguishing the different types of fluctuations at all length scale: both for undulations of the interface and for bulk density fluctuations. They first described an interface as a continuous variation of the density, which was caused by density fluctuations within the bulk phase, but with no undulations of the interfacial position. In the second step, the undulation predicted by the capillary wave theory was taken into account. Recently, the density functional theory was checked experimentally by means of grazing incidence X-ray scattering measurements at an air/water interface [29]. Fradin *et al.* [29] predicted that the surface was rougher than that expected from the simple capillary model at small length scales. Such a density functional theory is not contradicted by our interfacial model: the water/DCE interface is thin (~1 nm), but is rough with respect to the spatial resolution of the excitation energy transfer quenching method (~7 nm, Figure 12.7). Furthermore, no clear relationship between the interfacial tension and interfacial roughness could be obtained in this study (Table 12.3). The results suggest that molecular-scale interfacial roughness would be governed by not only the interfacial tension (as a macroscopic property), but also by the chemical/physical nature of an organic solvent itself. Since liquid/liquid interfacial structures would be governed by several factors such as the dipole moment, the dielectric constant, the hydrogen-bonding ability, the density, the molecular shape of a liquid and so forth, interfacial roughness and nature of a liquid could be correlated in a very complex manner. Further experimental and theoretical studies including molecular dynamic simulations will reveal detailed characteristics at a liquid/liquid interface.

12.6. A RELATIONSHIP BETWEEN INTERFACIAL STRUCTURE AND POLARITY AT A LIQUID/LIQUID INTERFACE

The polarity at a water/oil interface sometimes plays important roles in deciding heterogeneous reaction kinetics [31]. Such results suggest that solvent environments at the interface are different from those in bulk media. Recently, surface-selective spectroscopic techniques have been applied to studying the polarity at a water/oil interface. Wang *et al.* reported an SHG (second harmonic generation) spectroscopic study on the polarities of water/DCE and water/chlorobenzene (CB) interfaces by using N,N-diethyl-p-nitroaniline (DEPNA) as a probe [32]. According to their study, the interfacial polarity ($P_{A/B}$) is equal to the arithmetic average of the polarities of the adjoining bulk phase (P_A and P_B) [Equation (19)].

$$P_{A/B} = \frac{P_A + P_B}{2} \tag{19}$$

They explained the results by dominance of long-range solute–solvent interactions in determining the difference in the excited- and ground-state solvation energies around DEPNA at the interface, but not by local interfacial interactions. On the other hand, Michael and Benjamin reported molecular dynamics computer simulations of an electronic spectrum of DEPNA at a water/DCE interface [33] and predicted that the interfacial polarity was influenced by both interfacial roughness and the position of the probe molecule at the interface [21]. In addition to the long-range solute–solvent interactions at the interface, they also demonstrated that short-range solute–solvent interactions were also important in determining the difference in the excited- and ground-state solvation energies around the probe molecule at the interface. Although such studies are very important for advances in the relevant research fields, the number of both experimental and theoretical studies on the polarity at a water/oil interface is still limited [5,34,35]. In particular, the applicability of Equation (19) is worth studying further in a wide range of a solvent polarity, and the role of interfacial roughness in determining the polarity should be clarified experimentally.

In order to obtain a clearer picture on the interfacial polarity, we conducted picosecond TIR fluorescence spectroscopy by using sulforhodamine B (SRB) as a polarity probe molecule [5]. On the basis of fluorescence dynamic measurements of SRB adsorbed on a water/oil interface, we studied a relationship between thickness/roughness and the polarity at the interface.

The photophysical properties of a xanthene dye have been extensively studied. In order to explain a relationship between the non-radiative decay rate constant (k_{nr}) of the dye and a solvent polarity parameter ($E_T(30)$), Quitevis *et al.* proposed a two-state model [36,37]. According to the model, k_{nr} is given by

$$k_{nr} \propto \exp\{-(\beta/RT + \kappa)(E_T(30) - 30)\} \exp(-\Delta G_{A^*B^*}{}^0/RT) \tag{20}$$

where β and κ are constants and $\Delta G_{A^*B^*}{}^0$ is the Gibbs free energy difference between the two excited-states of A* and B* in a non-polar solvent. If Equation (20) holds, a plot of ln k_{nr} vs. $E_T(30)$ should be linear, with the slope value being equal to $-(\beta/RT + \kappa)$. The physical meaning of Equation (20) is not straightforward and it will be better to consider this as an empirical equation. Nonetheless, the relation is useful to explain solvent effects on the spectroscopic properties of a xanthene dye.

The fluorescence lifetimes (τ determined at 580 nm) and quantum yields (Φ) of SRB were determined in water–dioxane mixtures and a series of alcohols at 25°C. The k_{nr} value varied with the medium in the range of (4.1–0.7) $\times$ 10^8 s^{-1}, whereas the radiative decay rate constant (k_r) was rather insensitive to the medium properties: (2.8–1.7) $\times$ 10^8 s^{-1}. The relationship between ln k_{nr} and $E_T(30)$ fall on a straight line and the slope value of the plot was 0.074 $\pm$ 0.01. Therefore, the photophysical properties of SRB and Equation (20) are applicable to probing the polarity at a water/oil interface.

It is noteworthy that the TIR fluorescence decay curve observed at each water/oil interface is reasonably fitted by a single-exponential function, and its lifetime is always longer than that in an aqueous phase (1.5 ns). The interfacial polarity $E_T(30)_{int}$ in each water/oil system was then estimated on the basis of the relevant k_{nr} value and the relationship between k_{nr} and $E_T(30)$ determined by bulk measurements. The results are summarized in Figure 12.8, together with the interfacial polarity $E_T(30)_{calc}$ calculated by Equation (19). These data demonstrate that $E_T(30)_{int}$ observed always takes an intermediate value between $E_T(30)$ of water and that of the organic phase. In the case of a relatively low polarity solvent (cyclohexane, CCl_4 or toluene), $E_T(30)_{int}$ agreed very well with $E_T(30)_{calc}$ as predicted by Equation (19). In these interfacial systems, our fluorescence data indicate that the interface is thin and two-dimension-like. When a water/oil interface is thin and sharp, therefore, it is concluded that the interfacial polarity is well predicted by the arithmetic average of the polarities of the water and organic phases [Equation (19)]. In the case of a relatively high polarity solvent (CB, *o*-DCB or DCE), on the other hand, $E_T(30)_{int}$ was always lower than $E_T(30)_{calc}$. It is worth noting that these water/oil interfaces are thin ($\sim$1 nm) but rough (in the spatial resolution of R_0) as estimated by the fluorescence dynamic spectroscopies: dynamic anisotropy and energy transfer method. The results demonstrate that an interfacial polarity deviates from $E_T(30)_{calc}$ when the interface is rough. We suppose that the origin of the present results

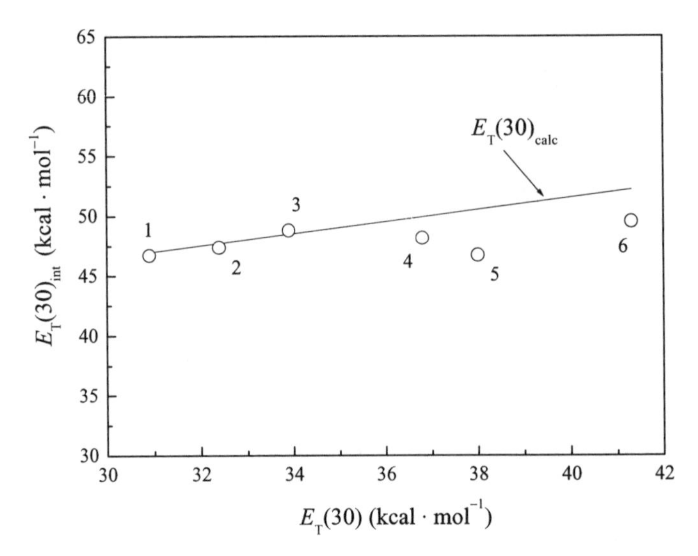

FIGURE 12.8. Relationship between $E_T(30)_{int}$ and $E_T(30)$ of the oil phase. Open circles and solid curve represent $E_T(30)_{int}$ and $E_T(30)_{calc}$, respectively. 1: cyclohexane, 2: carbon tetrachloride, 3: toluene, 4: chlorobenzene, 5: *o*-dichlorobenzene, 6: 1,2-dichloroethane.

would be due to interfacial roughness and orientations of SRB adsorbed on the interface, as discussed elsewhere [5].

12.7. PHOTOCHEMICAL OBSERVATION OF MOLECULAR RECOGNITION AT A LIQUID/LIQUID INTERFACE

Molecular recognition in biological systems proceeds at microscopic interfaces such as cell and protein surfaces in aqueous media [38]. Therefore, hydrogen-bonding interactions at a liquid/liquid interface have received current interests as a model of molecular/ion recognition in biological systems. Nonetheless, owing to the experimental difficulties in observing directly molecular recognition at a liquid/liquid interface, there have appeared only a few spectroscopic studies on molecular recognition at a water/oil interface [39–41].

It is anticipated that orientational dynamics of a host–guest complex produced by molecular recognition are different from those of a host or guest. Therefore, time-resolved fluorescence spectroscopy under TIR conditions has a possibility to sense molecular recognition at a water/oil interface. The system studied is a model of flavoenzymes as shown in Figure 12.9.

Since riboflavin (RF) is surface-active and possesses a high fluorescence quantum yield, the dye is very suitable for TIR fluorescence measurements at a water/oil interface. For TIR experiments, an aqueous RF solution was poured onto a CCl_4 solution containing a diamino-*s*-triazine derivative (DTT), which acted as a flavin receptor via triple hydrogen bonds. On the basis of fluorescence dynamic measurements of RF adsorbed on a water/oil

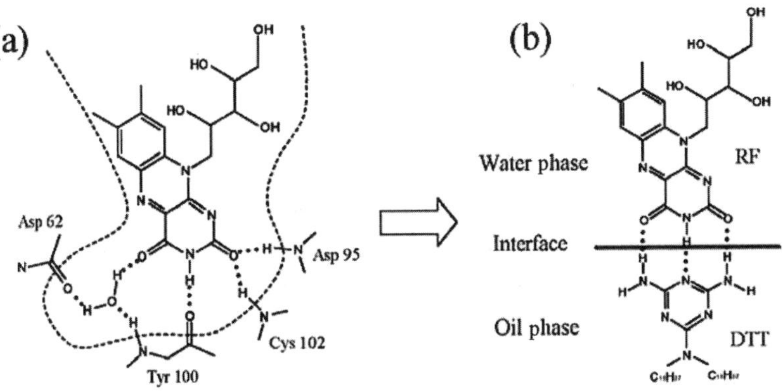

FIGURE 12.9. A schematic illustration of molecular recognition mediated by hydrogen-bonding interactions at a liquid/liquid interface.

interface, one can follow molecular recognition mediated by complementary hydrogen bonding at the interface.

In the absence of DTT in the CCl_4 phase, the rotational reorientation time of RF at the interface was 210 ps, while fast (160–220 ps) and slow (670–750 ps) rotational reorientation times were observed in the presence of DTT in the CCl_4 phase. This slow rotational reorientation time was assigned to that of the RF-DDT complex at the water/CCl_4 interface. These results indicate that molecular recognition mediated by complementary hydrogen bonding takes place effectively even at the artificial water/oil interface. Fluorescence dynamics spectroscopy was shown to be powerful enough to elucidate molecular recognition at the interface as well [42].

12.8. CONCLUDING REMARKS

In order to obtain structural information about water/oil interfaces at a molecular level, two different approaches were introduced to the study. One is a magic angle dependence of the TIR fluorescence decay profile of SR101 adsorbed on a water/oil interface. The other approach was a structural dimension analysis of excitation energy transfer dynamics between dye molecules adsorbed on water/oil interfaces. The latter method was shown to have high potential to elucidate the thickness and roughness of the interfacial layer in the spatial resolution of the critical energy transfer distance (~7 nm) between an energy donor and an acceptor. By using these two methods, a water/CCl_4 interface was shown to be very sharp with respect to molecular dimension of SR101, while a water/DCE interface was concluded to be rougher than a water/CCl_4 interface. It is worth noting that the present results are the first experimental proof of the structural differences between the water/CCl_4 and water/DCE interfaces. These approaches are very meaningful in respect to comparative discussion on the same systems by both experiments and available computer simulation data by Benjamin and co-workers [9, 21]. Furthermore, it has been suggested that roughness of the water/DCE interface is caused by thermal fluctuation of the sharp (~1 nm) interface in the range of ~7 nm.

These methodologies were also extended to study other water/oil interfacial systems with the properties of the oil being varied. Such a systematic study on water/oil interfaces has rarely been explored till now and, therefore, the present study contributes to further advances in the relevant researches.

Systematic investigations on water/oil interfaces are the first step to reveal factors governing structural and physical characteristics of the interfaces. Usually, interfacial roughness is explained in terms of the thermal capillary waves propagating at the interface, which is related to the interfacial tension of the system. However, the thermal capillary wave theory derived from macroscopic properties at a liquid/liquid interface is not necessarily sufficient to explain interfacial roughness at a molecular level. In practice, no clear relationship between the interfacial tension and interfacial roughness was obtained in the present study. The results clearly suggest that molecular-scale interfacial roughness would be governed by not only the interfacial tension (as a macroscopic property), but also the chemical/physical nature of an organic solvent itself.

Various chemical processes *at* or *across* a liquid/liquid interface are governed by the molecular level structures of the interface. In this review, we discussed the relationship between interfacial thickness/roughness and the nature of an oil phase. Also, the present study suggested that interfacial roughness would govern various reaction processes across a liquid/liquid interface. In separation sciences, such processes are very important with the typical examples being solvent extraction and liquid chromatography. Although the present study is focused on flat water/oil interfaces, the results and methodologies introduced by the study will contribute to further researches in the chemistry at various liquid/liquid interfaces, including oil-in-water or water-in-oil emulsion systems.

ACKNOWLEDGEMENTS

The authors are grateful for a Grant-in-Aid from the Ministry of Education, Science, Sports and Culture (Nos. 13740416 and 13129201 (Priority Research Area on "Nano-Chemistry at Liquid–Liquid Interfaces") to S.I. and No. 14050001 to N.K.) for support of the research.

REFERENCES

1. S. Ishizaka and N. Kitamura, Bull. Chem. Soc. Jpn. **74**, 1983 (2001).
2. K. Nakatani, S. Ishizaka and N. Kitamura, Anal. Sci. **12**, 701 (1996).
3. S. Ishizaka, K. Nakatani, S. Habuchi and N. Kitamura, Anal. Chem. **71**, 419 (1999).
4. S. Ishizaka, S. Habuchi, H.-B. Kim and N. Kitamura, Anal. Chem. **71**, 3382 (1999).
5. S. Ishizaka, H.-B. Kim and N. Kitamura, Anal. Chem. **73**, 2421 (2001).
6. W.WM. Wendlandt and H.G. Hecht, *Reflectance Spectroscopy*, John Wiley and Sons, New York, 1966, p. 21.
7. H. Masuhara, F.C. De Schryver, N. Kitamura and N. Tamai, *Microchemistry; Spectroscopy and Chemistry in Small Domains*, North-Holland, Amsterdam, 1994, p. 257.
8. S. Hamai, N. Tamai and H. Masuhara, J. Phys. Chem. **99**, 4980 (1995).
9. I. Benjamin, J. Chem. Phys. **97**, 1432 (1992).
10. D.V. OConnor and D. Phillips, *Time-Correlated Single Photon Counting*, Academic Press, New York, 1986, Chap. 8.

11. J.M. Kovaleski and M.J. Wirth, J. Phys. Chem. **99**, 4091 (1995).
12. M.J. Wirth and J.D. Burbage, Anal. Chem. **63**, 1311 (1991).
13. R.L. Christensen, R.C. Drake and D. Phillips, J. Phys. Chem. **90**, 5960 (1986).
14. Th. Förster, Discuss. Faraday Soc. **27**, 7 (1959).
15. N.J. Turro, *Modern Molecular Photochemistry*, University Science Books, Sansalito, CA, 1991, Chap. 9.
16. J. Klafter and A. Blumen, J. Chem. Phys. **80**, 875 (1984).
17. J. Klafter and A. Blumen, J. Lumin. **34**, 77 (1985).
18. N. Tamai, T. Yamazaki and I. Yamazaki, Chem. Phys. Lett. **147**, 25 (1988).
19. I. Yamazaki, N. Tamai and T. Yamazaki, J. Phys. Chem. **94**, 516 (1990).
20. A.G. Volkov and D.W. Deamer, Eds, *Liquid–Liquid interfaces*, CRC Press, Boca Raton, FL, 1996, Chap. 9.
21. D. Michael and I. Benjamin, J. Chem. Phys. **107**, 5684 (1997).
22. F.P. Buff, R.A. Lovett and F.H. Stillinger, Jr., Phys. Rev. Lett. **15**, 621 (1965).
23. A. Braslau, P.S. Pershan, G. Swislow, B.M. Ocko and J. Als-Nielsen, Phys. Rev. A **38**, 2457 (1988).
24. P.S. Pershan, Faraday Discuss. Chem. Soc. **89**, 231 (1990).
25. G.J. Simpson and K.L. Rowlen, Chem. Phys. Lett. **309**, 117 (1999).
26. J.A. Riddick and W.B. Bunger, *Techniques of Chemistry: Organic Solvents*, Vol. II, Wiley-Interscience, New York, 1970, Chap. 3.
27. J.N. Israelachvili, *Intermolecular and Surface Forces*, Academic Press, London, 1992, Chap. 15.
28. M.J. Wirth and J.D. Burbage, J. Phys. Chem. **96**, 9022 (1992).
29. C. Fradin, A. Braslau, D. Luzet, D. Smilgies, M. Alba, N. Boudet, K. Mecke and J. Daillant, Nature **403**, 871 (2000).
30. K.R. Mecke and S. Dietrich, Phys. Rev. E **59**, 6766 (1999).
31. J.M. Perera, G.W. Stevens and F. Grieser, Colloids Surf. A **95**, 185 (1995).
32. H. Wang, E. Borguet and K.B. Eisenthal, J. Phys. Chem. B **102**, 4927 (1998).
33. D. Michael and I. Benjamin, J. Phys. Chem. B **102**, 5145 (1998).
34. W.H. Steel, F. Damkaci, R. Nolan and R.A. Walker, J. Am. Chem. Soc. **124**, 4824 (2002).
35. W.H. Steel and R.A. Walker, J. Am. Chem. Soc. **125**, 1132 (2003).
36. K.G. Casey and E.L. Quitevis, J. Phys. Chem. **92**, 6590 (1988).
37. K.G. Casey, Y. Onganer and E.L. Quitevis, J. Photochem. Photobiol. A: Chem. **64**, 307 (1992).
38. H. Gohlke and G. Klebe, Angew. Chem. Int. Ed. **41**, 2644 (2002).
39. K. Nochi, A. Yamaguchi, T. Hayashita, T. Uchida and N. Teramae, J. Phys. Chem. B **106**, 9906 (2002).
40. M.J. Crawford, J.G. Frey, T.J. VanderNoot and Y. Zhao, J. Chem. Soc., Faraday Trans. **92**, 1369 (1996).
41. T. Kakiuchi, K. Ono, Y. Takasu, J. Bourson and B. Valeur, Anal. Chem. **70**, 4152 (1998).
42. S. Ishizaka, S. Kinoshita, Y. Nishijima and N. Kitamura, Anal. Chem. **75**, 6035 (2003).

13

Development of Surfactant-Type Catalysts for Organic Synthesis in Water

Kei Manabe and Shū Kobayashi
Graduate School of Pharmaceutical Sciences, The University of Tokyo, Hongo, Bunkyo-ku, Tokyo 113-0033, Japan

13.1. INTRODUCTION

Most chemical reactions of organic substances conducted in the laboratory as well as in industry need organic solvents as reaction media, in spite of the fact that water is safe, benign, environmentally friendly and cheap compared with organic solvents. Although todayŝ environmental consciousness imposes the use of water as a solvent on both industrial and academic chemists [1,2], organic solvents are still used instead of water for mainly two reasons. First, most organic substances are insoluble in water and, as a result, water does not function as reaction media. Second, many reactive substrates, reagents and catalysts are decomposed or deactivated by water. Our goal is to develop a novel catalytic system that enables the use of water as a solvent for a wide range of reactions of organic materials.

The first drawback in the use of water (the solubility problem) may be overcome by using surfactants, which solubilize organic materials or form emulsions with them in water. Indeed, surfactants have been occasionally used in organic synthesis [3–6]. A successful example is emulsion polymerization [7]. Some late transition metal-catalysed reactions in water have also been conducted in the presence of surfactants or surfactant-like ligands [8–15]. In many other cases, however, large quantities of surfactant molecules compared with the reaction substrates are needed for the desired reactions to proceed efficiently, and thus, the systems are impractical even if water can be used as a solvent. From the viewpoints of practicability and applicability, the surfactant-aided organic synthesis is still at the preliminary stage.

In the course of our investigations to circumvent the second drawback in the use of water (the decomposition problem), we have found that some metal salts such as rare earth metal triflates (triflate = trifluoromethanesulfonate) can be used as water-compatible Lewis acids [16,17]. Lewis acid catalysis has attracted much attention in organic synthesis [18]. Although various kinds of Lewis acids have been developed and many have been applied in industry, these Lewis acids must be generally used under strictly anhydrous conditions. The presence of even a small amount of water stops the reactions, because most Lewis acids immediately react with water rather than substrates. In addition, recovery and reuse of the conventional Lewis acids are formidable tasks. These disadvantages have restricted the use of Lewis acids in organic synthesis.

On the other hand, we have found that the Mukaiyama aldol reaction of benzaldehyde with silyl enol ether **1** was catalysed by ytterbium triflate ($Yb(OTf)_3$) in water–THF (1/4) to give the corresponding aldol adduct in high yield (Equation (1)). When this reaction was carried out in dry THF (without water), the yield of the aldol adduct was very low (ca. 10%). Thus, this catalyst is not only compatible with water but also activated by water probably because of dissociation of the counteranions from the Lewis acidic metal. Furthermore, these catalysts can be easily recovered and reused.

PhCHO + 1 ($OSiMe_3$) → [$Yb(OTf)_3$ (10 mol%), H_2O–THF (1/4), rt, 20 h] → aldol adduct (Ph, OH, O) 91% (1)

These findings prompted us to investigate Lewis acid catalysis in aqueous media in further detail, because reactions in such media have the following advantages compared with reactions under anhydrous conditions: (1) It is not necessary to dry solvents and substrates for the reactions in aqueous media. This means that aqueous solutions of substrates or hydrated substrates can be directly used without further drying. (2) From the viewpoint of recent environmental consciousness, it is desirable to use water instead of organic solvents as a reaction solvent. Therefore, development of organic reactions in water will contribute to progress of green chemistry [19]. (3) Water has unique physical and chemical properties such as high dielectric constant and cohesive energy density compared with most organic solvents. This unique nature of water is also essential for most enzymatic reactions in living systems. Many enzymes catalyse desired reactions with high efficiency and excellent stereoselectivity under mild conditions in water, and this effectiveness is often regarded as a goal for synthetic chemists. Although many researchers have developed synthetic mimics of active sites of enzymes to realize the enzymatic activity, we have focused our attention on the medium of enzymatic reactions—water, which plays major roles in the reactions. By utilizing the unique nature of water, it would be possible to develop reaction systems that cannot be attained in dry organic solvents [20,21].

We have been also interested in Brøsted acid-catalysed reactions in water. In the course of the investigations on Lewis acid-catalysed reactions in water, we have found unique reaction systems using a surfactant-type Brøsted acid. Here we present our recent investigations on these Lewis and Brøsted acid-catalysed reactions in water.

13.2. SURFACTANT-TYPE LEWIS ACIDS FOR REACTIONS IN WATER

While the Lewis acid-catalysed aldol reactions in water-containing solvents described above were catalysed by several metal salts, a certain amount of organic solvents such as THF and ethanol still had to be combined with water to dissolve organic substrates and promote the reactions efficiently. However, it is desirable to avoid the use of harmful organic solvents. Therefore, we initiated investigations to develop a new system for Lewis acid-catalysed reactions in water without using any organic solvents.

The main drawback in the use of water (low solubility of most organic substances in water) would be overcome by using surfactants, which solubilize organic materials or form emulsions with them in water. To address this solubility issue, therefore, we planned to use surfactants, hopefully small amounts of them, for the Lewis acid-catalysed reactions in water.

The surfactant-aided Lewis acid catalysis was first demonstrated in the model reaction shown in Table 13.1 [22]. While the reaction proceeded sluggishly in the presence of 10 mol% scandium triflate ($Sc(OTf)_3$) in water, a remarkable enhancement of the reactivity was observed when the reaction was carried out in the presence of 10 mol% $Sc(OTf)_3$ in an aqueous solution of sodium dodecyl sulfate (SDS, 20 mol%, 35 mM), and the corresponding aldol adduct was obtained in high yield. It was found that the type of surfactant influenced the yield, and that Triton X-100, a non-ionic surfactant, was also effective in the aldol reaction (but required longer reaction time), while only a trace amount of the adduct was detected when using a representative cationic surfactant, cetyltrimethylammonium bromide (CTAB). The effectiveness of the anionic surfactant is attributed to high local concentration of scandium cation on the surfaces of dispersed organic phases, which are surrounded by the surfactant molecules.

The results mentioned above prompted us to synthesize a more simplified catalyst, scandium tris(dodecyl sulfate) ($Sc(DS)_3$) [23,24]. This new type of catalyst, "Lewis acid–surfactant-combined catalyst (LASC)", was expected to act *both* as a Lewis acid to activate the substrate molecules *and* as a surfactant to form emulsions in water. Engberts and co-workers also reported a surfactant-type Lewis acid, copper bis(dodecyl sulfate) ($Cu(DS)_2$) [25]. Although they studied detailed mechanistic aspects of Diels–Alder

TABLE 13.1. Effect of surfactants on aldol reaction in water.

PhCHO + 2 (1.5 eq.) (OSiMe$_3$, Ph) → [$Sc(OTf)_3$ (10 mol%), surfactant (20 mol%), H_2O, rt] → aldol adduct (OH, O, Ph, Ph)

Surfactant	Time (h)	Yield (%)
–	4	3
SDS	4	88
Triton X-100	60	89
CTAB	4	Trace

TABLE 13.2. Effect of solvents on LASC-catalysed aldol reaction.

PhCHO + 2 (1.5 eq.) → [$Sc(DS)_3$ (10 mol%), H_2O, rt, 4 h]

Solvent	Yield (%)	Solvent	Yield (%)
H_2O	92	THF	Trace
MeOH	4	Et_2O	Trace
DMF	14	Toluene	Trace
DMSO	9	Hexane	4
MeCN	3	–(neat)	31
CH_2Cl_2	3		

reactions in water, the reactions need excess of the catalyst compared with the reaction substrates and have severe substrate limitations. On the other hand, a catalytic amount of $Sc(DS)_3$ or another LASC **3** efficiently promoted the aldol reaction of benzaldehyde with silyl enol ether **2** in water (Equation (2)). While $Sc(DS)_3$ and **3** are only slightly soluble in water, stable emulsions were formed upon addition of the aldehyde with stirring or vigorous mixing. Addition of **2**, followed by stirring at room temperature for 4 hours, afforded the desired aldol product in high yields. It should be noted that hydrolysis of the silyl enol ether is not a severe problem under the reaction conditions in spite of the water-labile nature of silyl enol ethers under acidic conditions.

We also found that $Sc(DS)_3$ worked well only in water rather than in organic solvents (Table 13.2). A kinetic study on the initial rate of the aldol reaction revealed that the reaction in water was about 100 times faster than that in dichloromethane. In addition, the reaction under neat conditions was slower than that in water and resulted in lower yield (31%), showing the advantage of the use of water in this reaction. This advantageous effect of water is attributed to the following factors: (1) hydrophobic interactions in water to concentrate the catalyst and the substrates; (2) aggregation of the substrates through the hydrophobic interactions that squeeze water molecules out of the organic substrate phase, leading to reducing the rate of hydrolysis of the silyl enol ethers; (3) hydration of Sc(III) ion and the counteranion by water molecules to form highly Lewis acidic species such as $[Sc(H_2O)_n]^{3+}$; (4) rapid hydrolysis of the initially formed scandium aldolate to secure fast catalytic turnover.

PhCHO + 2 (1.5 eq.) → [LASC (10 mol%), H_2O, rt, 4 h]

LASC = $Sc(DS)_3$: 92%
3: 83% (2)

$Sc(DS)_3$

3

Various substrates have been successfully used in the present LASC-catalysed aldol reaction. Aromatic as well as aliphatic, α, β-unsaturated and heterocyclic aldehydes worked well. As for silicon enolates, silyl enol ethers derived from ketones as well as ketene silyl acetals derived from thioesters and esters reacted well to give the corresponding adducts in high yields. It is noted that highly water-sensitive ketene silyl acetals reacted smoothly in water under these conditions.

In the work-up procedure for the aldol reactions stated above, the crude products were extracted with ethyl acetate after quenching the reactions. The addition of ethyl acetate in this procedure facilitates the phase separation between the organic and aqueous phase and makes the separation of organic products facile. It is more desirable, however, to develop a work-up procedure without using any organic solvents such as ethyl acetate. In addition, the development of a protocol for recovery and reuse of the catalysts is indispensable to apply the LASC system to large-scale syntheses. Therefore, it is worthy to mention that centrifugation of the reaction mixture of a LASC-catalysed aldol reaction led to phase separation without addition of organic solvents. After centrifugation at 3500 rpm for 20 minutes, the emulsion mixture became a tri-phasic system where the LASC was deposited between a transparent water phase and an organic product phase. It is noted that this procedure enables, in principle, the recovery and reuse of LASCs and the separation of the organic products without using organic solvents.

In the LASC-catalysed reactions, the formation of stable emulsions seemed to be essential for the efficient catalysis. We thus undertook the observation of the emulsions by means of several tools. Optical microscopic observations of the emulsions revealed the formation of spherical emulsion droplets in water (Figure 13.1). The average size of the droplets formed from **3** in the presence of benzaldehyde in water was measured by dynamic light scattering, and proved to be ca. 1.1 μm in diameter. The shape and size of the emulsion droplets were also confirmed by transmission electron microscopy and atomic force microscopy.

Michael reactions are one of the most useful carbon–carbon bond-forming reactions in organic synthesis, and Lewis acid-catalysed versions have been developed to solve problems which are often observed in traditional, base-catalysed Michael reactions. $Sc(DS)_3$ can be also applied to Michael reactions in water as shown in Equation (3) [26]. Compared with $Yb(OTf)_3$-catalysed Michael reactions in water [27],

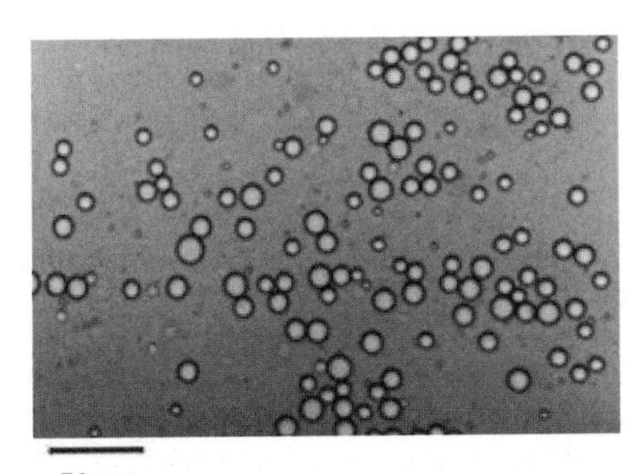

FIGURE 13.1. Optical micrograph of a mixture of **3** and benzaldehyde in water (**3**/PhCHO = 1:20).

the $Sc(DS)_3$-catalysed reaction was found to be very fast.

$Sc(DS)_3$ (10 mol%), H_2O, 30°C, 20 h; (3.0 eq.); 96% (3)

Similarly, reactions of indoles with electron-deficient olefins were catalysed by $Sc(DS)_3$ (Equation (4)) [28]. As for electron-deficient olefins, not only enones but also β-nitrostyrene was successfully used. It is noteworthy that solid substrates such as indole and β-nitrostyrene as well as liquid ones reacted smoothly. This is the first example of Lewis acid-catalysed Friedel–Crafts-type reactions of aromatic compounds in water.

$Sc(DS)_3$ (2.5 mol%), H_2O, 30°C, 6.5 h; (3.0 eq.); 91% (4)

Catalytic asymmetric aldol reactions in water have been attained by a combination of $Cu(DS)_2$ and chiral bis(oxazoline) ligand **4**. In this case, addition of a Brøsted acid, especially a carboxylic acid such as lauric acid, is essential for good yield and enantioselectivity (Equation (5)) [29]. This is the first example of Lewis acid-catalysed asymmetric aldol reactions in water without using organic solvents. Although the yield and the selectivities have not yet been optimized, it is noted that this enantioselectivity has been achieved at ambient temperature in water.

PhCHO + $OSiMe_3$ (1.5 eq.); **4** (24 mol%), $Cu(DS)_2$ (20 mol%), H_2O, 23°C, 20 h (5)

Without lauric acid: 23%, *syn*/*anti* = 76/24, 58% ee (*syn*)
With lauric acid (10 mol %): 76%, *syn*/*anti* = 74/26, 69% ee (*syn*)

We have also developed a new catalytic system for aldol reactions in water. The reaction of benzaldehyde with silyl enol ether **2** was catalysed by a combination of diphenylborinic acid (Ph_2BOH), benzoic acid and SDS in water to give the aldol adduct in high yield with high *syn/anti* ratio (Equation (6)) [30,31]. All of the three components in the catalyst combination were indispensable for the high yield and selectivity. This

TABLE 13.3. Diastereoselective aldol reaction in water.

PhCHO + (silicon enolate, R, 3.0 eq.) → [Ph_2BOH (10 mol%), $PhCO_2H$ (1 mol%), SDS (10 mol%), H_2O, 0°C] → Ph–CH(OH)–CH(CH_3)–C(=O)–R

Silicon enolate	Time (h)	Yield (%)	*syn/anti*
$OSiMe_3$ (*E*/*Z* = <1/>99)	8[a]	60	96/4
$OSiMe_3$, S*t*-Bu (*E*/*Z* = 98/2)	18	62	96/4
$OSiMe_3$, S*t*-Bu (*E*/*Z* = 3/97)	72	84	39/61

[a] At 30°C.

high diastereoselectivity was unexpected for us, because lower diastereoselectivity had been obtained in the previous Lewis acid-catalysed aldol reactions in water (in the case of $Sc(DS)_3$, *syn/anti* = ca. 1/1). Furthermore, the distereoselectivity was found to depend on the olefin geometry of the silyl enol ethers used as shown in Table 13.3.

PhCHO + **2** (1.5 eq.) ($OSiMe_3$, Ph) → [Ph_2BOH (10 mol%), $PhCO_2H$ (1 mol%), SDS (10 mol%), H_2O, 0°C, 24 h] → Ph–CH(OH)–CH(CH_3)–C(=O)–Ph (6)

93%, *syn*/*anti* = 94/6

A kinetic study of the Ph_2BOH-catalysed reactions of several aldehydes with **2** revealed that the rate of the disappearance of **2** followed first-order kinetics and was independent from the reactivity of the aldehydes used. Taking into account this result, we have proposed the reaction mechanism in which a silyl enol ether is transformed to the corresponding diphenylboryl enolate before the aldol addition step takes place (Scheme 13.1). The high diastereoselectivity is consistent with the mechanism, in which the aldol step proceeds via a chair-like six-membered transition state. The opposite diastereoselectivity in the reaction with the geometrical isomers of the thioketene silyl acetal shown in Table 13.3 also supports the mechanism via the boron enolate, because this trend was also observed in the classical boron enolate-mediated reactions in dry organic solvents. Although we have not yet observed the boron enolates directly under the reaction conditions, this mechanism can explain all of the experimental data obtained and is considered as the most reasonable one. As far as we know, this is the first example of

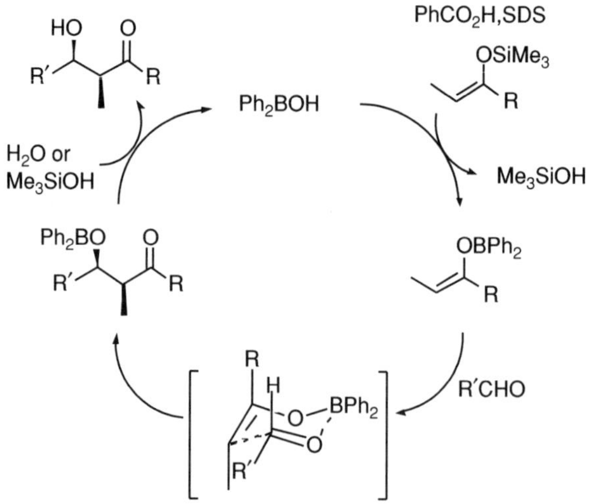

SCHEME 13.1. Mechanism of the Ph_2BOH-catalysed reaction.

the formation of boron enolates using a catalytic amount of a boron source. In addition, it is noted that these catalytic boron-enolate reactions have been attained in water, in which boron enolates had been believed to be too water-sensitive to use as substrates for aqueous reactions. These results are expected to open a new field for development of aqueous systems for water-sensitive compounds.

The concept of surfactant-type catalysts described above was also found to be applicable to catalytic systems other than Lewis acid-catalysed reactions. For example, we have developed palladium-catalysed allylic substitution reactions using a combination of $Pd(PPh_3)_4$ and a non-ionic surfactant, Triton X-100 [32].

13.3. SURFACTANT-TYPE BRØSTED ACIDS FOR REACTIONS IN WATER

As an extension of the studies on the LASC systems, we planned to develop a "Brøsted acid–surfactant-combined catalyst (BASC)", composed of a Brøsted acidic group and a hydrophobic moiety, and found that a BASC was an efficient catalyst for Mannich-type reactions in water.

Mannich-type reactions provide useful routes for the synthesis of optically active β-amino ketones or esters, which are versatile building blocks for the preparation of many nitrogen-containing biologically important compounds. We searched an efficient catalyst for Mannich-type reactions of silicon enolates with imines generated in situ from aldehydes and amines in water, and selected results are summarized in Table 13.4 [33,34]. Although the reaction in the presence of SDS alone afforded the desired β-amino ketone in a very low yield (entry 1), addition of a catalytic amount of HCl slightly improved the yield (entry 2). This result suggests that a combination of a Brøsted acid and an anionic surfactant leads to an effective catalyst for this Mannich-type reaction. We then tested *p*-dodecylbenzenesulfonic acid (DBSA) as a BASC, which was expected

TABLE 13.4. Mannich-type reactions in the presence of various catalysts in water.

PhCHO (1.0 eq.) + 2-methoxyaniline (1.0 eq.) + $CH_2=C(OSiMe_3)Ph$ (1.5 eq.) → (Catalyst, H_2O, 23°C) → Mannich product

Entry	Catalyst (mol%)	Time (min)	Yield (%)
1	SDS (30)	20	8
2	SDS (30) + HCl (10)	20	29
3	$Sc(DS)_3$ (10)	20	38
4	DBSA (10)	20	69
5	DBSA (10)	120	83
6	TsOH (10)	120	Trace

to behave both as a Brøsted acid and as a surfactant. Indeed, DBSA (10 mol%) was found to be a good catalyst for the reaction (entry 4), and interestingly, even better than a surfactant-type Lewis acid, $Sc(DS)_3$ (entry 3).

It is noteworthy that hydrolysis of the silicon enolate was not a severe problem even in the presence of the Brøsted acid. Akiyama and co-workers also developed Brøsted acid-catalysed Mannich-type reactions in water in the presence of a surfactant [35].

Although the reaction system stated above has extended the substrate applicability in Mannich reactions in water, there is still a drawback that the silicon enolates, which are prepared from the corresponding carbonyl compounds usually under anhydrous conditions, have to be used. From atom-economical and practical points of view, it is desirable to develop an efficient system for Mannich-type reactions in which the parent carbonyl compounds are directly used. Along this line, we next investigated three-component Mannich-type reactions in water using ketones, instead of silicon enolates, as nucleophilic components, and found that DBSA was also an effective catalyst [36]. An example is shown in Equation (7), where only 1 mol% DBSA was sufficient to give the desired product.

$$\text{PhCHO (1.0 eq.)} + \text{PhNH}_2 \text{ (1.0 eq.)} + \text{cyclohexanone (5.0 eq.)} \xrightarrow[\text{H}_2\text{O, 23°C}]{\text{DBSA (1 mol\%)}} \text{2-[Ph(PhNH)CH]cyclohexanone, 97\%} \quad (7)$$

Acid-catalysed direct esterification of carboxylic acids with alcohols is a fundamental and useful reaction in organic synthesis [37]. Generally, direct esterification is carried out in organic solvents and needs either of two methods to shift the equilibrium between reactants and products. One is removal (azeotropically or using dehydrating agents) of water generated as the reactions proceed, and the other is use of large excess amounts of one of the reactants. On the other hand, our idea is that the esterification would be realized even in water without using large excess reactants. The concept is shown in Figure 13.2. The surfactant-type Brøsted acid and organic substrates (carboxylic acids

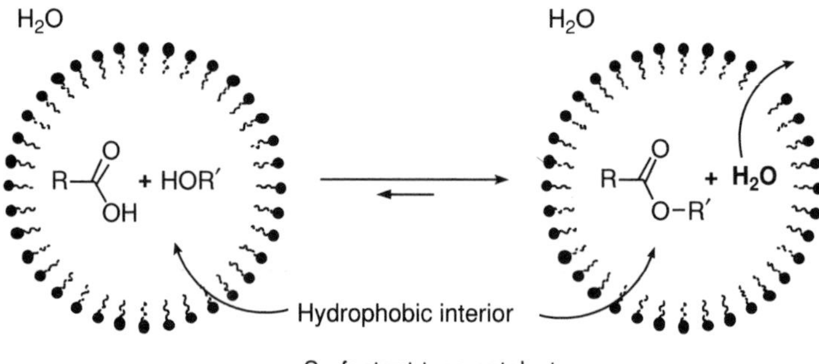

FIGURE 13.2. Illustration of direct esterification by dehydration in the presence of a surfactant-type catalyst in water.

and alcohols) in water would form the droplets whose interior is hydrophobic. The surfactants would concentrate a catalytic species such as proton onto the droplets'surfaces, at which the reaction takes place, and then enhance the rate to reach the equilibrium. For lipophilic substrates, the equilibrium position between the substrates and the products (esters) would lie at the ester side, because water molecules would be expelled out of the droplets owing to the hydrophobic nature of their interior. As a result, the dehydration reactions would efficiently proceed even in the presence of a large amount of water as a solvent. Although dehydration in water is really unusual, we have found that DBSA-catalysed esterification of hydrophobic carboxylic acids and alcohols indeed proceeds even in the presence of a large amount of water to afford the corresponding esters in high yields [38,39]. Representative examples are shown in Table 13.5.

Figure 13.3 shows the profiles of the DBSA-catalysed reaction of lauric acid with 3-phenyl-1-propanol (1:1) at 40°C in water (closed circle). The reaction reached its maximum yield of 84% in 170 hours. We also conducted hydrolysis of the corresponding ester (open square). Both esterification and hydrolysis finally led to the same composition of the reaction mixture, indicating that the reaction reached its equilibrium position.

TABLE 13.5. DBSA-catalysed esterification in water.

$$\underset{(1.0\text{ eq.})}{RCO_2H} + \underset{(2.0\text{ eq.})}{HOR'} \xrightarrow[H_2O,\ 40°C,\ 48\text{ h}]{DBSA\ (10\ mol\%)} RCO_2R'$$

R	R′	Yield (%)
$CH_3(CH_2)_{10}$–	–$(CH_2)_3Ph$	89
$CH_3(CH_2)_{10}$–	–$(CH_2)_{11}CH_3$	97
$CH_3(CH_2)_{10}$–	–$(CH_2)_{13}CH_3$	>99
$Ph(CH_2)_4$–	–$(CH_2)_{13}CH_3$	97
$CH_3(CH_2)_7$CH=CH$(CH_2)_7$–	–$(CH_2)_{13}CH_3$	97
c-Hex	–$(CH_2)_{13}CH_3$	95

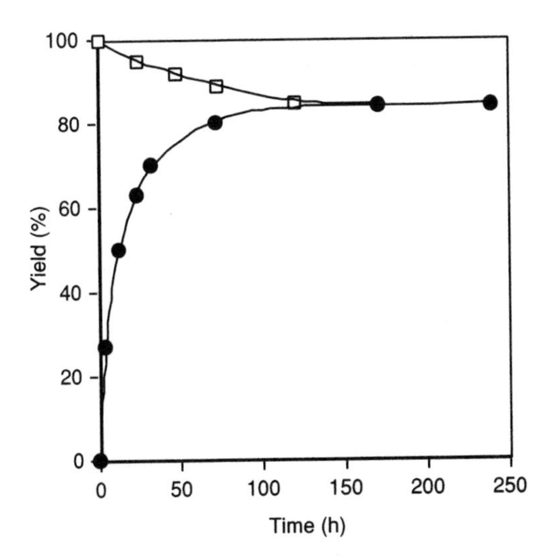

FIGURE 13.3. Reaction profiles for DBSA-catalysed reactions in water. Closed circle: esterification of lauric acid with 3-phenyl-1-propanol (1:1). Open square: hydrolysis of 3-phenyl-1-propyl laurate.

The high yield at the equilibrium position should be attributed to the formation of a hydrophobic area in water as shown in Figure 13.2.

The amounts of DBSA used were also found to affect the equilibrium position (Table 13.6). Each equilibrium position was confirmed by conducting both esterification of the carboxylic acid with the alcohol and hydrolysis of the ester. Table 13.6 clearly shows that increase of the amount of DBSA resulted in decrease of the yield of the ester at the equilibrium position. This result may be attributable to the size difference of the emulsion droplets that were formed by the hydrophobic substrates and the surfactant in water. As the amount of the surfactant-type catalyst increases, the size of each droplet may decrease, because the emulsion system may become a microemulsion system where the substrates are solubilized in water by a large amount of the surfactant. In fact, while 10 mol% DBSA gave the white turbid mixture, the reaction mixture was almost clear in the presence of 200 mol% DBSA, indicating that the size of the droplets became smaller. The smaller the droplets, the larger the sum of surface area of the droplets. As a result,

TABLE 13.6. Yield at the equilibrium position in the esterification in the presence of various amounts of DBSA in water.[a]

Entry	Amount of DBSA (mol%)	Yield (%)
1	10	84
2	50	71
3	100	58
4	200	32

[a] Conditions: lauric acid/3-phenyl-1-propanol = 1:1, at 40°C in water (3 ml/0.5 mmol of lauric acid).

TABLE 13.7. DBSA-catalysed selective esterification in water.

$$R^1CO_2H + R^2CO_2H + HOR \xrightarrow[H_2O,\ 40°C,\ 48\ h]{DBSA\ (10\ mol\%)} R^1CO_2R\ (\mathbf{A}) + R^2CO_2R\ (\mathbf{B})$$

(1:1:*x*)

Entry	R^1	R^2	R (x equiv)	Yield (%)[a] A	Yield (%)[a] B
1[b]	$CH_3(CH_2)_{10}$–	CH_3–	–$(CH_2)_{11}CH_3$ (1)	63	35
2	$CH_3(CH_2)_{10}$–	CH_3–	–$(CH_2)_{11}CH_3$ (1)	81	4
3	$CH_3(CH_2)_{10}$–	CH_3–	–$(CH_2)_{11}CH_3$ (2)	>99	16
4[c]	$CH_3(CH_2)_{10}$–	CH_3–	–$(CH_2)_3Ph$ (1)	78	trace
5	*c*-Hex	CH_3–	–$(CH_2)_{11}CH_3$ (1)	65	5
6	*c*-Hex	CH_3–	–$(CH_2)_{11}CH_3$ (2)	90	13
7	*c*-Hex	Ph	–$(CH_2)_{11}CH_3$ (2)	83	5
8	*c*-Hex	Pr(CH=CH)–	–$(CH_2)_{11}CH_3$ (2)	84	14
9	$Ph(CH_2)_2$–	Ph(CH=CH)–	–$(CH_2)_{11}CH_3$ (1)	78	trace

[a] NMR yield.
[b] Under neat conditions.
[c] 120 h.

the organic phase and water can come into contact with each other, and the hydrolysis of the ester becomes a favourable process. These results suggest that a large hydrophobic area inside the emulsion droplets is important to attain high yields at the equilibrium position.

We next carried out selective esterification of two substrates in this reaction system. When a 1:1 mixture of lauric acid and acetic acid was esterified with dodecanol in the presence of DBSA under neat conditions at 40°C for 48 h, the laurate ester and the acetate ester were obtained in 63% and 35% yields, respectively (Table 13.7, entry 1). On the other hand, when the same reaction was conducted in water, the laurate ester was predominantly obtained in 81% yield, and the yield of the acetate was only 4% (entry 2). Similar selective esterification of lauric acid over acetic acid was also observed in the reaction of another alcohol (entry 4). Furthermore, even cyclohexanecarboxylic acid, which is an α-disubstituted acid, was preferentially esterified in the presence of acetic acid (entries 5 and 6). These selectivities are attributed to the hydrophobic nature of lauric acid and cyclohexanecarboxylic acid as well as to the high hydrophilicity of acetic acid. These unique selectivities became possible by using water as a solvent. Selective esterification based on the difference in hydrophobicity was also attained in the reaction of two alcohols, one of which is hydrophobic and the other water-soluble.

DBSA is also applicable to other reactions in water. Ether formation from two alcohols is such an example [40]. We tried formation of symmetric ethers from benzylic alcohols in water using 10 mol% of DBSA as a catalyst. The reactions were found to proceed smoothly in water to afford the corresponding symmetric ethers in high yields (Table 13.8, entries 1 and 2). It should be noted that the etherification of the substrate shown in entry 1 in the presence of TsOH instead of DBSA gave only a trace amount

TABLE 13.8. DBSA-catalysed etherification in water.

$$\mathrm{ROH} + \mathrm{R'OH} \xrightarrow[\substack{H_2O \\ 24\ h}]{\text{DBSA(10 mol\%)}} \mathrm{ROR'}$$

Entry	R	R′	Yield (%)
1[a]	$H_3CO-C_6H_4-CH(Bu)-$	$H_3CO-C_6H_4-CH(Bu)-$	90
2[b]	Ph_2CH-	Ph_2CH-	91
3[b,c]	$CH_3(CH_2)_{11}-$	Ph_2CH-	89
4[c,d]	$CH_3(CH_2)_{11}-$	$Ph(CH_3)_2C-$	83
5[b,c]	$CH_3(CH_2)_{11}-$	$H_3CO-C_6H_4-CH_2-$	77

[a] For 7 h at 30°C.
[b] At 80°C.
[c] 2 equiv of R′OH was used.
[d] At 50°C.

of the product. This result indicates that the long alkyl chain of DBSA, which leads to the formation of emulsion droplets, is essential for the efficient catalysis as in the case of the dehydrative esterification. We then carried out unsymmetrical etherification of two alcohols, a primary alcohol and a benzylic alcohol (2 equiv). The reactions also proceeded smoothly to give the desired ethers in high yields (entries 3–5).

Dithioacetals are useful in organic synthesis as protective groups for carbonyl compounds, as precursors of acyl carbanion equivalents or as electrophiles under Lewis acidic conditions. The DBSA-catalysed system was also found to be applicable to dithioacetalization in water. In addition, easy work-up has been realized without the use of organic solvents when the products are solid and insoluble in water. In fact, the dithioacetalization of cinnamaldehyde on 10 mmol-scale with 1 mol% of DBSA proceeded smoothly to deposit crystals. The pure product was obtained in excellent yield after the crystals were filtered and washed with water (Equation (8)). This simple procedure is one of the advantages of the present reaction system.

$$\mathrm{PhCH{=}CHCHO} + \mathrm{HS(CH_2)_2SH}\ (1:1.1) \xrightarrow[\text{2) Filtered, then washed with } H_2O]{\text{1) DBSA (1 mol\%)},\ H_2O,\ 40^\circ C,\ 4\ h} \text{2-styryl-1,3-dithiolane}\ \ 96\% \quad (8)$$

With several types of DBSA-catalysed dehydration in hand, we tried chemospecific reactions of substrates having two reactive functional groups. As shown in Equation (9), mercaptoalcohols such as 2- mercaptoethanol and 6-mercapto-1-hexanol were subjected to reactions in the presence of both triphenylmethanol and lauric acid in an aqueous solution of DBSA. These three-component reactions predominantly gave tritylthioalkyl laurates in high yields, accompanied with small amounts of tritylthioalcohols. It is noteworthy that the thiol parts of the mercaptoalcohols were selectively transformed to the thioethers, and the alcohol parts to the esters.

$$\mathrm{CH_3(CH_2)_{10}CO_2H} + \mathrm{HO}\!\!\left(\!\!\right)_{\!n}\!\!\mathrm{SH} + \mathrm{HOCPh_3} \quad (2:1:1) \xrightarrow[\mathrm{H_2O,\ 80°C,\ 24\ h}\ \text{then}\ 40°C,\ 24\ h]{\mathrm{DBSA\ (10\ mol\%)}} \mathrm{CH_3(CH_2)_{10}CO_2}\!\!\left(\!\!\right)_{\!n}\!\!\mathrm{SCPh_3} \tag{9}$$

$n = 2$: 78%
$n = 6$: 91%

13.4. CONCLUDING REMARKS

We have developed several useful organic reactions in water. The key is the use of newly developed surfactant-type catalysts such as LASC and BASC. While most organic substrates are not soluble in water, emulsion droplets are formed immediately by combining the catalysts and organic materials in water, and catalytic organic reactions proceed smoothly. Although several acid-catalysed reactions have been demonstrated, the idea of the surfactant-type catalysts is not limited to acid catalysis, and may be applied to various types of catalytic reactions in water. In the light of the increased demand for reduction of organic solvents in industry, new methods of catalysis in water should be urgently developed. We believe that the surfactant-type catalysts described in this chapter will contribute to progress in green chemical processes.

ACKNOWLEDGEMENTS

This work was partially supported by CREST and SORST, Japan Science and Technology Corporation (JST) and a Grant-in Aid for Scientific Research from Japan Society of the Promotion of Science.

REFERENCES

1. P.A. Grieco, Ed., *Organic Synthesis in Water*, Blacky Academic and Professional, London, 1998.
2. C.-J. Li and T.-H. Chan, *Organic Reactions in Aqueous Media*, John Wiley & Sons, New York, 1997.
3. J.H. Fendler and E.J. Fendler, *Catalysis in Micellar and Macromolecular Systems*, Academic Press, London, 1975.
4. P.M. Holland and D.N. Rubingh, Eds, *Mixed Surfactant Systems*, American Chemical Society, Washington, DC, 1992.
5. C.J. Cramer and D.G. Truhlar, Eds, *Structure and Reactivity in Aqueous Solution*, American Chemical Society, Washington, DC, 1994.
6. D.A. Sabatini, R.C. Knox and J.H. Harwell, Eds, *Surfactant-Enhanced Subsurface Remediation*, American Chemical Society, Washington, DC, 1994.
7. W.D. Harkins, J. Am. Chem. Soc. **69**, 1428 (1947).
8. F.M. Menger, J.U. Rhee and H.K. Rhee, J. Org. Chem. **40**, 3803 (1975).
9. C. Larpent, E. Bernard, F.B. Menn and H. Patin, J. Mol. Cat. A: Chem. **116**, 277 (1997).

10. T. Dwars, U. Schmidt, C. Fischer, I. Grassert, R. Kempe, R. Fröhlich, K. Drauz and G. Oehme, Angew. Chem. Int. Ed. **37**, 2851, (1998).
11. I. Grassert, U. Schmidt, S. Ziegler, C. Fischer and G. Oehme, Tetrahedron: Asymmetry **9**, 4193 (1998).
12. R. Selke, J. Holz, A. Riepe and A. Börner, Chem. Eur. J. **4**, 769 (1998).
13. K. Yonehara, T. Hashizume, K. Mori, K. Ohe and S. Uemura, J. Org. Chem. **64**, 5593 (1999).
14. K. Yonehara, K. Ohe and S. Uemura, J. Org. Chem. **64**, 9381 (1999).
15. M.S. Goedheijt, B.E. Hanson, J.N.H. Reek, P.C.J. Kamer and P.W.N.M. van Leeuwen, J. Am. Chem. Soc. **122**, 1650 (2000).
16. S. Kobayashi, Chem. Lett. 2187 (1991).
17. S. Kobayashi and I. Hachiya, J. Org. Chem. **59**, 3590 (1994).
18. H. Yamamoto, Ed., *Lewis Acids in Organic Synthesis*, Wiley-VCH, Weinheim, 2000.
19. P. Anastas and J.C. Warner, *Green Chemistry: Theory and Practice*, Oxford University Press, Oxford, UK, 1998.
20. R. Breslow, Acc. Chem. Soc. **24**, 159 (1991).
21. K. Manabe and S. Kobayashi, Chem. Eur. J. **8**, 4094 (2002).
22. S. Kobayashi, T. Wakabayashi, S. Nagayama and H. Oyamada, Tetrahedron Lett. **38**, 4559 (1997).
23. S. Kobayashi and T. Wakabayashi, Tetrahedron Lett. **39**, 5389 (1998).
24. K. Manabe, Y. Mori, T. Wakabayashi, S. Nagayama and S. Kobayashi, J. Am. Chem. Soc. **122**, 7202 (2000).
25. S. Otto, J.B.F.N. Engberts and J.C.T. Kwak, J. Am. Chem. Soc. **120**, 9517 (1998).
26. Y. Mori, K. Kakumoto, K. Manabe and S. Kobayashi, Tetrahedron Lett. **41**, 3107 (2000).
27. E. Keller and B.L. Feringa, Tetrahedron Lett. **37**, 1879 (1996).
28. K. Manabe, N. Aoyama and S. Kobayashi, Adv. Synth. Catal. **343**, 174 (2001).
29. S. Kobayashi, Y. Mori, S. Nagayama and K. Manabe, Green Chem. **1**, 175 (1999).
30. Y. Mori, K. Manabe and S. Kobayashi, Angew. Chem. Int. Ed. **40**, 2815 (2001).
31. Y. Mori, J. Kobayashi, K. Manabe and S. Kobayashi, Tetrahedron **58**, 8263 (2002).
32. S. Kobayashi, W.W.-L. Lam and K. Manabe, Tetrahedron Lett. **41**, 6115 (2000).
33. K. Manabe, Y. Mori and S. Kobayashi, S. Synlett 1401 (1999).
34. K. Manabe, Y. Mori and S. Kobayashi, Tetrahedron **57**, 2537 (2001).
35. T. Akiyama, J. Takaya and H. Kagoshima, Synlett 1426 (1999).
36. K. Manabe and S. Kobayashi, Org. Lett. **1**, 1965 (1999).
37. R.C. Larock, *Comprehensive Organic Transformations*, John Wiley & Sons, New York, 1999.
38. K. Manabe, X.-M. Sun and S. Kobayashi, J. Am. Chem. Soc. **123**, 10101 (2001).
39. K. Manabe, S. Iimura, X.-M. Sun and S. Kobayashi, J. Am. Chem. Soc. **124**, 11971 (2002).
40. S. Kobayashi, S. Iimura and K. Manabe, Chem. Lett. 10 (2002).

14

Bioseparation Through Liquid–Liquid Interfaces

Tsutomu Ono and Masahiro Goto
Kyushu University, Fukuoka 812-8581, Japan

14.1. INTRODUCTION

Surfactant self-assemblies, such as reverse micelles or microemulsions, create unique environments in organic media. These aggregates display a range of interesting physicochemical properties that have brought about great potential in modern biotechnology because they are capable of solubilizing biomolecules (e.g., proteins, peptides, amino acids) in their aqueous interior with their intrinsic activity. Since a reverse micelle is a spontaneous nano-aggregate of surfactant molecules dispersed in an organic solvent, the solution is optically transparent and thermodynamically stable. Moreover, the hydrophilic headgroups of the surfactants face inward (towards each other) and provide aqueous nano-scale droplets, called "water pools", in non-polar solvents.

Molecular recognition using a molecular assembly is biologically important for the relationship between biomolecules and amphiphiles, and these interactions reflect the biomolecule–surface interaction in vivo, for example, protein–membrane interactions [1] and protein targeting through a cell membrane. Fundamentally, the biological membrane is characterized by three structural principles: molecular order, recognition and mobility. Therefore, reverse micellar recognition for biomolecules is conducted through the liquid–liquid interface utilizing specific interactions. In this chapter, we focus on the bioseparation of biomolecules through a liquid–liquid interface using reversed micelles.

14.2. REVERSE MICELLAR PROTEIN EXTRACTION

Luisi and co-workers were the first to mention the possibility of the reverse micellar solubilization of proteins for protein separation [2]. After that, the research groups of

Dekker [3] and Hatton [4] developed this process systematically for practical uses. They also demonstrated the factors affecting the protein separation based on the reverse micellar extraction. A recent review summarized the factors clarified so far [5].

Here, we review the key roles of the interaction between proteins and surfactants in liquid–liquid protein extraction. These assist in the understanding of the protein transfer mechanism through a water/oil interface and in the design of reverse micelles for protein separation processes.

14.2.1. Concept

Reverse micelles are self-organized aggregates of amphiphilic molecules that provide a hydrophilic nano-scale droplet in apolar solvents. This polar core accommodates some hydrophilic biomolecules stabilized by a surfactant shell layer. Furthermore, reverse micellar solutions can extract proteins from aqueous bulk solutions through a water–oil interface. Such a liquid–liquid extraction technique is easy to scale up without a loss in resolution capability, complex equipment design, economic limitations and the impossibility of a continuous mode of operation. Therefore, reverse micellar protein extraction has great potential in facilitating large-scale protein recovery processes from fermentation broths for effective protein production.

For separating and purifying proteins, a forward extraction operation facilitates the selective transfer of a target protein from an aqueous solution containing some kinds of proteins into an organic phase, and the extracted proteins are quantitatively recovered into a fresh aqueous solution by the subsequent back extraction. The transfer selectivity is based on the interaction between surfactants, which form reverse micelles, and the protein surface. On the other hand, the quantitative recovery of an objective protein from reverse micelles is accomplished by severing proteins from the enclosure with a surfactant layer.

Both processes are controllable by varying the operational conditions to alter the charge properties of the proteins and the surfactants.

14.2.2. Forward Transfer into Reverse Micelles

Proteins solubilized in aqueous solution interact more or less with hydrophilic groups of surfactants at the oil–water interface. Therefore, the type of hydrophilic group is strongly influenced by the protein extraction efficiency. Anionic and cationic surfactants interact with charged protein surfaces more strongly than non-ionic surfactants. This feature also means that the non-ionic surfactants are favourable for protein stabilization in water droplets because of the not-so-hard interaction between the protein and the surfactant. In protein extraction, such an electrostatic interaction between proteins and surfactants is the main driving force in protein transfer.

Aqueous pH alters the protein charge property and affects the extraction efficiency. Haemoglobin (Mw: 64,500, pI: 6.8) is a difficult protein in terms of being able to completely extract it into reverse micelles. The representative anionic surfactant, di-2-ethylhexyl sulfosuccinate (AOT), cannot extract it, and gives rise to an interfacial precipitate. In contrast, we succeeded in the complete extraction of haemoglobin using synthetic anionic surfactants, dioleyl phosphoric acid (DOLPA), as seen in

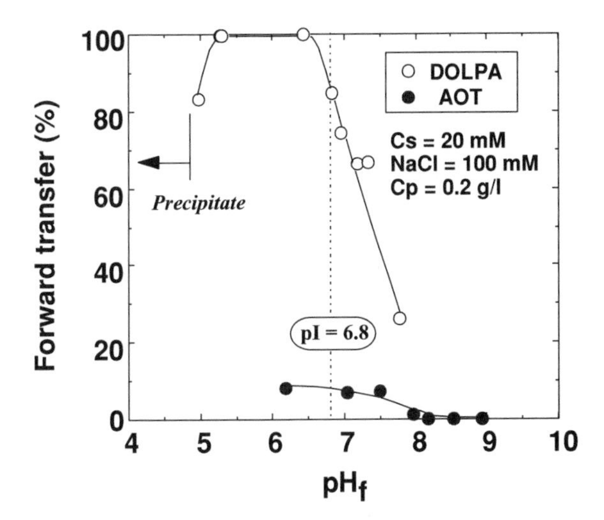

FIGURE 14.1. Effect of aqueous pH on the forward transfer of haemoglobin [6].

Figure 14.1 [6]. Only for pHs below the isoelectric point (pI) of the protein should the protein be extracted with anionic surfactants such as DOLPA or AOT, whereas for pHs above the pI, extraction into the reverse micellar phase should be inhibited because of the unfavourable electric repulsion. This result indicates that electrostatic interactions are required to extract haemoglobin efficiently into reverse micelles.

Although AOT is also an anionic surfactant of the same type as DOLPA, haemoglobin cannot be transferred into the AOT reverse micellar phase, and most haemoglobin can be seen at the oil–water interface as a red precipitate. Adachi and Harada have reported that cytochrome *c* precipitated as a cytochrome *c*–AOT complex at low concentrations of AOT [7]. It was found that this precipitate is likewise the AOT–haemoglobin complex (AOT/haemoglobin = 120:1) from the results of elemental analysis [8]. These results indicate that the difference in the extraction ability of DOLPA and AOT might depend on the hydrophobicity of the surfactants provided to the hydrophilic proteins.

Since most of the studies concerning protein extraction with reverse micelles have used a representative surfactant, AOT and trioctyl methyl ammonium chloride (TOMAC), to form reverse micelles, the protein extraction studies using a novel surfactant shed light on the important factors in the design of reverse micelles suitable for protein extraction. In particular, a comparative investigation concerning protein extraction efficiency with a variety of synthetic surfactants clarified the role of the hydrophobic tails of the surfactant on protein extraction [9].

Figure 14.2 summarizes the structures of the surfactants employed in this report. These are all anionic surfactants with phosphoric or sulfosuccinate groups and different types of hydrophobic chains.

We carried out the extraction of cytochrome *c* (Mw: 12,300, pI = 10.1) and haemoglobin using these surfactants. The effects of surfactant structure on the extraction efficiency of cytochrome *c* and haemoglobin are seen in Figures 14.3a and b, respectively.

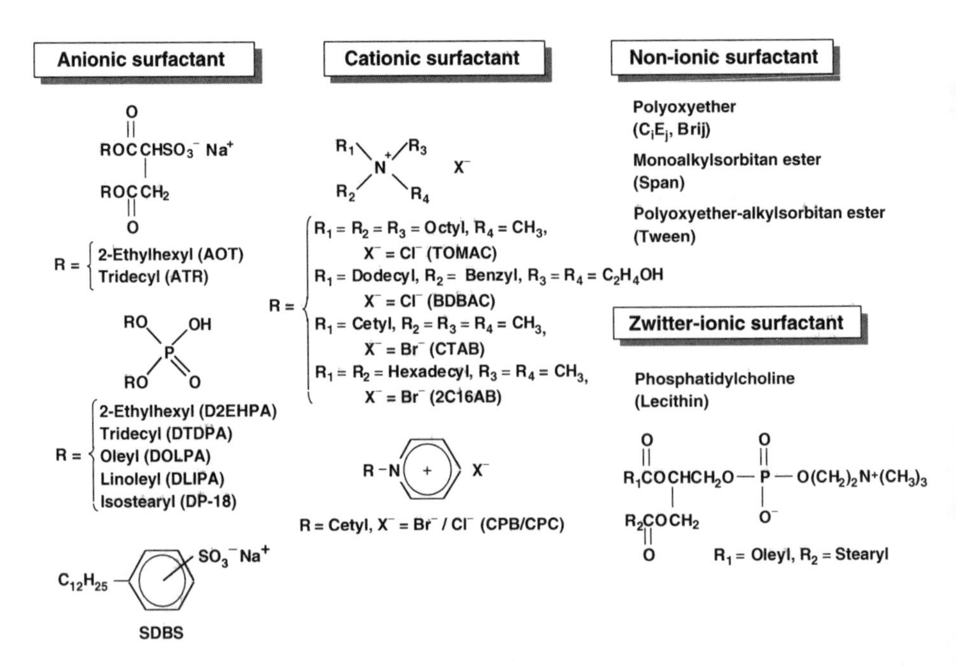

FIGURE 14.2. Structures and abbreviations of the representative surfactants used in the protein extraction studies.

These show that alkyl chains play an important role in the protein extraction ability. Bulky two-tailed surfactants are favourable in the protein transfer from an aqueous phase into an organic phase, while the other surfactants, having saturated-straight alkyl chains, dioctyl phosphoric acid (DOPA) and dioctyl sulfosuccinate (AOC), or a single alkyl chain, monooleyl phosphoric acid (MOLPA), are limited by their solubility into an organic solvent or their ability to form reverse micelles. A tridecyl group in di-tridecyl phosphoric acid (DTDPA) and di-tridecyl sulfosuccinate (ATR) comes from the tridecyl alcohol which includes several isomers, and therefore these surfactants also have bulky tails.

The results of surfactant-dependency on protein transfer indicate that protein extraction reverse micelles not only provide a hydrophilic droplet in a non-aqueous solvent to facilitate protein partition, but also make proteins sufficiently hydrophobic to solubilize into an organic solvent by coating the protein surface. Consequently, we suggest that proteins in the aqueous phase are extracted through the formation of an interfacial complex, a surfactant-coated protein and that the hydrophobic property dominates the extraction efficiency of the proteins, as seen in Figure 14.4. The unsaturated or branched alkyl chain may contribute to the formation of a soluble protein–surfactant complex into a non-aqueous solvent.

From the results of Figure 14.3, interestingly, the protein extraction behaviour by dilinoleyl phosphoric acid (DLIPA) is distinct from the result obtained by DOLPA, in spite of the similar hydrophobic chain. The oleyl and linoleyl groups are unsaturated C18 chains. The former has one unsaturated cis-double bond at the position between C9 and C10, while the latter has two at the position between C9 and C10 and between C12 and C13. The two double bonds in the hydrophobic part of the surfactant DLIPA provide some inflexibility in the chain, which can lead to steric hindrance and to solvation by the organic phase (Figure 14.5).

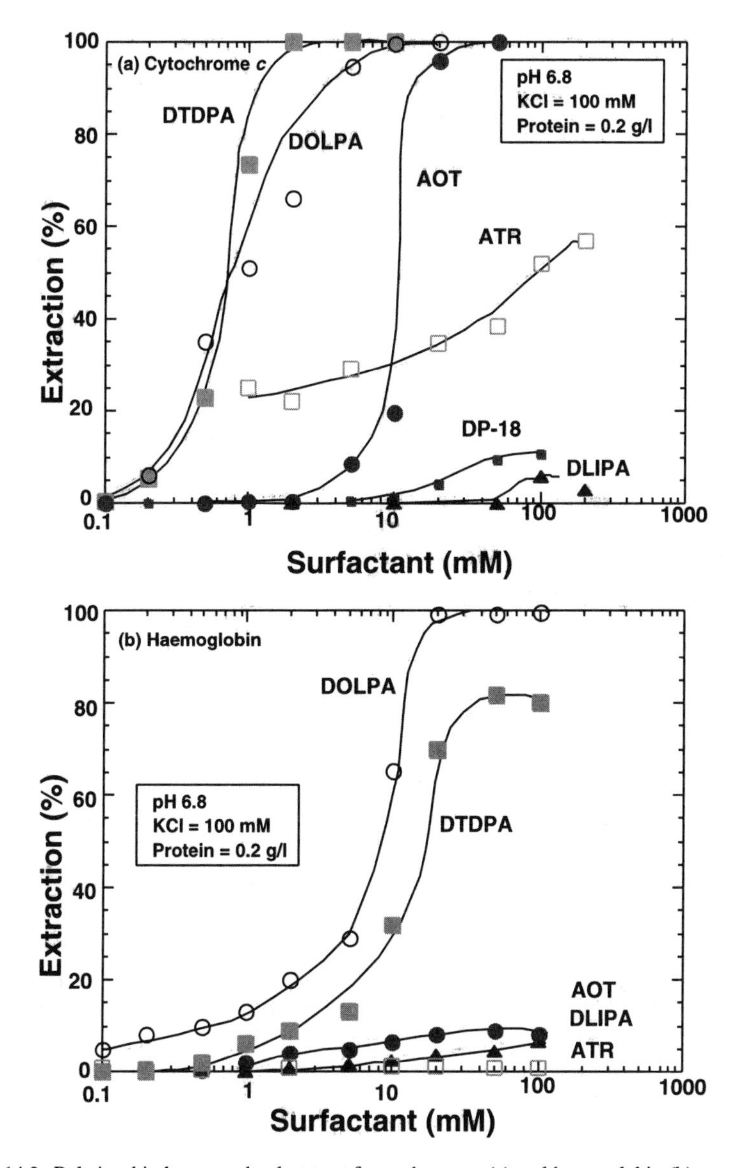

FIGURE 14.3. Relationship between the degrees of cytochrome *c* (a) and haemoglobin (b) extraction and surfactants [9].

To obtain the high extraction efficiency of a target protein, we need to design reverse micelles having sufficient hydrophobicity to extract the protein and prepare the conditions for enhancing the protein–surfactant interaction to form a protein–surfactant complex, which is a crucial intermediate for reverse micellar protein extraction.

14.2.3. Protein Recovery from Reverse Micelles

Since protein extraction is based on spontaneous phenomena, the protein release from thermodynamically stable reverse micelles with protein cores is difficult. Protein

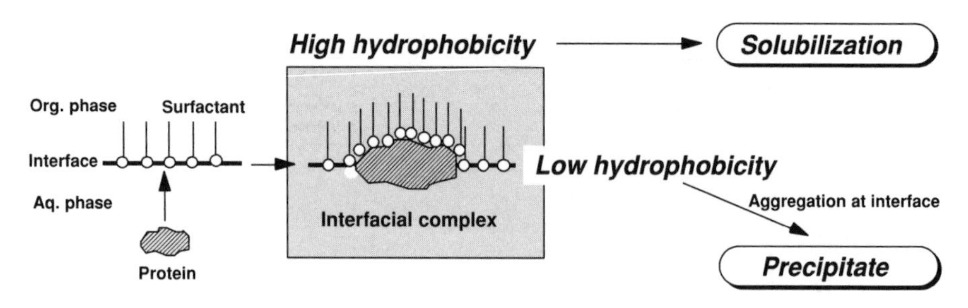

FIGURE 14.4. Schematic illustration of protein extraction by surfactants [9].

recovery issues in protein separation and purification using reverse micelles remain open. In particular, DOLPA reverse micelles form stable reverse micelles incorporating proteins, and result in no protein release from reverse micellar solutions. In general, the back transfer of proteins from anionic reverse micellar solutions is facilitated by adjusting the second aqueous solution not to extract proteins: higher pH than pI and a high salt concentration.

The basic pH brings about an electrostatic repulsion between the protein surface and anionic surfactants. On the other hand, the high salt concentration avoids the electrostatic adsorption of the surfactants. Both factors modify the electrostatic interaction between the proteins solubilized in reverse micelles and the surfactant layer. Indeed, however, in the back extraction process, the proteins solubilized in the droplets cannot suffer such a modification because of the little degree of coalescence of the reverse micelles containing proteins with a bulk aqueous solution or vacant reverse micelles. This is one of the reasons why the back extraction process is much slower than the forward extraction process.

Carlson and Nagarajan have reported that the addition of alcohol improved the back extraction efficiency [10]. It is well known that alcohol is a representative cosurfactant so far. A long-chain alcohol stabilizes the reverse micelles, while a short-chain alcohol sometimes inhibits the formation of reverse micelles. In haemoglobin back extraction, the addition of isopropyl alcohol or ethanol is significant in facilitating haemoglobin release from DOLPA reverse micelles (Figure 14.6) [6].

This shows that haemoglobin is recovered from a reverse micellar solution only at higher pH values than pI and high salt concentrations to inhibit the electrostatic

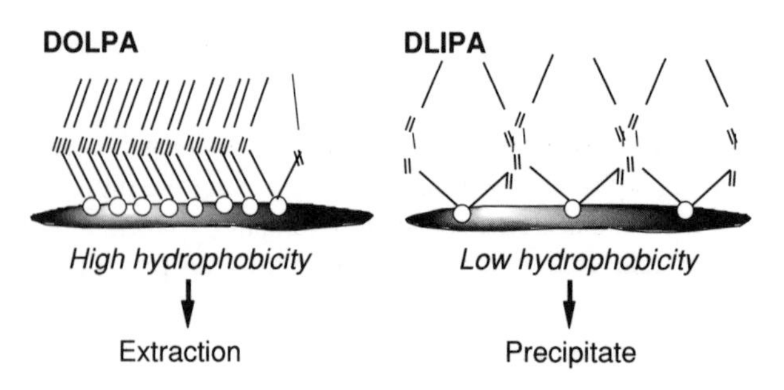

FIGURE 14.5. Schematic illustration of the difference in the extraction modes of DOLPA and DLIPA [9].

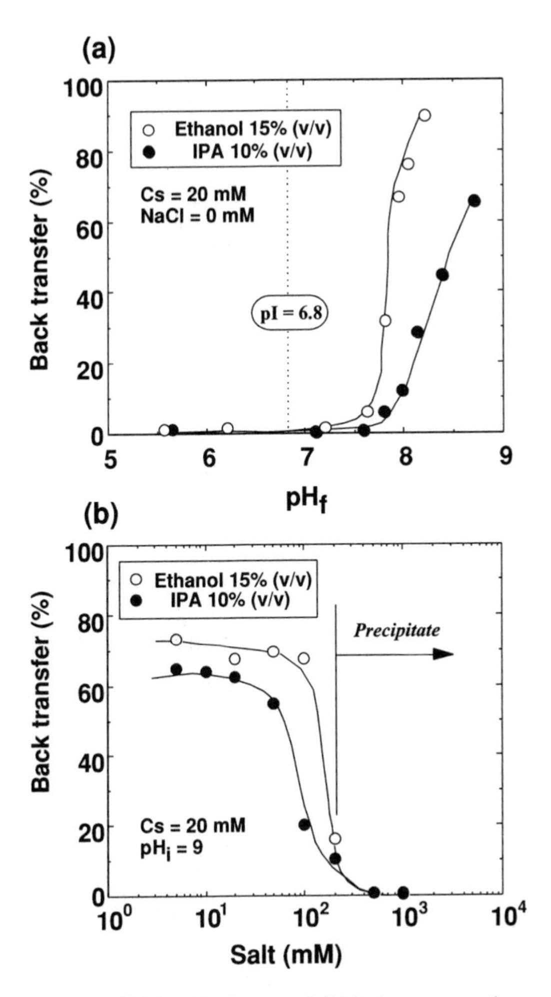

FIGURE 14.6. Effect of aqueous pH (a) and ionic strength (b) in the recovery phase on the back-transfer of haemoglobin [6].

interaction between the protein and the surfactant in the presence of alcohol. If not, DOLPA reverse micelles do not release any haemoglobin, even in an alkali pH, 1 N NaOH, and a high salt concentration, 2 M NaCl solution. These facts imply that alcohol promotes the coalescence between reverse micellar droplets and bulk aqueous solutions. To elucidate the effect of alcohol on the coalescence, we measured a lithium ion transfer from reverse micellar droplets to bulk water, which also means there is a solute leakage from reverse micellar droplets. Figure 14.7 demonstrates that the addition of alcohol facilitates a lithium ion leakage of about 20%.

Moreover, the alcohol addition during the back-extraction process was involved in the swelling of the reverse micelles. This phenomenon is strongly dependent on the salt concentrations in the water droplets and the recovery of the aqueous phase. Table 14.1 summarizes the droplet swelling ratios using recovery aqueous phases with different salt concentrations. In the presence of alcohol, as the salt concentration in the aqueous

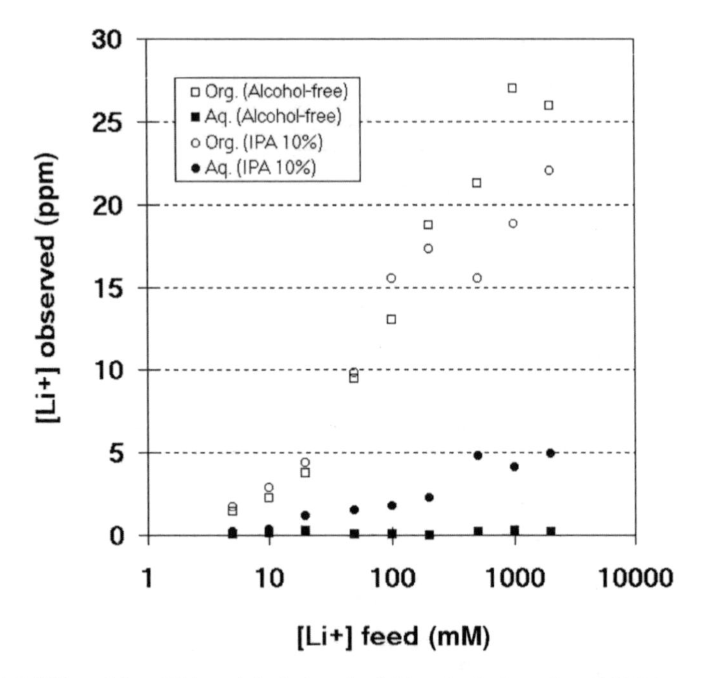

FIGURE 14.7. Effect of the addition of alcohol on the lithium ion leakage from DOLPA reverse micelles during back extraction. A back-extraction operation was performed by contacting the reverse micellar solution extracting lithium in advance and the fresh recovery of the aqueous solution. The observed amounts of lithium ion in each phase were measured using an ICP analyser.

TABLE 14.1. The swelling ratio of the reverse micellar droplet in alcohol-based back extraction.

		Swelling ratio (–) [b]	
$[NaCl]_{ini}$ [a]	$[NaCl]_{rec}$ [a]	Alcohol-free	Isopropyl alcohol 10%
100	0	1.52	4.70
	50	1.38	3.07
	100	1.40	2.30
	500	1.25	1.15
	1000	1.08	1.09
	2000	0.91	1.01
1000	0	1.34	5.01
	50	1.73	4.14
	100	1.82	4.03
	500	1.52	1.93
	1000	1.19	1.25
	2000	0.85	0.65

[a] The subscripts ini and rec indicate the initial and recovery aqueous phases, respectively.

[b] The ratio of the water content in the reverse micellar solution after back extraction to that before back extraction.

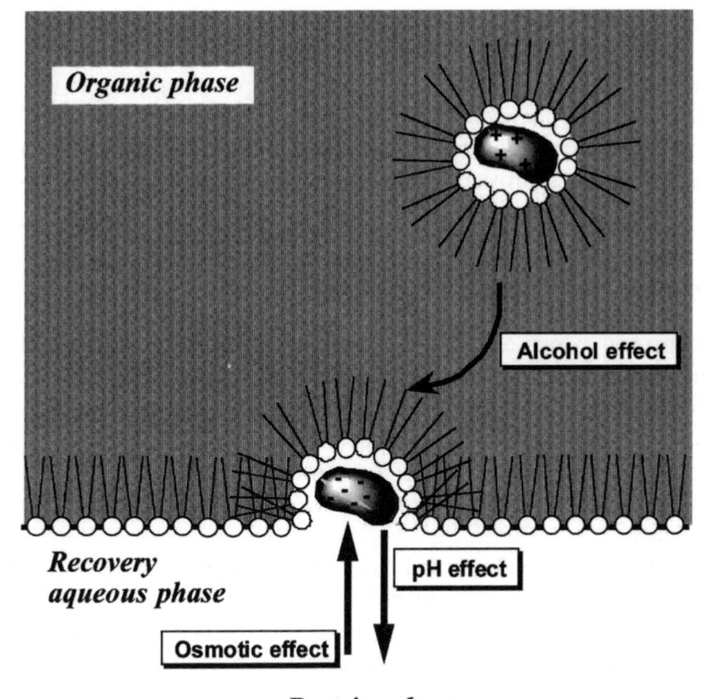

FIGURE 14.8. Schematic diagram of the back-transfer of proteins from the DOLPA reverse micellar phase by the addition of alcohol [6].

phase is decreased, the droplets became swollen. Also, when the salt concentration in the droplet was increased, the swelling ratio enhanced. The actual swelling of the droplets was supported experimentally by dynamic light scattering measurements; the droplet diameters were 4–6 nm after the alcohol-free back-extraction operation as well as before, whereas in the presence of isopropyl alcohol (10%), the diameters grew to 15–19 nm ($[LiCl]_{ini} = 10-100$ mM and $[LiCl]_{rec} = 0$ mM). These results mean that this swelling depends on an osmotic difference between the reverse micellar droplets and the bulk aqueous phase.

On the basis of the experimental evidence of back extraction, there are three dominant factors, as shown in Figure 14.8.

It is reasonable and widely accepted that the back-transfer process of proteins and other solutes is governed by an interfacial process and by a coalescence of the reverse micelles at the oil–water interface. According to the previous report, alcohol promotes the fusion/fission of the reverse micelles [11]. Such a modification in the dynamic property of reverse micellar droplets also affects the coalescence of the droplets and the bulk aqueous solution, and in this study results in an assistance in the release of proteins from the droplets. However, besides the alcohol addition, the appropriate pH and salt concentration in the recovery aqueous phase is required for protein release from the droplets into the recovery phase. The salt concentration leads to an osmotic effect and results in a swelling of the droplets in the presence of alcohol. The swelling droplets would be more fragile and facilitate the mass transfer between the droplets and the bulk solution. Finally, an electrostatic repulsion between the protein-charged surface and the surfactant headgroups is involved at solution pH values of over pI, which leads to the completion of the protein release from reverse micellar droplets. It is concluded that a lowering in the interfacial resistance for reverse micelles-containing proteins and an enhancement of the protein–surfactant electrostatic interaction is necessary for recovering the objective proteins extracted by reverse micelles.

14.2.4. Molecular Recognition in Protein Extraction

On the protein extraction behaviour using reverse micelles, the protein–surfactant interaction between charged protein surfaces and surfactant headgroups is a dominant factor in distinguishing the target proteins. In particular, some researchers have suggested that a protein can be extracted as a hydrophobic ion complex between a protein and surfactant molecules [6,7,12–15]. Therefore, an intrinsic factor of proteins also gives considerable modulation in the extraction behaviour, in which the environmental factors were maintained.

Recent studies have been carried out to alter protein solubilization behaviour by modifying the intrinsic properties of proteins. The liquid–liquid extraction of native cytochrome b_5 and mutants obtained by site-directed mutagenesis were achieved [16]. The mutant with a substitution of glutamic acid, negative-charged residue at moderate pH, for lysine, positive-charged residue at moderate pH, at three different sites along the sequence of cytochrome b_5 leads to an increase in the solubility in the AOT reverse micellar phase. Forney and Glatz took advantage of charged fusion proteins to improve the selectivity during the protein separation process [17,18]. The negatively charged fusion glucoamylase, which has amino-terminal fusions containing several additional asparatic acid residues, was efficiently extracted into cationic reverse micelles consisting of TOMAC or CTAB (cetyl trimethyl ammonium bromide). Although these approaches of utilizing genetic engineering techniques increase the electrostatic sites on the protein–surface accessible by surfactant headgroups, they change only to a few limited residues and give rise to alterations in the original charge distribution and asymmetry of the protein molecules.

Chemical modification facilitates the replacement of the residues on the protein surface without denaturation. In particular, chemical modifications without altering the electrical charge of the residues are useful for contrasting the protein–surfactant interaction because the pI of the modified protein is similar to that of the native one. A guanidylation reaction entails replacing the amino groups of lysine (Lys) residues with a guanidinium moiety, such as arginine (Arg) residues.

Guanidylated cytochrome *c* (G-cytc) was synthesized from native cytochrome *c* (N-cytc) and *O*-methylisourea (Figure 14.9) [19]. Almost all of the lysine residues of cytochrome *c* were substituted with a homoarginine moiety as seen in Table 14.2, but

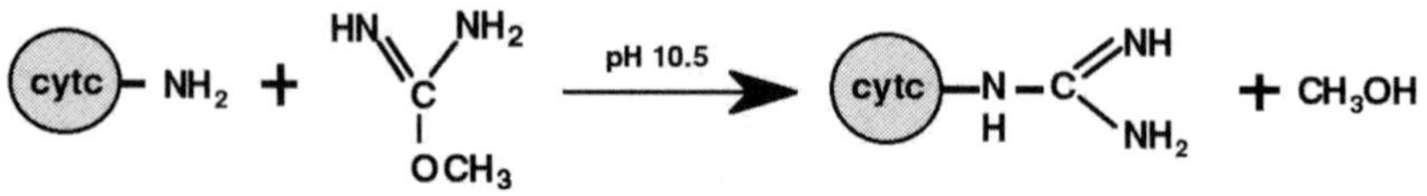

cytc–NH_2 + HN=C(NH_2)–OCH_3 —(pH 10.5)→ cytc–NH–C(=NH)–NH_2 + CH_3OH

FIGURE 14.9. Synthesis scheme of guanidylated cytochrome *c* (G-cytc).

TABLE 14.2. Cationic residues in the amino acid sequence of lysozyme and cytochrome *c* [19].

Cationic residue	Lysozyme	Cytochrome *c* Native	Cytochrome *c* Guanidylated
Arg	11	2	19.8 (2 + 17.8)[a]
Lys	6	19	1.2
His	1	3	3

[a] Homoarginine-containing groups modified by guanidylation.

this modification does not involve the disappearance of the cationic charge on the protein surface.

Figure 14.10 shows the extraction behaviours of N-cytc and G-cytc by DOLPA reverse micelles.

This shows that the cytochrome *c* extraction into an organic phase is carried out at a very low DOLPA concentration, in which reverse micelles cannot be formed. Also, the molar ratio required for the complete protein extraction was approximately 20, which corresponds to the number of cationic charged residues available for the electrostatic interaction with an anionic surfactant in one cytochrome *c* molecule. These results support the concept that proteins are extracted by electrostatic interactions with surfactant molecules, and that the existence of reverse micelles is not necessary for causing protein transfer, as mentioned above.

In addition, Figure 14.10 indicates that almost all of the G-cytc readily vanished from the aqueous feed solution at a relatively low DOLPA concentration in the organic

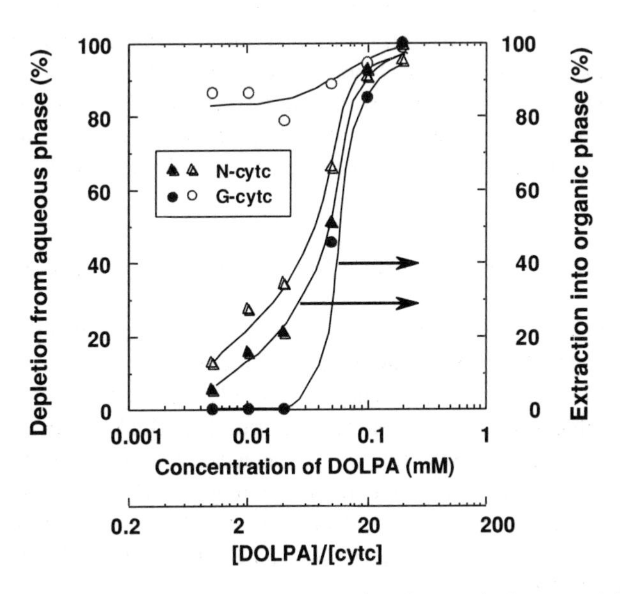

FIGURE 14.10. Extraction behaviour of N-cytc and G-cytc into the organic phase containing DOLPA. The amount of depleted proteins from aqueous solution (open symbols) and extracted proteins in the surfactant solution (closed symbols) were estimated by the decrease and increase in the protein concentration during the extraction operation, respectively [19].

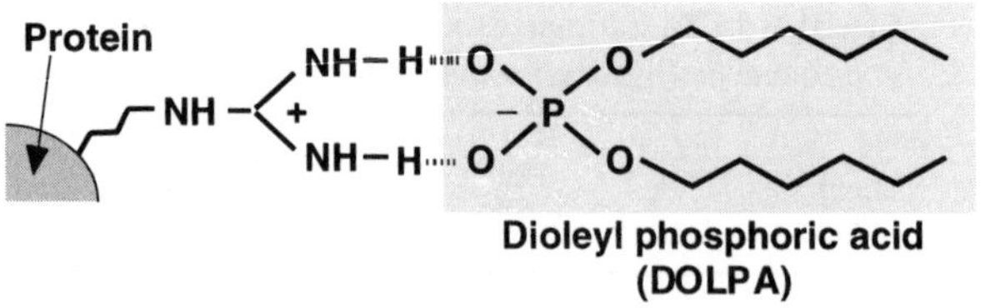

FIGURE 14.11. Schematic illustration of the guanidinium-phosphate interaction. This specific interaction is made up of electrostatic interactions and hydrogen bonding [19].

phase. The unbalanced amount of G-cytc after the extraction operation corresponds to the amount of interfacial precipitate, a protein–surfactant complex.

Although native cytochrome *c* is a Lys-rich protein, G-cytc altered by the guanidylation of N-cytc is a Arg-rich protein as well as lysozyme (Mw: 14,300, pI = 12.1). A previous study concerning the extraction behaviour of cytochrome c and lysozyme using DOLPA reverse micelles showed that the extraction efficiency of lysozyme is higher than that of cytochrome *c* [19]. Similarities in the molecular weight and the isoelectric point of lysozyme and cytochrome *c* highlight the influence of the surface charge distribution of proteins on the protein extraction efficiency. Therefore, it is clear that there is a specific interaction between the surface of the lysozyme, the Arg-rich protein, and the DOLPA headgroups.

A guanidinium group specifically interacts with the carboxylate or phosphate groups through hydrogen bonding along with electrostatic interactions [20]. In this study, it seems likely that the specific interaction between the phosphate groups of DOLPA and the guanidinium moiety on the cytochrome *c* surface enhanced the formation of the protein–surfactant complex at the oil–water interface, as seen in Figure 14.11.

Hence DOLPA reverse micelles recognize the Arg-rich proteins through specific phosphate-guanidinium groups. Such noncovalent bonds of surfactant molecules lead to alterations in the hydrophilic protein surface into sufficiently hydrophobic surfaces to be solubilized in a nonpolar solvent. Molecular recognition on the protein surface facilitates protein transfer, a significant characteristic for the specific separation of biomolecules.

14.3. EXTRACTION OF DNA THROUGH LIQUID–LIQUID INTERFACES

A transfer of DNA from an aqueous to organic phase was also investigated using reverse micelles. DNA is a significantly large biomolecule and possesses a double helix structure. To date there are a few instances where DNA has been able to be dissolved in an organic solvent. Imre and Luisi first reported that DNA can be solubilized in an organic solution by directly injecting a DNA aqueous solution [21]. It was shown that DNA maintains its double helix structure after being entrapped in the reverse micelles. A cationic surfactant forms an ionic complex with DNA, because DNA is negatively charged in a moderate pH condition. However, the extraction of DNA from an aqueous solution has yet to be achieved.

In this section, we will introduce the extraction behaviour of DNAs from an aqueous solution into an organic phase using reverse micelles, and show some important operational parameters for achieving the DNA transfer. We can control the DNA transfer from an aqueous to organic phase and vice versa. The key parameters for successfully transferring DNAs without the conformational change are discussed.

Sodium salt DNA from salmon testes (D-1626) was employed. The molecular weight of the purified DNA was confirmed to be around 300 kDa by electrophoresis with agarose gel. An extraction experiment of DNA was conducted in a 10-ml test tube. Eight surfactants, sodium di-2-ethylhexyl sulfosuccinate (AOT), cetyl trimethyl ammonium bromide (CTAB), trioctyl methyl ammonium chloride (TOMAC) and five di-alkyl dimethyl ammonium bromides which have different alkyl-chain lengths ($2C_{10}$, $2C_{12}$, $2C_{14}$, $2C_{16}$ and $2C_{18}$QA), were used without purification. The concentration of DNA in the aqueous and isooctane phases was detected at 260 nm by a UV-vis spectrophotometer.

The degree of extraction E is defined by the following equation:

$$E = [C_{\mathrm{DNA,org}}]_{\mathrm{eq}}/[C_{\mathrm{DNA,aq}}]_{\mathrm{init}} \tag{1}$$

The DNA concentration in the organic phase can generally be evaluated by the difference in the initial and residual concentrations of DNAs in the feed aqueous solution. However, it should be noted that the amount of DNA extracted into the organic phase is sometimes not in agreement with the reduced amount of DNA from the material source because of the formation of precipitate at the aqueous–organic interface. Thus, in this study, the percent extraction was determined by the direct measurement of the DNA concentration in the organic phase [22].

In the protein extraction by reverse micelles, it is well known that the selection of surfactants is very important and often determines the success of the protein extraction operation [9]. In the initial stage of our experiments, we chose the cationic surfactant TOMAC. It is easy to deduce that cationic surfactants are the best candidates for DNA extraction because DNA is negatively charged and behaves as a polyanion in an intermediate pH solution. Meanwhile TOMAC is a typical and effective cationic surfactant for protein extraction by reverse micelles [23]. However, the reverse micelles that are formed by TOMAC cannot extract DNA in the organic solvent, although the concentration of DNA in the aqueous phase is reduced. A similar tendency is observed when CTAB is employed. The effect of the surfactants on the DNA extraction is summarized in Table 14.3.

TABLE 14.3. Effect of surfactants on DNA extraction [22].

Surfactant	Alcohol content (%)	Percent DNA extraction (%)
AOT	0	0
CTAB	13	2.1
TOMAC	0.5	6.8
$2C_{10}$QA	1.0	38.6
$2C_{12}$QA	1.5	82.1
$2C_{14}$QA	2.0	98.6
$2C_{16}$QA	3.5	98.8
$2C_{18}$QA	5.2	98.4

DNA concentration: 50 g/ml, surfactant concentration: 10 mM, aqueous pH: 8.0, organic solvent: isooctane, temperature: 298 K alcohol: 2-ethylhexyl alcohol.

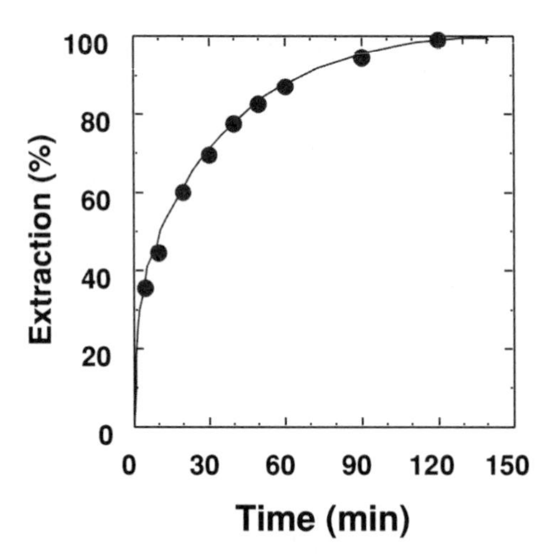

FIGURE 14.12. Time course of DNA extraction into an isooctane solution containing dimethyl dioctadecyl ammonium bromide (2C18QA). The extraction was performed under the following conditions: 10 mM Tris–HCl buffer, pH 8, DNA 50 g/ml, salt-free, 2C18QA 10 mM and 2-ethylhexylalcohol 5.2% (v/v) [22].

Since the quaternary ammonium salts of cationic surfactants are not directly soluble in hydrophobic organic solvents such as isooctane, we added 2-ethylhexyl alcohol to improve the solubility. The alcohol contents given in Table 14.3 are the minimum concentration for dissolving each 10 mM surfactant in isooctane. These results show that quaternary ammonium surfactants with double hydrophobic chains are the best surfactants for DNA extraction. When the anionic surfactant AOT is used, DNA transfer does not occur at all. These results demonstrate that the driving force in the DNA extraction is an electrostatic interaction between the negatively charged DNA and cationic surfactants. Another outstanding factor for achieving DNA transfer is the hydrophobicity of the cationic surfactants. The cationic surfactants are considered to form an ion complex with negatively charged DNA at the oil–water interface. However, in order to release the complex from the interface and dissolve it in the organic solvent, a considerably high hydrophobicity is required for the surfactant–DNA complex. Furthermore, since the negative charges of the phosphoric groups of DNA are closely packed on the surface of the DNA, a straight chain might be favourable for the hydrophobic part of the surfactants to prevent steric hindrance when interacting [6].

Figure 14.12 shows the time course of the DNA transfer from the aqueous phase to the reverse micellar phase. When the hydrophobicity of the surfactant–DNA complex is sufficiently high for dissolving in isooctane, the DNA transfer proceeds smoothly in 2 hours without the formation of precipitates. Figure 14.13 shows the effect of the surfactant concentration on the DNA extraction. The residual DNA concentration in the feed of the aqueous solution decreases with increasing surfactant concentration. In low concentrations of below 1 mM, however, the degree of DNA extraction does not increase because precipitate is formed at the interface. The DNA in the aqueous phase is completely transferred to the organic phase at 5 mM surfactant. In the protein extraction by reverse micelles, there are two important parameters: pH and ionic strength. Therefore,

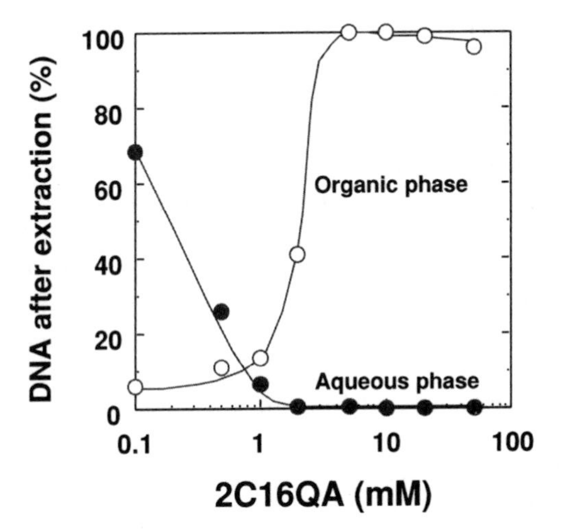

FIGURE 14.13. Effect of surfactant concentration (2C16QA) on the degree of extraction of DNA from the aqueous phase into the organic phase. The extraction was performed under the following conditions: 10 mM Tris–HCl buffer, pH 8, DNA 50 μg/ml, salt-free and 2-ethylhexylalcohol 3.8% (v/v) [22].

we investigated the effect of pH and ionic strength in the material solution. Although the degree of extraction is reduced to around 80% at pH 2, a complete extraction is achieved at an intermediate pH of between 6 and 8 (data not shown). Furthermore, the ionic strength also significantly affects the DNA extraction (Figure 14.14). The DNA transfer is strictly inhibited by the salt, and the divalent ion Ca^{2+} has a strong effect compared to the monovalent ion Na^+, which is also observed in protein extraction by reverse micelles [6]. This result suggests that the main driving force in DNA extraction is the electrostatic interaction.

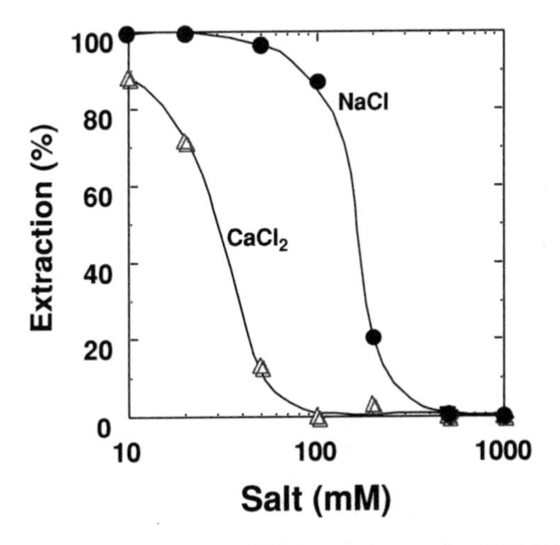

FIGURE 14.14. The effect of salt concentration on the degree of extraction of DNA from the aqueous phase into the organic phase. The extraction was performed under the following conditions: 10 mM Tris–HCl buffer, pH 8, DNA 50 μ g/ml, 2C16QA 10 mM and 2-ethylhexylalcohol 5% (v/v) [22].

14.4. CONCLUDING REMARKS

The functionalization of the reverse micelles will create a novel application in bioseparation processes in the analytical and medical sciences. It is therefore important to reveal the recognition mechanism of proteins at the liquid–liquid interface in reversed micellar solutions. DNA is also successfully extracted in a few hours by reversed micelles formed by cationic surfactants in isooctane. The driving force of the DNA transfer is the electrostatic interaction between the cationic surfactants and the negatively charged DNA. Another important factor is the hydrophobicity of the cationic surfactants. Double-chain type cationic surfactants are found to be one of the best surfactants ensuring the efficient extraction of DNA. These results have shown that reverse micellar solutions will become a useful tool not only for protein separation, but also for DNA separation.

ACKNOWLEDGEMENTS

This research was supported by a Grant-in-Aid for Science Research (No. 13129205) from the Ministry of Education, Science Sports & Culture of Japan.

REFERENCES

1. M.N. Jones, Chem. Soc. Rev. **1992**, 127 (1992).
2. P.L. Luisi, F.J. Bonner, A. Pellegrini, P. Wiget and R. Wolf, Helv. Chim. Acta **62**, 740 (1979).
3. M. Dekker, K. vant Riet, S.R. Weijers, J.W.A. Baltussen, C. Laane and B.H. Bijsterbosch, Chem. Eng. J. **33**, B27 (1986).
4. K.E. Goklen and T.A. Hatton, Biotechnol. Prog. **1**, 69 (1985).
5. M.J. Pires, M.R. Aires-Barros and J.M.S. Cabral, Biotechnol. Prog. **12**, 290 (1996).
6. T. Ono, M. Goto, F. Nakashio and T.A. Hatton, Biotechnol. Prog. **12**, 793 (1996).
7. M. Adachi and M. Harada, J. Phys. Chem. **97**, 3631 (1993).
8. M. Kuroki, T. Ono, M. Goto and F. Nakashio, Sep. Sci. Technol. **30**, 89 (1995).
9. M. Goto, T. Ono, F. Nakashio and T.A. Hatton, Biotechnol. Bioeng. **54**, 26 (1997).
10. A. Carlson and R. Nagarajan, Biotechnol. Prog. **8**, 85 (1992).
11. L. Mukhopadhyay, P.K. Bhattacharya and S.P. Moulik, Colloids Surf. **50**, 295 (1990).
12. V.M. Paradkar and J.S. Dordick, Biotechnol. Bioeng. **43**, 529 (1994).
13. M.C. Manning, J.E. Matuura, B.S. Kendrick, J.D. Meyer, J.J. Doemish, M. Vrkljan, J.R. Rush, J.F. Carpenter and E. Shefter, Biotechnol. Bioeng. **48**, 506 (1995).
14. T. Ono, T. Miyazaki, M. Goto and F. Nakashio, Solvent Extr. Res. Dev. Jpn. **3**, 1 (1996).
15. K. Kawakami, M. Harada, M. Adachi and A. Shioi, Colloids Surf. A **109**, 217 (1996).
16. M.J. Pires, P. Martel, A. Baptista, S.B. Petersen, R.C. Willson and J.M.S. Cabral, Biotechnol. Bioeng. **44**, 773 (1994).
17. C.E. Forney and C.E. Glatz, Biotechnol. Prog. **10**, 499 (1994).
18. C.E. Forney and C.E. Glatz, Biotechnol. Prog. **11**, 260 (1995).
19. T. Ono and M. Goto, Biotechnol. Prog. **14**, 903 (1998).
20. B. Spring and P. Haake, Bioorg. Chem. **6**, 181 (1977).
21. V.E. Imre and P.L. Luisi, Biochem. Biophys. Res. Commun. **107**, 538 (1982).
22. M. Goto, A. Horiuchi, T. Ono and S. Furusaki, J. Chem. Eng. Jpn. **32**, 123 (1999).
23. M. Dekker, K. vant Riet, B.H. Bijsterbosch, P. Fijneman and R. Hilhorst, Chem. Eng. Sci **45**, 2949 (1990).

Index